Reviews of

77 Physiology, Biochemistry and Pharmacology

formerly
Ergebnisse der Physiologie, biologischen
Chemie und experimentellen Pharmakologie

Editors

R. H. Adrian, Cambridge · E. Helmreich, Würzburg
H. Holzer, Freiburg · R. Jung, Freiburg
K. Kramer, München · O. Krayer, Boston
R. J. Linden, Leeds · F. Lynen, München
P. A. Miescher, Genève · J. Piiper, Göttingen
H. Rasmussen, Philadelphia · A. E. Renold, Genève
U. Trendelenburg, Würzburg · K. Ullrich, Frankfurt/M.
W. Vogt, Göttingen · A. Weber, Philadelphia

With 38 Figures

Springer-Verlag Berlin Heidelberg GmbH 1977

ISBN 978-3-662-30971-1   ISBN 978-3-540-37998-0 (eBook)
DOI 10.1007/978-3-540-37998-0

Library of Congress-Catalog-Card Number 74-3674

© by Springer-Verlag Berlin Heidelberg 1977
Originally published by Springer-Verlag Berlin Heidelberg New York in 1977
Softcover reprint of the hardcover 1st edition 1977

Contents

List of Contributors

BECKER, ELMER, Dr., Department of Pathology, University of Connecticut, Health Center, Farmington, Connecticut/USA

GRUNEWALD, W.A., Prof. Dr. Dr., Physiologisches Institut der Universität, Regensburg/Federal Republic of Germany

SOWA, W., Dr., Physiologisches Institut der Universität, Regensburg/Federal Republic of Germany

STARKE, KLAUS, Prof. Dr., Pharmakologisches Institut, Universitätsklinikum, Essen/Federal Republic of Germany

WARD, PETER, A., Prof. Dr., Department of Pathology, University of Connecticut, Health Center, Farmington, Connecticut/USA

Rev. Physiol. Biochem. Pharmacol., Vol. 77
© by Springer-Verlag 1977

Regulation of Noradrenaline Release by Presynaptic Receptor Systems*

K. STARKE

Contents

---

* Abbreviations not defined in text or tables: ATP: adenosine triphosphate; COMT: catechol-O-methyltransferase; cyclic AMP. cyclic adenosine 3′,5′-monophosphate; cyclic GMP: cyclic guanosine 3′,5′-monophosphate; DMPP: 1,1-dimethyl-4-phenylpiperazinium iodide; DOPEG: 3,4-dihydroxy-phenylglycol; e.j.p.: excitatory junction potential; MAO: monoamine oxidase; PGA_2, PGB_2, PGE_1, PGE_2, $PGF_{1\alpha}$, $PGF_{2\alpha}$: the respective prostaglandins.

1. Introduction, Scope, Definitions

The release of transmitter from noradrenergic nerve endings per unit time depends on the rate of impulse flow on the one hand and the release per impulse
on the other hand. Excitatory and inhibitory neurotransmitters and perhaps
circulating agents, which act on receptors in the soma-dendritic part of the
neurone, regulate the frequency of the action potentials travelling down the
axon. The quantum of noradrenaline released by an impulse is far from being
constant. The causes for its variation are only partly understood. However,
one mechanism has been recognized over the last ten years. It is very likely
that several substances naturally occurring in the body are able to increase
or decrease the release per pulse, and that they act on receptors in the terminal
varicose part of the axon (the "noradrenergic nerve terminal"; Haefely, 1972)

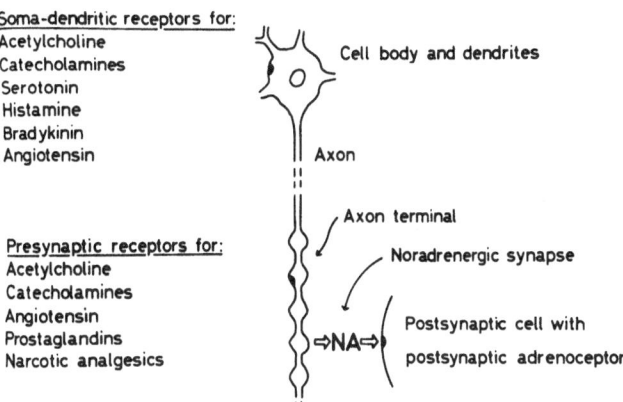

Fig. 1. Receptor systems of postganglionic sympathetic neurones. Not all systems indicated may be detectable in a particular neurone. NA, noradrenaline

where release is assumed to take place. There is thus a dual receptor-mediated regulation (Fig. 1). Soma-dendritic receptors control the generation of impulses, i.e., the electrical signal. Presynaptic receptors control impulse-evoked secretion, i.e., the chemical signal.

This review covers presynaptic effects of angiotensin, acetylcholine, prostaglandins, and catecholamines. Narcotic analgesics are also discussed because of the recent evidence that they are agonists related to endogenous compounds. The numerous foreign substances that affect release, such as local anesthetics, reserpine, false transmitters, scorpion toxins, and various cations, are not included. Evidence for modulation of release that is derived from measurement of the outflow of transmitter from the tissue is presented in detail. Evidence derived from changes in the response of effector cells is considered only to a limited extent. In whole animals, variations of impulse traffic and blood supply may obscure presynaptic effects. Therefore, studies on more or less isolated preparations are discussed preferentially.

The term "release" is used for any passage of noradrenaline across the neuronal membrane into the synaptic cleft, irrespective of the underlying mechanism. Unless defined further, it indicates secretion evoked by normal orthodromic action potentials, triggered for instance by electrical stimulation. Release can also be induced by drugs. Traces of noradrenaline pass out from the nerve terminals even in the absence of chemical or physical stimuli ("spontaneous release"). — "Overflow" or "outflow" describes the diffusion of noradrenaline or its metabolites from the tissue into the perfusion or incubation fluid. Release elicited by electrical pulses gives rise to "stimulation-evoked overflow". It will often be necessary to distinguish between the effect of a drug on the stimulation-evoked overflow and its effect on the overflow in the absence of electrical stimulation. In this connexion, the overflow in the absence of electrical stimulation will be called "basal outflow". An increase of basal outflow caused by a drug may, but does not necessarily, reflect drug-induced release (p. 10).

2. Basic Mechanisms in Noradrenergic Synaptic Transmission

Basic events occurring in catecholamine release have been reviewed by SMITH and WINKLER (1972), BALDESSARINI (1975) and STJÄRNE (1975a). A review by IVERSEN (1975) and a book edited by PATON (1976) deal with uptake mechanisms for biogenic amines, and articles by KOPIN (1972), LANGER (1974a) and MUSAC-CHIO (1975) analyze the metabolic fate of catecholamines. Noradrenergic receptor mechanisms are discussed by FURCHGOTT (1972) and BLOOM (1975). Only selected aspects are considered here.

2.1. Exocytosis

The varicosities of noradrenergic axon terminals are equipped with the biochemical machinery for the synthesis, storage, and inactivation of the transmitter and appear to be the structures from which it is secreted. Within the varicosities, noradrenaline is stored in granular vesicles together with ATP and at least two soluble proteins, chromogranin A and dopamine-β-hydroxylase. When action potentials arrive, release is initiated by depolarization of the varicosity membrane, which leads to an influx of calcium. Calcium is indispensable for release by nerve impulses, but it is not known at which stage it is involved in electrosecretory coupling. It seems likely that at least a large part of the transmitter is secreted by exocytosis. The vesicle and plasma membranes fuse and form an opening through which the vesicle contents can diffuse into the extracellular space. In agreement with the exocytosis hypothesis, release of noradrenaline is accompanied by release of chromogranin A and dopamine-β-hydroxylase. Released noradrenaline triggers the response of the postsynaptic cell by interaction with postsynaptic adrenoceptors.

2.2. Facilitation and Depression in Trains of Pulses

When noradrenergic fibers are stimulated by trains of pulses, preceding pulses modify release by subsequent ones. Electrophysiological and biochemical studies indicate that release *increases* with successive shocks in the guinea-pig uterine artery and vas deferens[1] as well as the rabbit ear artery, portal vein and vas deferens[1] (BURNSTOCK et al., 1964; BELL, 1972; RAND et al., 1973; HUGHES and ROTH, 1974). In contrast, release per shock *decreases* with successive shocks in guinea-pig atria (RAND et al., 1973). In the vas deferens of the mouse, the

---

[1] The nature of the postganglionic motor transmitter in the vas deferens is disputed. Some investigators believe it to be noradrenaline (SWEDIN, 1971; SJÖSTRAND, 1973b; BENNETT and MIDDLETON, 1975a), while others suppose that the rich noradrenergic innervation serves a presynaptic inhibitory role, modulating a motor transmission which is mediated by an unknown transmitter (AMBACHE and ZAR, 1971; AMBACHE et al., 1972, 1975; EULER and HEDQVIST, 1975). If authors interpret experiments in terms of a noradrenergic motor transmission, this review follows their view. However, a reservation should be made, especially when only the contractile response to nerve stimulation is measured.

amplitude of the e.j.p.s of smooth muscle cells increases exponentially when the sympathetic nerves are stimulated at less than 1 Hz. At higher frequencies, the e.j.p.s depart from exponential growth after the first few pulses and decline to a steady-state amplitude much smaller than that reached at low frequency (cf. Fig. 6a, p. 64; BENNETT, 1973a). If one assumes (BURNSTOCK et al., 1964; BENNETT, 1973a) that the e.j.p. amplitude is a measure of the quantum of noradrenaline released by a single impulse, then in the mouse vas deferens facilitation of release is soon followed by depression at frequencies higher than 1 Hz.

Facilitation and depression by preceding pulses probably account for the well-known frequency-dependence of release. In many tissues the average overflow per impulse *increases* with increasing stimulation frequency, presumably because facilitation becomes more pronounced, the shorter the interval between pulses (cat spleen: BROWN and GILLESPIE, 1957; CUBEDDU and WEINER, 1975a; guinea-pig uterine artery: BELL and VOGT, 1971; myenteric plexus-longitudinal muscle preparation: HENDERSON et al., 1972b, 1975; vas deferens: STJÄRNE, 1973k; rabbit portal vein and vas deferens: HUGHES, 1972). In the nictitating membrane of the cat and the vas deferens of the mouse the average overflow per pulse is constant over a wide range of frequencies (FARNEBO and MALMFORS, 1971; HENDERSON and HUGHES, 1974; HENDERSON et al., 1975). In the rat cerebral cortex and the rabbit superior cervical ganglion it *decreases* with increasing frequency (MONTEL et al., 1974a; NOON and ROTH, 1975). It seems likely that in these tissues facilitation is soon overcome by depression, the depression being more pronounced the shorter the interval between pulses.

Any theory of noradrenaline secretion has to account for these observations. Tentative explanations have been and can now be given for both facilitation and depression. On the other hand, the reason for the surprising tissue differences in facilitation and depression and in the slopes of frequency-overflow curves is obscure.

Facilitation has been explained by the calcium residue hypothesis first developed for cholinergic nerves (see YOUNKIN, 1974). Each impulse is assumed to leave some calcium bound to an active site. This calcium adds to the calcium entering during subsequent impulses, so that the release per impulse increases with successive shocks. The shorter the interval between pulses, the more calcium is left and the greater the facilitation, so that the average release per impulse increases with increasing frequency (BENNETT and FLORIN, 1975; STJÄRNE, 1975e). – The existence of presynaptic receptors opens up new aspects, which will be discussed in the appropriate chapters. There is evidence that at least part of the depression is mediated by presynaptic α-adrenoceptors (p. 64). Moreover, the possibility that presynaptic β-receptors contribute to facilitation merits consideration (p. 78).

The calcium residue hypothesis also helps to understand the frequency dependence of the effects of external calcium and many drugs on the release of noradrenaline. It seems likely that release increases rapidly with increasing intracellular calcium only within the range occurring during low rates of impulse flow. At high intraneuronal calcium concentrations, such as during high rates of impulse flow, the "release receptors" for calcium become fully occupied. *Variations of intraneuronal calcium near the saturation level will lead to mini-*

mal, if any, variation of release; a further rise may even be inhibitory (STJÄRNE, 1973e, 1975e; KIRPEKAR et al., 1975). This hypothetical relationship between frequency, intraneuronal calcium, and release may explain why modification of the *extracellular* calcium concentration (between 1.3 and 5.1 mM) has strong effects on the average release per pulse at low frequency, but only insignificant effects at high frequency (HENDERSON and HUGHES, 1974); at high frequency, any resultant change of the *intraneuronal* calcium level appears to remain within the saturation range. Moreover, the relationship may explain why the presynaptic modulators under review lose their influence on release when the frequency is high. Many modulators have been proposed to act by enhancing or impairing the stimulation-evoked rise of intraneuronal calcium. The consequences will be minimized when calcium approaches saturation levels during high rates of impulse flow.

2.3. Inactivation

Released as well as exogenous noradrenaline is removed from the region of the receptors (sometimes called the "biophase"; FURCHGOTT, 1972, p. 302) by three processes. Firstly, a fraction is taken up back into the neurone by a high affinity transport system located in the axonal membrane. The extent of recapture varies in different organs and may amount to 70–80% in densely innervated tissues. Amines chemically related to noradrenaline such as dopamine, metaraminol, and tyramine can be taken up by the same transfer mechanism. Neuronal uptake is blocked by relatively low concentrations of cocaine or desipramine. Secondly, another fraction enters extraneuronal cells via a low affinity membrane transport system. Extraneuronal uptake is blocked by relatively low doses of normetanephrine or steroids such as corticosterone and estradiol. Thirdly, the fraction that escapes uptake diffuses into the environment ("overflow"). A further mechanism for the uptake of exogenous noradrenaline (BLAKELEY et al., 1974) is discussed on p. 57. After neuronal uptake the fate of noradrenaline differs from that after extraneuronal uptake. After entry into the axoplasm, part of the noradrenaline is redistributed into the storage vesicles. The remainder of the axoplasmic noradrenaline and the bulk of the extraneuronal noradrenaline undergo catabolism.

It has been known for some time that part of the noradrenaline released by nerve impulses is metabolized (HERTTING and AXELROD, 1961). The traditional view that axoplasmic noradrenaline is mainly metabolized by MAO, whereas extraneuronal noradrenaline is mainly metabolized by COMT (KOPIN, 1972), has recently been refined (LANGER, 1974a). In many tissues the major metabolite formed from released noradrenaline is DOPEG (CUBEDDU et al., 1974a; LANGER and ENERO, 1974; LEVIN, 1974; BRAESTRUP and NIELSEN, 1975; LUCHELLI-FORTIS and LANGER, 1975). It appears to originate from intraneuronal deamination of recaptured noradrenaline, followed by reduction of the aldehyde to the alcohol. Inhibition of reuptake by cocaine decreases the formation of DOPEG. On the other hand, O-methylation to normetanephrine occurs at extraneuronal sites and is specifically prevented by drugs that inhibit extraneuronal uptake

(LUCHELLI-FORTIS and LANGER, 1975). Biotransformation of released noradrenaline to O-methylated deaminated products probably takes place intraneuronally in the nictitating membrane (LANGER and ENERO, 1974; LUCHELLI-FORTIS and LANGER, 1975), but reflects extraneuronal metabolism in the spleen of the cat (CUBEDDU et al., 1974a) and the rat brain (BRAESTRUP and NIELSEN, 1975). Changes in the pattern of metabolites may be used to recognize changes in neuronal and extraneuronal uptake.

2.4. Release by Other Stimuli

There are fundamental differences in the mechanisms by which drugs or other stimuli elicit release of noradrenaline. High *potassium* concentrations depolarize the varicosity membrane by reducing the gradient between the intracellular and extracellular spaces. Potassium-evoked release is calcium-dependent (KIRPE-KAR and WAKADE, 1968; SORIMACHI et al., 1973) and probably exocytotic, since dopamine-β-hydroxylase is released concomitantly (GARCIA and KIRPEKAR, 1975; THOA et al., 1975). Similarly, release by *nicotinic drugs* appears to involve membrane depolarization, calcium entry (LINDMAR et al., 1967a; HAEUSLER et al., 1968), and presumably exocytosis, though the release of dopamine-β-hydroxylase has not been reported. Release can also be induced by exposure of the tissue to a *sodium-deficient* medium; it is only partly calcium-dependent (GARCIA and KIRPEKAR, 1973; DUBEY et al., 1975) and has been proposed to be partly exocytotic (VARGAS et al., 1976; see, however, GARCIA and KIRPEKAR, 1975). The indirectly acting sympathomimetic amine *tyramine* does not depolarize noradrenergic nerve endings (CABRERA et al., 1966), does not require calcium for release (LINDMAR et al., 1967a) and does not release dopamine-β-hydroxylase (CHUBB et al., 1972). Tyramine is taken up into the neurone and displaces noradrenaline from the storage vesicles, which then diffuses from the cytosol across the plasma membrane into extracellular space (SCHÜMANN and PHILIPPU, 1961); the mechanism is clearly distinct from exocytosis. Similarly, *reserpine,* which impairs the vesicular storage of noradrenaline, releases the transmitter without dopamine-β-hydroxylase, and its effect is calcium-independent (THOA et al., 1975). Finally, the small *spontaneous* release in the absence of nerve impulses or other stimuli (for its mechanism see HENSELING et al., 1976) appears to be calcium-independent, in contrast to the spontaneous release of acetylcholine from motor nerve terminals (RUBIN, 1970).

3. Determination of Noradrenaline Release

The amount of noradrenaline released from the interior of the nerve terminals cannot be measured directly. All the available methods are indirect. In principle, two *groups of methods* are used. Firstly, the response of the postsynaptic cell is taken as a measure of release. Either the electrical or the mechanical response is evaluated (postsynaptic response methods). Secondly, the quantity of noradrenaline or of noradrenaline plus its metabolites that overflows into the perfusion or incubation medium during and after release is determined. Either the overflow of endogenous noradrenaline is measured chemically or by bioassay, or the tissue is pretreated with labeled noradrenaline or its precursors, and the overflow of labeled compounds is recorded (overflow methods). How useful are these methods for the investigation of drug effects oñ release by nerve impulses?

Drug-induced changes in the *postsynaptic response* to nerve impulses may reflect changes of release, but also changes of the inactivation mechanisms or of the sensitivity of the postsynaptic cells. In order to check the latter possibilities, the influence of the drug on the response to nerve stimulation can be compared with its influence on the response to exogenous noradrenaline. However, in many tissues the distribution and inactivation of exogenous noradrenaline on the one hand, and neurally released transmitter on the other hand, are quite different (BEVAN and SU, 1971; BRANDÃO and GUIMARÃES, 1974). Moreover, released and exogenous noradrenaline may act on different populations of postsynaptic receptors (HOTTA, 1969). Even if a drug selectively modulates the response to nerve impulses, it is difficult to exclude selective interference with the inactivation of, or the receptor mechanism for, released transmitter. It should also be noted that electrical stimulation may excite non-noradrenergic in addition to noradrenergic fibers. A drug may affect the non-noradrenergic component of the postsynaptic response.

The *overflow* of ńoradrenaline evoked by nerve impulses is the difference between the amount released and the amount recaptured by the neurones (and subsequently stored in the vesicles or metabolized) or taken up into extraneuronal cells (and subsequently metabolized). Drug-induced changes in overflow may reflect not only changes in release but also changes in uptake or biotransformation. The following methods help to distinguish between these possibilities.

1. The influence of the agent under study on MAO and COMT activities and uptake mechanisms can be tested directly. Several concentrations should be used. For instance, high concentrations of some α-adrenolytic drugs inhibit the neuronal and extraneuronal uptake of noradrenaline, whereas lower concentrations are sufficient to increase the stimulation-evoked overflow. Failure to determine dose–response curves has delayed the realization that low concentrations selectively facilitate release (7.2.).

2. The neuronal stores can be labeled with radioactive noradrenaline, and the influence of the drug on the stimulation-evoked overflow of both labeled noradrenaline and labeled metabolites can be determined. Effects on biotransformation will thus be disclosed. Exogenous noradrenaline previously taken up and the endogenous transmitter probably differ in their distribution and fate. However, it is unlikely that drugs affect exogenous and endogenous noradrenaline in a qualitatively different way. It should be noted that the various methods used for the separation of labeled noradrenaline and its metabolites have yielded inconsistent results. For instance, the stimulation-evoked overflow of total radioactivity from the cat nictitating membrane was reported to contain 70–75%

noradrenaline and less than 7% DOPEG (ENERO et al., 1972; LANGER et al., 1972). In contrast, more recent experiments from the same laboratory with an improved column chromatographic procedure (GRAEFE et al., 1973) have shown that noradrenaline accounts for only 23%, whereas the percentage of DOPEG is much higher (48%; LANGER and ENERO, 1974). In the original procedure, a large part of the labeled DOPEG contaminated the "noradrenaline" fraction. Another word of caution concerns nomenclature. Alumina has often been employed to separate catechol compounds from O-methylated metabolites. The adsorbed fraction contains noradrenaline, DOPEG, and 3,4-dihydroxymandelic acid, and under many conditions DOPEG is a substantial or even the major constituent. It may be misleading to term the adsorbed fraction "noradrenaline" or "catecholamines" (e.g., ZIMMERMAN et al., 1972; HUGHES, 1973). For instance, KIRPEKAR and PUIG (1971) preperfused cat spleens with ^{3}H-noradrenaline. The overflow of total tritium evoked by splenic nerve stimulation was almost quantitatively accounted for by what the authors called "^{3}H-noradrenaline", but what were in fact ^{3}H-catechols adsorbed on alumina. The conclusion was that appreciable metabolism of noradrenaline released by nerve stimulation did not occur. On the other hand, further separation of the ^{3}H-catechols in similar experiments by DUBOCOVICH and LANGER (1973) revealed that a large part was ^{3}H-DOPEG, indicating extensive metabolism[2]. — The separation of noradrenaline metabolites is time-consuming. Fortunately, even the determination of the stimulation-evoked overflow of total radioactivity alone is a valuable method for the study of noradrenaline release. The metabolites formed from released noradrenaline are not bound within the tissue and appear in the overflow. The outflow of 3,4-dihydroxymandelic acid and 3-methoxy-4-hydroxymandelic acid is slow in comparison with that of normetanephrine and the neutral metabolites (LEVIN, 1974); however, the acids are minor products in most tissues. Therefore, the stimulation-evoked overflow of total radioactivity approximates release better than the overflow of noradrenaline alone.

3. The influence of the drug on the stimulation-evoked overflow can be tested after neuronal and/or extraneuronal uptake mechanisms have been blocked by, e.g., cocaine or desipramine (neuronal uptake) and corticosterone or normetanephrine (extraneuronal uptake). Under these conditions, any change in overflow should reflect a change in release. It should be realized, however, that the uptake inhibitors themselves can affect release and that they can modify the effect of release-modulating agents. For instance, the increase of the biophase concentration of noradrenaline resulting from the blockade of uptake will enhance the activation of presynaptic α-adrenoceptors (7.); this in turn will lead to depression of release even though the overflow of noradrenaline may be augmented (FARNEBO and HAMBERGER, 1971a; ENERO et al., 1972); as a further consequence, the inhibitory effect of exogenous α-receptor agonists is reduced (STARKE 1972a; STARKE and MONTEL, 1973d; cf. 7.4.). Uptake inhibitors may also reduce release by local anesthesia; in rabbit hearts, cocaine at 2×10^{-6} and 2×10^{-5} M increases the stimulation-evoked overflow of noradrenaline,

---

[2] In the literature synopses of Tables 1–6 and 9 the nomenclature of the respective authors is used. Thus, the term "^{3}H-noradrenaline" is adopted for the tritium eluted from alumina even if the amine was probably only a minor constituent of the total radioactive material.

but at 2×10^{-4} M causes a decrease, presumably because suppression of impulse conduction prevails (STARKE et al., 1972). Finally, normetanephrine and desipramine are known to respectively activate and block α-receptors; in the nictitating membrane of the cat normetanephrine is one half as potent as noradrenaline (LANGER and RUBIO, 1973). When these drugs are used as uptake inhibitors it is advisable to exclude effects on the presynaptic α-receptors of the tissue under study.

4. During exocytosis, vesicular soluble proteins are secreted together with noradrenaline. There is no evidence that a mechanism for the removal of released dopamine-β-hydroxylase exists within the tissue. Therefore, parallel changes in overflow of noradrenaline and the enzyme support the view that they reflect changes in release.

Judicious use of the methods outlined above will in general make it possible to decide whether a drug modulates the release of noradrenaline. However, just how much noradrenaline is released by an impulse or a series of impulses under normal conditions, in the absence of drugs, remains impossible to determine at the present time. Newly recognized obstacles, namely presynaptic receptors in particular for the transmitter itself, have been added to the traditional ones, namely inactivation mechanisms within the tissue. Any procedure meant to eliminate inactivation mechanisms must be expected to interfere, either directly or indirectly, with presynaptic receptor systems.

The preceding considerations are also valid for the determination of release evoked by other stimuli. It should be noted that drugs may increase the basal outflow of noradrenaline from the tissue (in the absence of nerve stimulation) not only by triggering some release mechanism, but also by blocking MAO or COMT or the cellular uptake of spontaneously released noradrenaline. However, agents that inhibit uptake mechanisms or metabolizing enzymes produce only a small, if any, increase of basal outflow (LUCHELLI-FORTIS and LANGER, 1975; TAUBE et al., 1975; cf. Fig. 2a, p. 52); large increases, as for instance by nicotinic compounds, indicate drug-induced release. Analogous precautions are necessary when the effect of a drug on the uptake of noradrenaline is investigated. Methods commonly used to study uptake—i.e. transfer from the extracellular space across the cell membrane into the cytoplasm—include the accumulation of the amine in the tissue and its removal from the medium (see GRAEFE, 1976). Both accumulation and removal are reduced not only by uptake inhibitors but also by releasing agents. In order to avoid false classification, a releasing effect should be excluded before a drug is designated as an uptake inhibitor (cf. HEIKKILA et al., 1975).

4. Angiotensin

Angiotensin exerts a variety of effects on the central nervous system and peripheral autonomic nerves (reviews by KHAIRALLAH, 1972; REIT, 1972; ROTH, 1972; STARKE, 1972c; ZIMMERMAN et al., 1972; SEVERS and DANIELS-SEVERS, 1973;

REGOLI et al., 1974). In particular, soon after synthetic angiotensin became available it was found that the peptide enhanced noradrenergic neuroeffector transmission (ZIMMERMAN, 1962; McCUBBIN and PAGE, 1963; BENELLI et al., 1964). It seems likely that the enhancement is partly due to a direct action on the effector, as for instance smooth muscle cells, leading to an increase of their sensitivity to noradrenaline (THOENEN et al., 1965; PANISSET and BOURDOIS, 1968; DAY and MOORE, 1976); like other smooth muscle stimulants, angiotensin at low doses causes a partial depolarization of the cell membrane which brings the cells closer to the firing level of action potentials (SJÖSTRAND and SWEDIN, 1974). However, in addition to this postsynaptic component angiotensin has presynaptic actions which increase the biophase concentration of noradrenaline. Direct evidence was first presented by ZIMMERMAN and WHITMORE (1967) who showed that angiotensin augments the overflow of noradrenaline evoked by sympathetic nerve stimulation. There has been considerable controversy concerning the mechanism of this effect, the main postulates being that angiotensin inhibits the neuronal uptake of noradrenaline (see KHAIRALLAH, 1972), and that it facilitates the average release per impulse (McCUBBIN and PAGE, 1963; BENELLI et al., 1964; cf. STARKE, 1972c; ZIMMERMAN et al., 1972). A possible direct releasing effect and an enhancement of the biosynthesis of noradrenaline add to the somewhat confusing multitude of proposed actions on noradrenergic nerve endings. – The term "angiotensin" is used here for either of the four closely related octapeptides 1-Asp-5-Ile-angiotensin II, 1-Asn-5-Ile-angiotensin II, 1-Asp-5-Val-angiotensin II, and 1-Asn-5-Val-angiotensin II.

4.1. Angiotensin-Induced Release and Inhibition of Neuronal Uptake?

Since the mid-sixties much work has been devoted to the question of whether angiotensin releases noradrenaline and/or inhibits its neuronal uptake. Even now an unequivocal answer cannot be given, mainly for two reasons.

Firstly, many experimental results can be explained both by inhibition of uptake and by angiotensin-induced release (cf. p. 10). It has been shown that angiotensin, mostly at high concentrations (10^{-8} M or higher), slightly increases the basal outflow of noradrenaline from some tissues, and it was concluded that the peptide releases transmitter from sympathetic nerve endings. However, in the majority of investigations the possibility that the effect reflected inhibition of the re-uptake of spontaneously released noradrenaline has not been ruled out (e.g., KIRAN and KHAIRALLAH, 1969). Unfortunately, many uptake studies lead to the converse dilemma. For instance, low concentrations of angiotensin (2×10^{-11} M and higher) have been reported to inhibit the accumulation of ^{3}H-noradrenaline in perfused rabbit hearts (PEACH et al., 1969; DAVILA and KHAIRALLAH, 1970). Since data on a possible releasing effect were not presented it cannot be decided whether the reduction of accumulation was due to inhibition of uptake or to an angiotensin-induced release. This distinction, of course, is of crucial importance for a drug to which either action has been attributed.

Secondly, studies concerning the effect of angiotensin on the uptake of noradrenaline have yielded contradictory results even in the same organ. As mentioned above, the peptide reduced the accumulation of noradrenaline in the rabbit heart (PEACH et al., 1969; DAVILA and KHAIRALLAH, 1970). On the other hand, similar concentrations failed to diminish the removal of noradrenaline from the medium perfusing isolated rabbit hearts (SCHÜMANN et al., 1970b; STARKE, 1971b; for similar discrepancies in the rat heart see, on the one hand, DAVILA and KHAIRALLAH, 1970, and KHAIRALLAH, 1972; on the other hand, HUGHES and ROTH, 1969, 1971, and CHEVILLARD and ALEXANDRE, 1970).

When one excludes reports which for one of these reasons are difficult to interpret, the following conclusions remain. Firstly, very high concentrations of angiotensin (10^{-5} M and higher) in fact slightly decrease the neuronal uptake of noradrenaline (SCHÜMANN et al., 1970b; STARKE, 1971b; JANOWSKY et al., 1972). In agreement with this view, 10^{-5} M angiotensin also reduces the — probably intraneuronal — formation of ^{14}C-DOPEG + ^{14}C-3,4-dihydroxymandelic acid in rabbit hearts perfused with medium containing ^{14}C-noradrenaline (STARKE, 1971b). An appreciable inhibitory effect of lower concentrations is doubtful (cf. THOENEN et al., 1965; GOMER and ZIMMERMAN, 1973; DAY and MOORE, 1976). One argument stems from the interaction of angiotensin and tyramine. Agents which block neuronal uptake are known to diminish the release of noradrenaline by, and the postsynaptic response to, indirectly acting sympathomimetic amines. However, up to 5×10^{-6} M angiotensin fails to reduce the release of noradrenaline by tyramine (STARKE, 1971b; CHEVILLARD and ALEXANDRE, 1972). The peptide enhances rather than reduces postsynaptic effects of tyramine (MCCUBBIN and PAGE, 1963; KANEKO et al., 1966; BLUMBERG et al., 1975a). — Secondly, in the rabbit coeliac artery preincubated with ^{3}H-noradrenaline as little as 10^{-10} M angiotensin accelerates the basal outflow of tritium. Since the uptake of noradrenaline was unchanged, a releasing effect seems possible (HUGHES and ROTH, 1971). However, the coeliac artery appears to be an exception. In other tissues, much higher concentrations of angiotensin are required to obtain a slight increase, if any, of basal noradrenaline outflow, and the mechanism remains unknown. Any angiotensin-induced release and inhibition of neuronal uptake probably plays a minor role in comparison with facilitation of release evoked by nerve impulses.

4.2. Effect on Release by Nerve Impulses

Angiotensin enhances the stimulation-evoked overflow of noradrenaline from the hind paw and kidney of the dog and from the rabbit heart, coeliac and pulmonary arteries, portal vein, and vas deferens, but not from several other organs (Table 1). The increase of overflow reflects facilitation of release rather than inhibition of the inactivation of released noradrenaline within the tissue, for the following reasons (cf. Table 1). (1) Low concentrations are required, which probably do not interfere with neuronal uptake. Moreover, angiotensin inhibits neither the extraneuronal uptake of noradrenaline (STARKE, 1971b; SALT, 1972), nor MAO or COMT (BARTH, 1970). (2) In tissues prelabeled with radioactive noradrenaline, angiotensin enhances the stimulation-evoked overflow of total radioactivity, i.e., of the sum of labeled transmitter plus labeled metabolites. Therefore, the increase in the overflow of noradrenaline does not result from diminished metabolic degradation. In the rabbit heart, angiotensin even *increases* the stimulation-evoked overflow of O-methylated products, probably because of an augmented supply of noradrenaline to extraneuronal COMT (STARKE, 1971b). (3) The effect if fully retained after the neuronal, or both the neuronal and extraneuronal uptake have been blocked. One finding appears inconsistent at first sight: Angiotensin does not further increase overflow after treatment with phenoxybenzamine. However, phenoxybenzamine itself strongly facilitates release in addition to blocking uptake mechanisms. The upper limit for release per pulse (cf. p. 5) is probably reached by phenoxybenzamine alone (STARKE and SCHÜMANN, 1972). (4) In rabbit atria, angiotensin enhances the overflow of both noradrenaline and dopamine-β-hydroxylase, indicating an increase of exocytotic release (BLUMBERG et al., 1975b; ACKERLY et al., 1976).

The facilitation is mediated by specific receptors. Two other peptides, namely vasopressin (STARKE et al., 1970; HUGHES and ROTH, 1971) and substance P (TAUBE et al., 1976) do not increase release in the rabbit heart, pulmonary artery, and portal vein. In the pulmonary artery, angiotensin I augments the stimulation-evoked overflow of noradrenaline. The concentrations required are 10 times higher than those of angiotensin II (STARKE et al., 1975c). Since the artery is rich in converting enzyme, at least part of the effect of the decapeptide is probably mediated by angiotensin II. 1-Sar-8-Ile-angiotensin II and 1-Sar-8-Ala-angiotensin II cause no or only a minimal increase of stimulation-evoked overflow. Either peptide strongly counteracts the facilitatory effect of angiotensin (STARKE et al., 1975c; ACKERLY et al., 1976; STARKE, unpublished). The results supplement postsynaptic response studies indicating that both antagonists block neuronal as well as smooth muscle effects of angiotensin (SWEET et al., 1973; ZIMMERMAN, 1973). The presynaptic angiotensin receptors may slightly differ from myocardial and smooth muscle receptors (BLUMBERG et al., 1975a; see, however, ZIMMERMAN, 1973).

The mechanism of the facilitatory effect is not known. Some information can be derived from the following observations. (1) The facilitation probably does not result from an action on noradrenaline storage vesicles, since angiotensin does not affect granular noradrenaline uptake and release (SCHÜMANN, 1970, p. 210). (2) Angiotensin promotes the biosynthesis of noradrenaline. However, its effect on release is not secondary to enhanced synthesis, since the release of exogenous, previously taken up noradrenaline is also increased. (3) Prostaglandins depress noradrenaline secretion. The effect of angiotensin is not due to suppression of prostaglandin formation, since the peptide *enhances* their biosynthesis (p. 16), and since it enhances release even after the production of prostaglandins has been blocked by indometacin (STARKE, unpublished). (4) Presynaptic α- and β-adrenoceptors are not involved. The effect of angiotensin persists in the presence of high concentrations of propranolol or the α-adrenergic agonist oxymetazoline, which prevents the facilitatory effect of α-adrenolytic drugs (STARKE, unpublished). (5) Angiotensin enhances release by nerve impulses at doses which do not affect basal outflow (Table 1). Moreover, it does not increase release by tyramine or amphetamine (STARKE, 1971b; CHEVILLARD and ALEXANDRE, 1972). Its influence on release by other stimuli is not

Table 1. Effect of angiotensin on transmitter overflow evoked by electrical stimulation of noradrenergic nerves

Species	Tissue	Concentration of angiotensin (M)	Effect	Comment	References
Dog	Hind paw Gracilis muscle	5×10^{-8} 4×10^{-8}	+ −	Blood-perfused tissue. Effect on overflow of endogenous noradrenaline evoked by stimulation of the lumbar sympathetic chain at 20 Hz. Results obtained in muscle difficult to interpret, since controls are lacking	ZIMMERMAN and WHITMORE, 1967
	Kidney	3×10^{-9}–10^{-8}	+	Blood-perfused kidney. Effect on overflow of endogenous noradrenaline or previously stored ^{3}H-noradrenaline or ^{3}H-metaraminol evoked by renal nerve stimulation at 2–5 Hz. No effect at 10 Hz. No effect on basal outflow	ZIMMERMAN and GISSLEN, 1968; ZIMMERMAN et al., 1972; GOMER and ZIMMERMAN, 1973
Cat	Spleen	10^{-9} Injection of 25–50 ng	0 0	Perfused spleen. Effect on overflow of endogenous noradrenaline or previously stored ^{3}H-noradrenaline evoked by splenic nerve stimulation at 6 or 30 Hz. No effect on basal outflow	THOENEN et al., 1965; HERTTING and SUKO, 1966
	Nictitating membrane	2×10^{-9}–10^{-5}	0	Tissue preincubated with ^{3}H-noradrenaline. Effect on overflow of total ^{3}H evoked by field stimulation at 0.2–20 Hz. 10 μM angiotensin increased basal outflow	CERVONI and REIT, 1975
Rat	Cerebral cortex Hypothalamus	10^{-10}–10^{-6}	0	Slices preincubated with ^{3}H-noradrenaline. Effect on overflow of total ^{3}H evoked by field stimulation at 3 Hz. No effect on basal outflow	STARKE et al., 1975c; STARKE, unpublished
Mouse	Vas deferens	10^{-6}	0	Incubated tissue. Effect on overflow of endogenous noradrenaline evoked by field stimulation at 0.2–15 Hz. No details	HENDERSON and HUGHES, 1974

Rabbit				Comment	Reference
	Heart	$10^{-10}-10^{-5}$	+	Perfused heart or heart preperfused with ^{14}C-noradrenaline. Effect on overflow of endogenous noradrenaline, ^{14}C-noradrenaline, or total ^{14}C evoked by accelerans nerve stimulation at 5 Hz. After overflow had been raised by cocaine, pronethalol, desipramine, protriptyline, or propranolol, angiotensin caused a further increase; after overflow had been raised by phenoxybenzamine, angiotensin had no further effect. No effect on basal outflow	Starke et al., 1969, 1970; Schümann et al., 1970a; Starke, 1970, 1971b; Starke and Schümann, 1972
	Heart	10^{-10}	+	Incubated atria. Effect on overflow evoked by field stimulation. No details	Blumberg et al., 1975b
	Coeliac artery Portal vein	$10^{-10}-10^{-7}$ $10^{-8}-2\times10^{-6}$	+ +	Strips preincubated with ^{3}H-noradrenaline. Effect on overflow of total ^{3}H evoked by field stimulation at 0.5–4 Hz. No effect at or above 5 Hz. After overflow had been raised by cocaine, angiotensin caused a further increase. Basal outflow from coeliac artery was also increased, that from portal vein only irregularly at high concentrations	Hughes and Roth, 1969, 1971
	Pulmonary artery	$10^{-10}-10^{-6}$	+	Strips preincubated with ^{3}H-noradrenaline. Effect on overflow of total ^{3}H evoked by field stimulation at 2 or 4 Hz was tested in presence of cocaine + corticosterone. No effect on basal outflow	Starke et al., 1975c; Taube et al., 1976; Starke, unpublished
	Vas deferens	$2\times10^{-7}-10^{-6}$	+	Incubated tissue. Effect on overflow of endogenous noradrenaline evoked by field stimulation at 0.5–5 Hz. No effect at higher frequencies. No details	Henderson and Hughes, 1974

+ = increase, − = decrease, 0 = no change. In each case, the effect measured is briefly defined under "Comment".

known. The available evidence is compatible with the view that angiotensin selectively enhances calcium-dependent secretion. (6) The enhancement is pronounced at low, but declines at high frequency of stimulation (HUGHES and ROTH, 1971; HENDERSON and HUGHES, 1974). (7) In an electrophysiological study on the vas deferens and uterine artery of the guinea pig, BELL (1972) found that angiotensin did not increase the e.j.p. evoked by a single nerve impulse. It did, however, increase the degree of facilitation of successive e.j.p.s during repetitive low frequency stimulation. If one assumes that facilitation in a train of pulses reflects increasing intraneuronal levels of calcium, until at high frequency the "release receptors" are saturated, then these results are compatible with the idea that angiotensin promotes the availability of calcium for the release process (STARKE, 1972c). (8) Catecholamines, acting on β-receptors, and angiotensin are the only endogenous compounds known to enhance release per impulse. It has been proposed that activation of presynaptic β-receptors leads to intraneuronal accumulation of cyclic AMP, the latter mediating the facilitation (ADLER-GRASCHINSKY and LANGER, 1975; cf. 9.2.1.). Angiotensin releases vasopressin possibly through stimulation of cyclic AMP formation (GAGNON et al., 1975). Could a presynaptic adenylate cyclase in noradrenergic nerves be involved in its facilitatory effect?

In several tissues angiotensin fails to increase noradrenaline release (Table 1). One possible reason for negative results is stimulation at high frequency. Moreover, angiotensin promotes the formation of prostaglandins for instance in the cat spleen (PESKAR and HERTTING, 1973). Prostaglandins inhibit noradrenaline secretion (6.1.). Under certain conditions they may mask the facilitation by angiotensin. Facilitation should then be revealed after blockade of prostaglandin biosynthesis. There may be other factors which prevent expression of the effect of angiotensin despite the existence of presynaptic receptors (11.4.). However, it is also possible that in some tissues the noradrenergic axon terminals lack angiotensin receptors.

Facilitation of per pulse release is probably the most important presynaptic component in angiotensin-induced enhancement of noradrenergic neuroeffector transmission, and may be of physiological significance (see STARKE, 1972c). The required concentrations occur in plasma, at least when renin secretion is high. There is evidence that the facilitatory effect of angiotensin generated by renin of renal origin assists in the circulatory compensation for acute hemorrhage (LIAO et al., 1975). Renin is probably also synthesized in blood vessel walls where, together with converting enzyme, it catalyzes the local production of angiotensin. Not only blood-borne but also locally formed angiotensin may presynaptically enhance noradrenergic transmission (MALIK and NASJLETTI, 1976). – The brain contains the components of the renin–angiotensin system (SEVERS and DANIELS-SEVERS, 1973). Noradrenergic neurones may participate in the central effects of angiotensin (SEVERS et al., 1971). A presynaptic effect on central noradrenergic fibers would therefore be of particular interest. The negative results obtained with slices of rat cerebral cortex and hypothalamus (Table 1) do not preclude that facilitation occurs under more physiologic conditions.

4.3. Effect on Biosynthesis

Angiotensin concentrations of 10^{-9} M and higher enhance the biosynthesis of ^{14}C-noradrenaline[3] from ^{14}C-tyrosine in isolated rat atria (BOADLE et al., 1969; DAVILA and KHAIRALLAH, 1971) and vasa deferentia, guinea-pig atria and portal veins, and rabbit portal veins (BOADLE-BIBER et al., 1972; ROTH, 1972). The increase caused by the peptide becomes significant after about 1 h of incubation and maximally amounts to about 50%. Angiotensin has no effect in guinea-pig or rabbit coeliac arteries and vasa deferentia and bovine splenic nerve trunks, and does not increase the synthesis of noradrenaline when dopa is used as precursor (DAVILA and KHAIRALLAH, 1971; BOADLE-BIBER et al., 1972). On the other hand, it accelerates the formation of ^{3}H-noradrenaline from ^{3}H-dopamine in rat heart slices (CHEVILLARD et al., 1971). It should be noted that all these experiments were performed without electrical nerve stimulation, so that the effects on synthesis cannot be secondary to facilitation of nerve impulse-induced release.

Angiotensin has been proposed to be a noradrenaline-releasing agent and/or to inhibit the neuronal uptake and re-uptake of noradrenaline. Therefore, one possible explanation for the increase of synthesis is that tyrosine hydroxylase is freed from end-product inhibition by axoplasmic noradrenaline. However, angiotensin does not augment synthesis in the rabbit coeliac artery, though it enhances basal noradrenaline outflow from this tissue at very low concentrations (p. 12). Moreover, in the experiments by BOADLE-BIBER et al. (1972) there was no correlation between increase in synthesis and increase in basal outflow. Finally, disinhibition of tyrosine hydroxylase would not account for the acceleration of synthesis from dopamine.

ROTH and HUGHES (1972) presented evidence that angiotensin stimulates the biosynthesis of enzyme(s) involved in noradrenaline formation. Increased formation of noradrenaline coincided with increased protein synthesis; dose–response curves for either effect were similar. Moreover, puromycin and cycloheximide had no effect on the basal rate of noradrenaline formation, but blocked the synthesis of new protein and abolished the angiotensin-induced *increase* in noradrenaline synthesis. Angiotensin may also enhance the conversion of dopamine to noradrenaline by promoting the synthesis of protein, probably dopamine-β-hydroxylase (CHEVILLARD et al., 1975). It appears to act at the translational level.

Angiotensin augments both the biosynthesis and the nerve impulse-evoked release of noradrenaline. It is not known whether the presynaptic receptors and the early biochemical events mediating the two effects are identical. Surprisingly, the peptide fails to increase biosynthesis in two tissues where it facilitates release (rabbit coeliac artery and vas deferens). If in a given tissue both synthesis and release are enhanced, the effects may cooperate. Increased biosynthesis may replenish the pool from which release is enhanced.

---

[3] In many experiments only total ^{14}C-catechols were determined.

4.4. Effect on Release of Other Substances

Angiotensin stimulates sympathetic ganglion cells and the homologous chromaffin cells of the adrenal medulla (reviews by TRENDELENBURG, 1967; HAEFELY, 1972; REIT, 1972; STARKE, 1972c). It depolarizes isolated chromaffin cells (DOUGLAS et al., 1967), and catecholamine release requires the presence of calcium (POISNER and DOUGLAS, 1966). Angiotensin acts on a specific receptor that is blocked by angiotensin antagonists. The receptor may differ from the angiotensin receptor in smooth muscle (PEACH and OBER, 1974).

From the scanty evidence available it appears that angiotensin has analogous effects on cholinergic and noradrenergic nerve endings. It enhances the stimulation-evoked, but not the basal overflow of acetylcholine from sympathetic and parasympathetic ganglia (PANISSET, 1968). The stimulation-evoked overflow of acetylcholine from the guinea-pig ileum (PANISSET, 1968) and the outflow from the cerebral cortex of cats (ELIE and PANISSET, 1970) are also increased; the site of action in the neuronal networks of these tissues is difficult to define.

4.5. Conclusion

The terminals of some noradrenergic neurones appear to be endowed with receptors for angiotensin. Activation of these presynaptic receptors by low concentrations of angiotensin (10^{-10}–10^{-9} M) triggers two parallel changes. On the one hand, the release of noradrenaline per impulse is increased; on the other hand, the biosynthesis of the transmitter is accelerated. Two further, often not sufficiently distinguished effects, namely a direct releasing action and inhibition of neuronal uptake, are probably of minor significance at low concentrations. Facilitation of nerve impulse-evoked release is a specific effect. It is not mediated by presynaptic prostaglandin, α- or β-receptors and is antagonized by peptide analogues lacking intrinsic activity. Enhancement of release and biosynthesis does not occur in all noradrenergically innervated tissues. In particular, attempts to demonstrate presynaptic angiotensin receptors on central noradrenergic neurones have so far been unsuccessful. The reason for the tissue differences is not known.

Present evidence is compatible with the view that angiotensin selectively facilitates calcium-dependent release processes, perhaps by making more calcium available for stimulus-secretion coupling. The increase of biosynthesis appears to be secondary to enhanced synthesis of new protein, including enzymes catalyzing noradrenaline formation. In tissues where both release and synthesis are promoted, the effects may cooperate.

Angiotensin augments the release and biosynthesis of noradrenaline at concentrations which occur in plasma in vivo, at least when renin secretion is high. Renin is secreted in response to a fall of blood pressure or volume. Under these conditions, enhancement of sympathetic neuro-effector transmission may help to ensure adequate circulation. However, evidence for a physiologic or pathophysiologic operation of the presynaptic effects of angiotensin is scanty.

5. Acetylcholine

According to present knowledge, cholinergic drugs exert two main effects on noradrenergic nerve endings: (1) by an action on nicotine receptors they release the transmitter, and (2) by an action on muscarine receptors they inhibit release evoked by nerve impulses (see reviews by FERRY, 1966; MUSCHOLL, 1970, 1973a and b; KOSTERLITZ and LEES, 1972).

5.1. Release Induced by Nicotinic Agonists

5.1.1. Basic Evidence

Early observations indicated that acetylcholine (mostly in the presence of atropine) and nicotine had effects on sympathetically innervated organs that mimicked the effect of catecholamines. Release of "an epinephrine-like substance" by acetylcholine and nicotine was first demonstrated by HOFFMANN et al. (1945) in the cat heart. Effects of cholinergic agonists on the basal outflow of noradrenaline are summarized in Table 2. Outflow is increased by nicotine and DMPP, drugs which predominantly activate nicotine receptors. It is not or minimally increased by methacholine and oxotremorine, drugs which predominantly activate muscarine receptors. Moreover, the overflow evoked by the nicotinic agonists is reduced by the antinicotinic agents hexamethonium, chlorisondamine, or pempidine; it is not changed or, in the case of acetylcholine, even enhanced by atropine. Clearly, the effect is mediated by nicotine receptors.

Some results appear inconsistent with this conclusion. High doses of atropine diminish the effect of DMPP and acetylcholine (HERTTING and WIDHALM, 1965; LINDMAR et al., 1968); however, at these doses atropine may block nicotine receptors (LINDMAR et al., 1968), or may interfere with a step following the nicotinic receptor activation. In guinea-pig atria, the overflow of noradrenaline evoked by nicotine is antagonized by hexamethonium, but overflow evoked by DMPP is not (BHAGAT et al., 1967); in some tissues, DMPP has a tyramine-like, hexamethonium-resistant effect (cf. MUSCHOLL, 1970). The catecholaminergic neurones of the chicken heart are unique in that neither acetylcholine (plus atropine) nor nicotine increases basal transmitter outflow (LINDMAR and DESANTIS, 1974; LÖFFELHOLZ, 1975).

The magnitude of the overflow evoked by nicotinic drugs precludes any possibility that it reflects inhibition of the inactivation of spontaneously released noradrenaline within the tissue. For instance, in rabbit hearts maximally effective concentrations of acetylcholine (plus atropine), nicotine, or DMPP induce overflow of about 20% of the total noradrenaline content within 2 min, so that the outflow per min rises up to 2,000-fold (LÖFFELHOLZ, 1970a). It may be added that most cholinergic drugs do not substantially inhibit the neuronal uptake of noradrenaline (LINDMAR et al., 1968; NEDERGAARD and BEVAN, 1969a; ALLEN et al., 1972b, 1973; WESTFALL and BRASTED, 1972; BALFOUR, 1973; GOODMAN, 1974). Moreover, isotope experiments indicate that the increase of the outflow of noradrenaline does not result from a decrease of its metabolic degradation. Nicotinic drugs release noradrenaline from the nerve endings.

5.1.2. Mechanism

BURN and GIBBONS (1964b) suggested that nicotine and acetylcholine release noradrenaline from nerve endings by promoting the entry of calcium. In fact release evoked by nicotinic agonists is reduced in low calcium media (Table 2). In guinea-pig atria, the effect of DMPP is not affected by changes in calcium concentration, confirming its tyramine-like mechanism of action in this organ (BHAGAT et al., 1967). For rat brain slices, both calcium-dependence (WESTFALL, 1974a) and calcium-independence (GOODMAN, 1974) of the effect of nicotine have been reported.

When nicotinic drugs are injected or infused into sympathetically innervated tissues, they induce antidromic impulses in postganglionic sympathetic fibers (FERRY, 1963; CABRERA et al., 1966; DAVEY et al., 1968; HAEUSLER et al., 1968, 1969a and b; KRAUSS et al., 1970; BEVAN and HAEUSLER, 1975). The firing is inhibited by antinicotinic drugs, but not by atropine. FERRY (1963) suggested that, if acetylcholine excited the postganglionic fibers somewhere near the nerve endings, then the impulses might propagate along the fibers in both directions, and the orthodromic pulses might cause the release of noradrenaline. Experiments with tetrodotoxin, which suppresses action potentials by inhibiting the sodium entry associated with their rising phase, partly confirm this view. In several tissues tetrodotoxin blocks the postsynaptic sympathomimetic effects of nicotine or acetylcholine (BELL, 1968; ENDOH et al., 1970).

However, the induction of action potentials is only one factor. In the cat heart, relatively low concentrations of acetylcholine trigger antidromic firing throughout the period of infusion, but the release of noradrenaline is small; at higher concentrations, or at low concentrations in the presence of atropine, the duration of discharges is limited to the first few seconds of infusion, yet release is markedly increased; moreover, the omission of calcium leads to a moderate increase in discharge amplitude, but abolishes release (HAEUSLER et al., 1968). In the cat heart and spleen, tetrodotoxin prevents acetylcholine from eliciting antidromic impulses, but not from releasing noradrenaline (HAEUSLER et al., 1968; KRAUSS et al., 1970; cf. analogous results by SU and BEVAN, 1970b; WESTFALL and BRASTED, 1972; FOZARD and MWALUKO, 1976). FURCHGOTT et al. (1975) have shown that tetrodotoxin strongly inhibits the sympathomimetic effect of low doses, but only slightly inhibits that of high doses of nicotine. The dose dependence may help to explain the contradictory findings with tetrodotoxin.

It seems, therefore, that two mechanisms contribute to the release of noradrenaline by nicotinic drugs, their relative importance depending on dose, tissue, or other experimental conditions (cf. FERRY, 1966, p. 426; KRAUSS et al., 1970; KOSTERLITZ and LEES, 1972, p. 778; FURCHGOTT et al., 1975). One is the induction of tetrodotoxin-sensitive action potentials, which lead to calcium entry and, ultimately, secretion. The second mechanism also leads to calcium entry and secretion, but does not depend on action potentials and persists in the presence of tetrodotoxin; its nature is not known, but can be tentatively inferred from analogies.

The action of nicotinic drugs on noradrenergic nerve endings is probably analogous to their effect on voluntary muscle endplates, autonomic ganglia

Table 2. Effect of cholinergic agonists on basal transmitter outflow from noradrenergically innervated tissues

Species	Tissue	Drug	Concentration (M)	Effect	Comment	References
Chicken	Heart	Acetylcholine (+atropine)	$2 \times 10^{-4} - 2 \times 10^{-3}$	0	Perfused heart. Effect on outflow of endogenous catecholamines	Lindmar and DeSantis, 1974; Löffelholz, 1975
		Nicotine	$10^{-4} - 10^{-3}$	0		
Dog	Arteries and veins	Acetylcholine	$6 \times 10^{-7} - 10^{-6}$	0	Strips preincubated with ^3H-noradrenaline. Effect on outflow of total ^3H	Vanhoutte et al., 1973; Vanhoutte, 1974
Cat	Heart	Acetylcholine	$5 \times 10^{-5} - 6 \times 10^{-3}$	+	Perfused heart. Effect on outflow of endogenous noradrenaline. Effect of acetylcholine was decreased by hexamethonium, pilocarpine, bretylium, tetracaine, or omission of calcium, not changed by tetrodotoxin, and enhanced by atropine	Haeusler et al., 1968, 1969a and b; Bevan and Haeusler, 1975
		DMPP	$3 \times 10^{-5} - 6 \times 10^{-5}$	+		
		Nicotine	$10^{-5} - 10^{-3}$	+		
	Spleen	Acetylcholine	Injection of 0.05–3 mg	+	Blood- or saline-perfused spleen. Effect on outflow of endogenous noradrenaline or previously stored ^3H-noradrenaline. Effect of acetylcholine or DMPP was not changed by atropine, except for a decrease at high concentrations, or tetrodotoxin, and was decreased by hexamethonium, chlorisondamine, pempidine, or high concentrations of bretylium	Brandon and Boyd, 1961; Blakeley et al., 1963; Hertting and Widhalm, 1965; Fischer et al., 1966; Davey et al., 1968; Krauss et al., 1970; Kirpekar et al., 1972a
		DMPP	Injection of 50–150 µg	+		
		Carbachol	$5 \times 10^{-6} - 5 \times 10^{-5}$	0		
	Nictitating membrane	Acetylcholine	3×10^{-5}	0	Tissue preincubated with ^3H-noradrenaline. Effect on outflow of total ^3H	Langer, 1970
Rat	Cerebral cortex Hypothalamus Striatum	Acetylcholine	$10^{-8} - 10^{-5}$	0	Slices preincubated with labeled noradrenaline. Effect on outflow of labeled noradrenaline or total radioactivity. Effect of nicotine was decreased by acetylcholine or hexamethonium	Hall and Turner, 1972; Goodman, 1974; Westfall, 1974a; Starke, unpublished
		Oxotremorine	$10^{-6} - 10^{-4}$	0		
		Nicotine	$3 \times 10^{-4} - 2 \times 10^{-2}$	+		

Table 2 (continued)

Species	Tissue	Drug	Concentration (M)	Effect	Comment	References
Rat	Hypothalamus Hippocampus	Carbachol	5×10^{-5}	−	Synaptosomes preincubated with ^{14}C-noradrenaline. Effect on loss of total ^{14}C from particles	BALFOUR, 1973
		Carbachol	5×10^{-4}	0		
		Nicotine	$5 \times 10^{-5}-5 \times 10^{-4}$	+		
Guinea pig	Heart	Acetylcholine	$10^{-6}-10^{-4}$	0	Perfused heart or heart preperfused with ^{3}H-noradrenaline. Effect on outflow of endogenous noradrenaline, ^{3}H-noradrenaline or total ^{3}H. Effect of nicotine was not changed by tetrodotoxin; it was decreased by acetylcholine, methacholine, hexamethonium, cocaine, lidocaine, piperocaine, bretylium, morphine, metanephrine, phenoxybenzamine, colchicine, cytochalasin B, prostaglandins E_1 and E_2, and omission of calcium	LINDMAR et al., 1968; WESTFALL and BRASTED, 1972, 1973, 1974; SORIMACHI et al., 1973; WESTFALL and HUNTER, 1974
		Acetylcholine (+atropine)	$2 \times 10^{-4}-10^{-3}$	+		
		Methacholine	$10^{-6}-10^{-3}$	+		
		Nicotine	10^{-5}	0		
		Nicotine	10^{-4}, or injection of 100 µg	+		
	Heart	Nicotine	6×10^{-5}	+	Atria preincubated with ^{3}H-noradrenaline. Effect on outflow of total ^{3}H or on loss of ^{3}H-noradrenaline from tissue. Effect of nicotine, but not that of DMPP, was enhanced in high calcium media and decreased by hexamethonium	BHAGAT et al., 1967; ALLEN et al., 1972b
		DMPP	$3 \times 10^{-5}-10^{-4}$	+		
	Taenia caeci	Nicotine	10^{-4}	+	Tissue preincubated with ^{3}H-noradrenaline. Effect on outflow of total ^{3}H	KUCHII et al., 1973
Rabbit	Heart	Acetylcholine	$6 \times 10^{-9}-6 \times 10^{-6}$	0	Perfused heart. Effect on outflow of endogenous noradrenaline and/or several false transmitters. The concentrations of acetylcholine (+atropine), nicotine and DMPP are those which caused a half-maximal overflow (LÖFFELHOLZ, 1970a). In some experiments, methacholine caused a small increase which was not changed by atropine (DUBEY et al., 1975). Effect of	RICHARDSON and WOODS, 1959; LINDMAR and MUSCHOLL, 1961; MUSCHOLL and MAÎTRE, 1963; LINDMAR et al., 1967a and b, 1968;
		Acetylcholine (+atropine)	$6 \times 10^{-5}-2 \times 10^{-3}$	+		
			10^{-4}	+		
		Methacholine	up to 10^{-4}	0		
		Carbachol	up to 4×10^{-5}	0		
		Pilocarpine	up to 10^{-3}	0		
		Furtrethonium	up to 2×10^{-4}	0		

				Comment	References
	Oxotremorine	up to 10^{-5}	0	acetylcholine, but not that of DMPP, was enhanced by atropine. Effect of DMPP was decreased by muscarinic drugs; this inhibition was reversed by atropine. Effect of DMPP was also decreased by hexamethonium, cocaine, tetracaine, psicaine, guanethidine, oxymetazoline, phentolamine, various inhalational anesthetics, alcohols, and omission of calcium	LÖFFELHOLZ, 1967, 1970a and b; LÖFFELHOLZ and MUSCHOLL, 1969; KILBINGER et al., 1971; FOZARD and MUSCHOLL 1972; MUSCHOLL, 1973b; GÖTHERT, 1974; STARKE and MONTEL, 1974; DUBEY et al., 1975; GÖTHERT et al., 1976
	Nicotine	2×10^{-5}	+		
	DMPP	2×10^{-5}	+		
Aorta	Acetylcholine	10^{-3}	+	Segments preincubated with ^{3}H-noradrenaline. Effect on outflow of total ^{3}H.	KIRAN and KHAIRALLAH, 1969
Ear artery	Acetylcholine	10^{-5}	+	Arteries preincubated with ^{3}H-noradrenaline. Effect on outflow of total ^{3}H	ALLEN et al., 1975
Pulmonary artery	Acetylcholine	10^{-7}–10^{-5}	0	Strips preincubated with ^{3}H-noradrenaline. Effect on outflow of ^{3}H-noradrenaline or total ^{3}H. Effect of nicotine was decreased by cocaine, desipramine, and phenoxybenzamine	SU and BEVAN, 1970b; NEDERGAARD and SCHROLD, 1973; STARKE, unpublished
	Nicotine	10^{-4}	+		
Portal vein	Acetylcholine	7×10^{-6}–7×10^{-5}	+	Strips preincubated with ^{3}H-noradrenaline. Effect on outflow of total ^{3}H. The increase was minimal	HUGHES and ROTH, 1971

+ = increase, − = decrease, 0 = no change. In each case, the effect measured is briefly defined under "Comment".

and, in particular, the adrenal medulla (SMITH and WINKLER, 1972). Both in chromaffin cells and, under certain conditions, in nerve endings, nicotinic drugs evoke secretion without the generation of action potentials. In chromaffin cells, the drug–receptor interaction elicits a nonselective increase in ion conductance. The ensuing depolarization enhances the inflow of calcium. However, there is little doubt that the depolarization causes only part of the calcium entry. In addition to the calcium gates opened by depolarization there are probably gates which are more directly opened by nicotinic agonists. In nerve endings, the sequence of events may be similar. In contrast to the adrenal medulla, the depolarization induced by nicotinic drugs sets up action potentials which contribute to the overall release of noradrenaline (first component). However, even when spike generation is prevented, the local, nonpropagated depolarization, and some specific membrane action of the drugs (HAEUSLER et al., 1968), lead to calcium entry and secretion (second component). It seems possible that either component of release is exocytotic.

5.1.3. Time Course

Postsynaptic sympathomimetic effects of nicotinic drugs are transient, disappearing despite continued exposure. The tissues are then refractory for some time both to the same drug and to other nicotinic agonists. On washout, sensitivity slowly recovers (e.g., STEINSLAND and FURCHGOTT, 1975). This time course of the effector response mirrors the time course of transmitter release. The outflow of noradrenaline caused by continuous exposure to acetylcholine, nicotine, or DMPP declines rapidly (LINDMAR et al., 1967a; LÖFFELHOLZ, 1967; SU and BEVAN, 1970b; NEDERGAARD and SCHROLD, 1973). LÖFFELHOLZ (1970a) presented evidence that in the perfused rabbit heart nicotinic drugs release noradrenaline in an "explosive" manner during the first 5–10 sec of their infusion only; the rapid decay could not be attributed to exhaustion of transmitter stores. On the other hand, there are divergent observations. Nicotine-induced release of noradrenaline from central neurones has been reported to be maintained (GOODMAN, 1974; WESTFALL, 1974a; see, however, HALL and TURNER, 1972). HAEUSLER et al. (1968) found a fairly constant outflow of noradrenaline from cat hearts throughout the 1-min periods of acetylcholine or DMPP infusion. It seems possible that in their experiments a slow washout of the extracellular space obscured the explosive nature of the secretion. However, as mentioned above, low concentrations of the agonists induce a persistent antidromic firing. It would be of interest to know whether this continuous electrical activity is accompanied by continuous release.

The refractory state has been termed "autoinhibition" or "desensitization" (LÖFFELHOLZ, 1970a; STEINSLAND and FURCHGOTT, 1975). Similar phenomena are known for the voluntary muscle endplate membrane and the membrane of autonomic ganglion cells (reviews by TRENDELENBURG, 1967; HAEFELY, 1972; HUBBARD, 1973; HUBBARD and QUASTEL, 1973). Both of these membranes become unresponsive to nicotinic agents during prolonged exposure. Early during this phase, the membranes are depolarized. However, after times varying from seconds to several minutes, they are repolarized or even hyperpolarized. When noradrenergic nerve endings are refractory to nicotinic agonists, the release of transmitter by, and the postsynaptic response to, electrical nerve stimulation are unchanged or even enhanced (LÖFFELHOLZ and MUSCHOLL, 1969; NEDER-GAARD and BEVAN, 1969 a and b with a discussion of earlier publications; LÖFFEL-HOLZ, 1970 b; SU and BEVAN, 1970 b; ROSS, 1973; STEINSLAND and FURCHGOTT, 1975). LÖFFELHOLZ (1970 b) concluded that "a marked depolarization persisting for more than one minute can be ruled out" (see, on the other hand, HAEUSLER et al., 1968; BEVAN and HAEUSLER, 1975).

It has been proposed that when the acetylcholine receptors of the skeletal muscle endplate are exposed to nicotinic drugs for a relatively long time, they are converted to an inactive form, or, alternatively, some essential step normally following the agonist-receptor interaction is inactivated (see HUBBARD, 1973; HUBBARD and QUASTEL, 1973). The membrane is then repolarized, but is unresponsive. In noradrenergic nerve endings, an analogous "receptor inactivation" may follow the initial activation (STEINSLAND and FURCHGOTT, 1975).

5.1.4. Drug Effects

As discussed above, the release of noradrenaline by nicotinic agonists is blocked by antinicotinic agents. It is inhibited by tetrodotoxin only under certain experimental conditions. Like other calcium-dependent release processes, release by nicotinic drugs is sensitive to some of the presynaptic modulators discussed in this review. It is decreased by muscarinic compounds (Table 2). When secretion is elicited by acetylcholine, atropine causes a large increase, since it blocks the muscarinic component; atropine does not augment secretion evoked by DMPP, which lacks muscarinic activity. Release by nicotine is also reduced by PGE_1 and PGE_2 (WESTFALL and BRASTED, 1974). In the rabbit heart, release by DMPP is diminished by the α-receptor agonist oxymetazoline; however, the degree of inhibition is greater than would be expected from the activation of presynaptic α-receptors. Moreover, phentolamine reduces rather than enhances release by DMPP. Apparently, the effect on presynaptic α-receptors is overshadowed by a "nonspecific" depression (STARKE and MONTEL, 1974).

Apart from these agents with a fairly well-known mode of action, an amazing variety of apparently unrelated drugs interfere with the release of noradrenaline by nicotinic agonists. Perhaps the earliest example is the finding that procaine or cocaine at 1/20 of their local anesthetic threshold concentration block the pilomotor response of the human skin to intracutaneous acetylcholine or nicotine (COON and ROTHMAN, 1940). Inhibitory agents include lidocaine, piperocaine, cocaine, tetracaine, psicaine, guanethidine, bretylium, desipramine, metanephrine, phenoxybenzamine, phentolamine, morphine, and various inhalation anesthetics and aliphatic alcohols (Table 2). Some of these drugs inhibit the neuronal uptake of noradrenaline at doses which interfere with nicotine-evoked release. Therefore, SU and BEVAN (1970 b) suggested "that the sympathomimetic action of nicotine ... depends upon an intact noradrenaline uptake mechanism" and may be linked to an entry of nicotine into the terminals "by a process related to the uptake mechanism for noradrenaline". BEVAN and SU (1972) later showed that, in rabbit aortic strips, nicotine is concentrated in the inner third of the adventitia, the region of the terminal sympathetic plexus, and that this accumulation is reduced by 10^{-4} M of cocaine, desipramine, guanethidine, or phenoxybenzamine, but not by d-tubocurarine, pentolinium, or hexamethonium. None of these agents blocked the small accumulation in the nerve-free media.

However, several considerations make it unlikely that uptake of nicotine by the neuronal noradrenaline transport mechanism is a prerequisite for release. Since nicotine is to a large part nonionized at the pH of the interstitial fluid, and since the nonionized form is very lipid-soluble, it penetrates readily into nerve cells. A specific uptake into noradrenergic nerve terminals is questionable (WESTFALL and BRASTED, 1972). The experiments of BEVAN and SU (1972) do not exclude a preferential *extraneuronal* binding of nicotine in the inner adventitia, which might also be reduced by the high concentrations of cocaine-like agents. The fact that in most tissues nicotine causes no or minimal inhibition of the uptake of noradrenaline (p. 19) argues against a common transport system. Moreover, under certain conditions phenoxybenzamine and metanephrine block the noradrenaline uptake mechanism without impairing the release of noradrenaline by nicotine (WESTFALL and BRASTED, 1972; FURCHGOTT et al., 1975).

Nicotine and its congeners release noradrenaline by the same mechanism. Thus, if nicotine acted at an intraneuronal site, then all nicotinic agents would be taken up. In this case, the antinicotinic drugs, which do not inhibit the uptake of nicotine, would also exert their antagonistic action intracellularly (BEVAN and SU, 1972). It seems rather unlikely that all nicotinic agonists enter nerve endings by the noradrenaline transfer system, and that drugs like cocaine and tetracaine inhibit release evoked by either acetylcholine, nicotine, and DMPP at the transport site. A strong argument against this possibility is that acetylcholine, like nicotine, does not cause any substantial inhibition of noradrenaline uptake (p. 19).

In conclusion, the mode of action of the "nonspecific" inhibitors—if they share a common mode—remains unknown. It seems possible that nicotinic drugs act on receptors on the surface of the nerve endings, and that the inhibitors interfere with some event that *follows* this interaction, as for instance the local depolarization and/or the entry of calcium (HAEUSLER et al., 1969b; WESTFALL and BRASTED, 1972; FOZARD and MUSCHOLL, 1974; STARKE and MONTEL, 1974; FURCHGOTT et al., 1975). In fact, such a possibility was not denied by BEVAN and SU (1972).

5.2. Effect on Release by Nerve Impulses

In 1935 BRÜCKE reported that subcutaneous injection of acetylcholine in the tail of the cat greatly diminished the pilomotor response of hairs at the injection site to sympathetic nerve stimulation; the response to adrenaline was unchanged. He concluded that „das Acetylcholin an den sympathischen Nervenendigungen angreift; vielleicht hemmt es dort die Sympathinproduktion, so daß die Haarmuskeln auf den nervösen Reiz nicht reagieren können". This is perhaps the first description and correct interpretation of muscarinic inhibition of noradrenaline release; the reservation "perhaps" is necessary, since BRÜCKE states, without giving details, that the inhibitory effect of acetylcholine was not abolished by atropine. The concept of, and unequivocal evidence for, presynaptic muscarine receptors originate from studies of LÖFFELHOLZ et al. (1967) and LINDMAR et al. (1968), who proceeded from the fortuitous finding that atropine enhances the overflow of noradrenaline elicited by infusion of acetylcholine into the rabbit heart. They proposed "that the peripheral adrenergic nerve fibre contains inhibitory muscarine receptors in addition to the well-known excitatory nicotinic receptors mediating noradrenaline release" (LINDMAR et al., 1968). In contrast to the extensive evidence for muscarinic inhibition (5.2.1.), evidence for a cholinergic facilitation of release evoked by nerve impulses is sparse (5.2.2.).

5.2.1. Muscarinic Inhibition

Muscarinic drugs reduce the stimulation-evoked overflow of noradrenaline from chicken heart, dog blood vessels, cat spleen and nictitating membrane, guinea-pig vas deferens, and rabbit heart, ear artery, and pulmonary artery (Table 3). However, there are some divergent results. Muscarinic inhibition has been found in the perfused heart of the guinea pig (LANGLEY and GARDIER, 1974; cf. muscarinic inhibition of release evoked by nicotinic drugs: LINDMAR et al., 1968; WESTFALL and HUNTER, 1974), but not in incubated guinea-pig atria (STORY et al., 1975). Acetylcholine does not change noradrenaline overflow from rat cerebral cortex slices evoked by electrical stimulation (STARKE et al., 1975c), though it reduces overflow from hypothalamic and cerebellar slices evoked by nicotine or high potassium concentrations (WESTFALL, 1974a and b). The reason for these differences is not known. It cannot be excluded that noradrenergic nerve endings in adjoining tissues, as for instance cardiac atria and ventricles, differ in their being equipped with muscarine receptors. Drugs that lack muscarinic activity mostly fail to reduce the stimulation-evoked overflow of noradrenaline (Table 3). The block that occurs after prolonged exposure to DMPP is not a cholinergic effect and resembles the block caused by guanethidine and bretylium (MUSCHOLL, 1970).

The decrease of overflow reflects inhibition of transmitter release rather than enhancement of its degradation or retention within the tissue, for the following reasons (cf. Table 3). (1) Muscarinic drugs do not increase the neuronal uptake of noradrenaline (see references on p. 19; there is one report that pilocarpine augments the uptake of noradrenaline into synaptosomes of the rat hippocampus, but not hypothalamus: BALFOUR, 1973). (2) In tissues treated with labeled noradrenaline, muscarinic drugs diminish the stimulation-evoked overflow of total radioactivity, i.e., of labeled noradrenaline plus labeled metabolites; they do not decrease the outflow of noradrenaline by promoting its biotransformation. (3) Inhibition is retained after neuronal and extraneuronal uptake have been blocked. (4) In perfused guinea-pig hearts, acetylcholine decreases the stimulation-evoked overflow of both noradrenaline and dopamine-β-hydroxylase, suggesting inhibition of exocytotic release (LANGLEY and GARDIER, 1974).

The receptors mediating the inhibition are muscarine receptors, since they are selectively activated by muscarinic agonists and blocked by atropine, but not by hexamethonium (Table 3). FOZARD and MUSCHOLL (1972) compared

Table 3. Effect of cholinergic agonists on transmitter overflow evoked by electrical stimulation of noradrenergic nerves

Species	Tissue	Drug	Concentration (M)	Effect	Comment	References
Chicken	Heart	Acetylcholine	$10^{-5}-10^{-4}$	−	Perfused heart. Effect on overflow of catecholamines evoked by accelerans nerve stimulation. Effect was antagonized by atropine	LÖFFELHOLZ, 1975
Dog	Arteries and veins	Acetylcholine	$3 \times 10^{-7}-3 \times 10^{-6}$	−	Strips preincubated with ^{3}H-noradrenaline. Effect on overflow of ^{3}H-noradrenaline and total ^{3}H evoked by field stimulation at 2 or 5 Hz	VANHOUTTE et al., 1973; VANHOUTTE, 1974
Cat	Spleen	Acetylcholine Carbachol	6×10^{-6} 5×10^{-6} 5×10^{-5}	− 0 −	Perfused spleen. Effect on overflow of endogenous noradrenaline or previously stored ^{3}H-noradrenaline evoked by splenic nerve stimulation at 1–30 Hz. Effect was inversely related to frequency and antagonized by atropine	KIRPEKAR et al., 1972a, 1975
	Nictitating membrane	Acetylcholine		−	Tissue preincubated with ^{3}H-noradrenaline. Effect on overflow of total ^{3}H evoked by field stimulation. Effect was antagonized by atropine	LANGER, personal communication
Rat	Cerebral cortex	Acetylcholine Oxotremorine	$10^{-8}-10^{-5}$ $10^{-6}-10^{-5}$	0 0	Slices preincubated with ^{3}H-noradrenaline. Effect on overflow of total ^{3}H evoked by field stimulation at 3 Hz.	STARKE et al., 1975c; STARKE, unpublished
Guinea pig	Heart	Acetylcholine		−	Perfused heart. Effect on overflow of noradrenaline evoked by accelerans nerve stimulation. No details	LANGLEY and GARDIER, 1974
	Heart	Acetylcholine	$10^{-10}-10^{-5}$	0	Atria preincubated with ^{3}H-noradrenaline. Effect on overflow of total ^{3}H evoked by field stimulation. No details	STORY et al., 1975
	Vas deferens	Acetylcholine	$10^{-6}-5 \times 10^{-6}$	−	Tissue preincubated with ^{3}H-noradrenaline. Effect on overflow of total ^{3}H evoked by field stimulation at 1 Hz was tested in presence of desipramine+normetanephrine. Effect was antagonized by atropine	STJÄRNE, 1975c

Species	Tissue	Substance	Concentration	Effect	Comment	Reference
Rabbit	Heart	Acetylcholine	6×10^{-11}–6×10^{-9}	0	Perfused heart. Effect on overflow of endogenous noradrenaline and/or previously stored α-methyladrenaline evoked by accelerans nerve stimulation at 3 or 10 Hz. Concentrations of methacholine, carbachol, pilocarpine, furtrethonium, and oxotremorine are those which caused a half-maximal inhibition (FOZARD and MUSCHOLL, 1972). Muscarinic inhibitory effects were antagonized by atropine. Effect of methacholine was not changed by indometacin. Acetylcholine (+atropine) had no effect when added 4–7 min before stimulation, but caused an increase when added 1 min before stimulation. DMPP had no effect when added 3 min before stimulation, but caused a decrease when added 7 min before stimulation	LÖFFELHOLZ and MUSCHOLL, 1969; LÖFFELHOLZ, 1970b; FOZARD and MUSCHOLL, 1972; MUSCHOLL, 1973a and b; FUDER and MUSCHOLL, 1974
		Methacholine	6×10^{-8}–6×10^{-5}	−		
		Carbachol	7×10^{-6}	−		
		Pilocarpine	10^{-5}	−		
		Furtrethonium	8×10^{-4}	−		
		Oxotremorine	8×10^{-5}	−		
			10^{-6}			
		Acetylcholine (+atropine)	6×10^{-5}–2×10^{-4}	0 or +		
		Nicotine	4×10^{-5}	0		
		DMPP	3×10^{-6}–3×10^{-5}	0 or −		
	Heart	Acetylcholine	10^{-12}–10^{-11}	+	Atria preincubated with ³H-noradrenaline. Effect on overflow of total ³H evoked by field stimulation. No details	STORY et al., 1975
			$>10^{-7}$	−		
	Ear artery	Acetylcholine	10^{-12}	0	Arteries preincubated with ³H-noradrenaline. Effect on overflow of total ³H evoked by periarterial nerve stimulation at 5 Hz. Inhibitory effect of higher concentrations was antagonized by atropine; increase caused by lower concentrations was not changed by atropine, hexamethonium, or neostigmine	ALLEN et al., 1972a, 1975; STORY et al., 1975
			10^{-11}–10^{-10}	+		
			10^{-9}–10^{-8}	0		
			3×10^{-8}–10^{-5}	−		
	Ear artery	Acetylcholine	2×10^{-6}	−	Arteries pre-superfused with ³H-noradrenaline. Effect on overflow of total ³H evoked by field stimulation at 10 Hz	STEINSLAND et al., 1973
	Pulmonary artery	Acetylcholine	10^{-11}–10^{-8}	0	Strips preincubated with ³H-noradrenaline. Effect on overflow of total ³H evoked by field stimulation at 2 Hz was tested either in absence or presence of cocaine +corticosterone. Inhibitory effect was antagonized by atropine	STARKE et al., 1975c; TAUBE et al., 1976; STARKE, unpublished
			10^{-7}–10^{-5}	−		

+ = increase, − = decrease, 0 = no change. In each case, the effect measured is briefly defined under "Comment".

pre- and postsynaptic effects of nine muscarinic agonists in the rabbit heart. They concluded that the muscarine receptors mediating inhibition of atrial tension development, ventricular rate, and noradrenaline release were similar. From a postsynaptic response study in the rabbit ear artery, Steinsland et al. (1973) arrived at an analogous conclusion. — McN-A-343 (4-(m-chlorophenylcarbamoyloxy)-2-butynyltrimethylammonium chloride) and AHR 602 (N-benzyl-3-pyrrolidyl acetate methobromide) are thought to activate selectively the depolarizing muscarine receptors of sympathetic ganglion cells (Trendelenburg, 1967; Haefely, 1972). In the rabbit heart, they increased the stimulation-evoked overflow of noradrenaline, probably through inhibition of re-uptake. However, even when allowance was made for the effect on uptake, no evidence for an agonist action on presynaptic muscarine receptors was found, indicating that these receptors do not correspond to the ganglionic excitatory muscarine receptors (Fozard and Muscholl, 1972, 1974). It should be noted, on the other hand, that McN-A-343 inhibits noradrenaline release in the rabbit ear artery (Allen et al., 1974).

The mechanism of muscarinic inhibition is not known. Some information can be derived from the following observations. (1) The inhibition probably does not reflect an action on noradrenaline storage granules, since acetylcholine does not change the outflow of noradrenaline from isolated granules of the splenic nerve trunk (Euler and Lishajko, 1961). (2) The inhibition does not result from a decrease of noradrenaline biosynthesis, since release of exogenous, previously taken up noradrenaline is also depressed. (3) The inhibition is not changed by indometacin at concentrations which block the formation of prostaglandins. Prostaglandins are not involved as a chemical link (Fuder and Muscholl, 1974). (4) Muscarinic compounds inhibit release evoked by electrical stimulation (Table 3), nicotinic drugs (Table 2), and high potassium concentrations (Muscholl, 1973a and b; Westfall, 1974b; Dubey et al., 1975; Vanhoutte and Verbeuren, 1975). In isolated rabbit hearts, secretion caused by high potassium-low sodium solution declines exponentially in two phases. Only the first phase requires calcium in the perfusion fluid; only this calcium-dependent component is inhibited by methacholine (Muscholl et al., 1975). Similarly, noradrenaline release evoked by sodium deprivation alone (at normal potassium) is partly calcium-dependent in the rabbit heart, and only the calcium-dependent fraction is methacholine-sensitive (Dubey et al., 1975). Muscarinic agonists affect neither basal outflow (Table 2), nor release evoked by tyramine (Löffelholz and Muscholl, 1969; Muscholl, 1973a and b; Vanhoutte et al., 1973; Vanhoutte, 1974). The results indicate that muscarinic agents selectively inhibit calcium-dependent release. (5) Lowering the calcium concentration enhances the inhibitory effect of methacholine (Muscholl, 1973c; Dubey et al., 1975). (6) Muscarinic inhibition is inversely related to the frequency of nerve stimulation (Kirpekar et al., 1975; postsynaptic response studies by Rand and Varma, 1970; Steinsland et al., 1973).

In conclusion, all these data are compatible with the idea that the activation of presynaptic muscarine receptors decreases the availability of calcium for stimulus-secretion coupling (Muscholl, 1973c; Dubey et al., 1975; Kirpekar et al., 1975). Both an increase of external calcium and intraneuronal accumulation of calcium during high frequency stimulation would overcome this inhibition. However, even if one accepts this hypothesis, the steps that lead to reduction of calcium availability remain unknown. It has been suggested that muscarinic drugs hyperpolarize the membrane of the nerve terminals (HAEUSLER et al., 1968; VANHOUTTE and VERBEUREN, 1975); this hyperpolarization might counteract the depolarizing effect of nicotinic drugs, but would not easily explain inhibition of release evoked by electrical stimulation (HAEFELY, 1972, p. 706). Another proposal is that muscarinic agonists depolarize the nerve terminals and thereby inactivate calcium channels (KIRPEKAR et al., 1972a). If so, depolarization by muscarinic agonists must differ from that caused by nicotinic compounds in time course and/or magnitude, since it does not trigger release.

5.2.2. Facilitation of Release?

By their effect on presynaptic muscarine receptors, cholinergic drugs are able to depress the response of effector cells to noradrenergic nerve impulses. Much less is known about the nature of potentiation phenomena which have been repeatedly described. They can perhaps be classified into three groups.

Postsynaptic synergism occurs in smooth muscles that are stimulated by both noradrenaline and muscarinic agonists. Muscarinic drugs may thus augment contractions elicited by sympathetic nerve impulses even if the release of noradrenaline is simultaneously depressed (SJÖSTRAND, 1973a; SJÖSTRAND and SWEDIN, 1974; VANHOUTTE, 1974).

On the other hand, enhancement of effector responses to sympathetic nerve stimulation by high doses of nicotine and DMPP may be presynaptic in origin (MALIK and LING, 1969a; NEDERGAARD and BEVAN, 1969a; SU and BEVAN, 1970b; STEINSLAND et al., 1973; STEINSLAND and FURCHGOTT, 1975). An analogous finding may be the increase of the stimulation-evoked overflow of noradrenaline from rabbit hearts caused by short perfusion with 6×10^{-5} M acetylcholine (plus atropine; LÖFFELHOLZ, 1970b). However, the influence of this treatment on the inactivation of noradrenaline within the heart has not been reported. Moreover, it is not known whether the increase of overflow is antagonized by antinicotinic drugs. Thus, the tempting idea that muscarinic inhibition has a counterpart in nicotinic facilitation requires further study.

Of particular interest is a third potentiation phenomenon, caused by very low doses of cholinergic drugs and above all by picomolar concentrations of acetylcholine (MALIK and LING, 1969b; RAND and VARMA, 1970; ALLEN et al., 1972a, 1975; STORY et al., 1975; see, however, HUME et al., 1972). In perfused rabbit ear arteries, acetylcholine at $10^{-11}-10^{-10}$ M enhances both vasoconstriction and overflow of transmitter evoked by sympathetic nerve stimulation, presumably through facilitation of release. The increase is not affected by atropine or hexamethonium (ALLEN et al., 1972a, 1975). Similar results have been obtained in rabbit atria (STORY et al., 1975). In contrast, the same concentrations of acetylcholine have no effect in perfused rabbit hearts (MUSCHOLL, 1973b), rabbit pulmonary arteries (TAUBE et al., 1976), and guinea-pig atria (STORY et al., 1975). More work is needed to clarify the occurrence and nature of this—apparently neither nicotinic nor muscarinic—facilitation.

5.3. Possible Significance

There is at present no good evidence that noradrenergic nerve endings in addition to noradrenaline contain stores of acetylcholine. On the other hand, several tissues receive both a noradrenergic and a cholinergic innervation, and cholinergic and noradrenergic terminals lie in close apposition (about 20 nm). The morphological arrangement leaves little doubt that cholinergic fibers can influence noradrenergic varicosities (see KOSTERLITZ and LEES, 1972; MUSCHOLL, 1973b).

Does the nicotinic releasing effect of acetylcholine normally occur in vivo? Much of the interest in cholinergic effects on noradrenergic nerve endings stems from the proposal by BURN and RAND (1959) that acetylcholine, thought to be contained in all postganglionic sympathetic fibers, plays an indispensable role in noradrenergic transmission; impulses passing down the axon first release acetylcholine, which then, probably from the outside of the fiber, releases noradrenaline. According to this postulate, the release of noradrenaline evoked by the nicotinic action of acetylcholine would be of the utmost importance. The arguments for and against the cholinergic link hypothesis have been repeatedly reviewed (FERRY, 1966; MUSCHOLL, 1970; KOSTERLITZ and LEES, 1972; BALDESSARINI, 1975). Most authors tend to reject the hypothesis and, thereby, a physiologic role of the nicotinic releasing effect of acetylcholine.

A second question concerns the significance of the modulation of release evoked by nerve impulses. There is no evidence for a physiological operation of either the possible nicotinic facilitatory effect of high concentrations of acetylcholine, or the hexamethonium- and atropine-resistant facilitatory effect of very low concentrations. On the other hand, two types of experiment suggest that muscarinic inhibition may normally occur.

In the first type of experiment, the influence of vagal nerve impulses on the release of noradrenaline evoked by sympathetic nerve impulses has been tested. In isolated rabbit atria and in dog hearts *in situ,* stimulation of the vagus nerves reduces the overflow of noradrenaline evoked by sympathetic nerve stimulation, indicating that released acetylcholine inhibits release from neighboring noradrenergic fibers (LÖFFELHOLZ and MUSCHOLL, 1970; LEVY and BLATTBERG, 1976). In contrast, stimulation of the vagus nerves has no effect on catecholamine release in the chicken heart (LÖFFELHOLZ, 1975).

Another approach is to study the effect of atropine on noradrenaline release evoked by autonomic nerve stimulation. If both acetylcholine and noradrenaline are secreted, the former depressing release of the latter, then atropine should enhance the release of noradrenaline. Only concentrations of atropine of up to about 10^{-6} M can be used in this type of experiment, since higher concentrations increase release by other than antimuscarinic effects, for instance by blockade of presynaptic α-adrenoceptors (STARKE et al., 1975c; TAUBE et al., 1976). Low concentrations of atropine have no effect on noradrenaline release in the rabbit heart (stimulation of postganglionic sympathetic nerve trunks; FOZARD and MUSCHOLL, 1972, 1974), ear artery (periarterial nerve stimulation; ALLEN et al., 1975), pulmonary artery (field stimulation; STARKE et al., 1975c), and the rat cerebral cortex (field stimulation; STARKE et al., 1975c), but increase release in the chicken heart (combined vagal and sympathetic nerve stimulation; LÖFFELHOLZ, 1975) and the guinea-pig vas deferens (field stimulation; STJÄRNE, 1975c).

Taken together, these results open up the possibility that in four organs, namely dog, rabbit, and chicken heart and guinea-pig vas deferens, endogenous acetylcholine slows down the release of noradrenaline. The source of acetylcholine in the chicken heart and guinea-pig vas deferens remains uncertain. In the chicken heart, the cholinergic fibers appear to run in the sympathetic rather than the vagal nerves, since vagal stimulation did not inhibit noradrenaline release (LÖFFELHOLZ, 1975). On the other hand, the experiments in rabbit atria and dog hearts strongly suggest a physiologically relevant interpretation (LÖFFELHOLZ and MUSCHOLL, 1970; LEVY and BLATTBERG, 1976). The two divisions of the autonomic nervous system seem to act antagonistically at two levels. Firstly, acetylcholine and noradrenaline have opposite effects on many effector cells: postsynaptic antagonism. Secondly, acetylcholine inhibits transmitter release from noradrenergic nerve endings (and noradrenaline inhibits release from cholinergic nerve endings, see p. 72): presynaptic antagonism.

5.4. Effect on Release of Other Substances

In analogy to sympathetic ganglion cell bodies, the chromaffin cells of the adrenal medulla contain nicotine and muscarine receptors. Their activation initiates hormone secretion (SMITH and WINKLER, 1972). — Nicotine increases the basal outflow of radioactive compounds from hypothalamic and striatal slices preincubated with labeled dopamine, presumably by releasing the transmitter from dopaminergic neurones (GOODMAN, 1974; WESTFALL, 1974a). The overflow evoked by nicotine is reduced by hexamethonium, acetylcholine, and methacholine. Acetylcholine also reduces the overflow of previously stored ^{3}H-dopamine evoked by electrical stimulation or high potassium concentrations. The results indicate that central dopaminergic nerve endings may contain both excitatory nicotine and inhibitory muscarine receptors (WESTFALL, 1974a and b; GIORGUIEFF et al., 1976).

Cholinergic agents affect cholinergic nerve endings. High doses release acetylcholine from motor and preganglionic autonomic nerve endings. However, the predominant effect at these sites is thought to be inhibition of release evoked by orthodromic action potentials, the presynaptic receptors being of the nicotine receptor type (see NISHI, 1970; HUBBARD and QUASTEL, 1973; KRNJEVIĆ, 1974, p. 495). Central cholinergic nerve endings are possibly endowed with inhibitory muscarine receptors, since release from brain slices evoked by potassium or electrical stimulation is inhibited by oxotremorine and, in the presence of cholinesterase inhibitors, enhanced by atropine (POLAK, 1971; SZERB and SOMOGYI, 1973; KATO et al., 1975). Oxotremorine also reduces the release of acetylcholine in the guinea-pig ileum, and the effect is reversed by atropine (KILBINGER and WAGNER, 1975). Though the evidence is far from complete, the results suggest that in motor, preganglionic autonomic, postganglionic parasympathetic, and central cholinergic nerve endings presynaptic cholinoceptors mediate a negative feedback mechanism in which released acetylcholine inhibits the release of more acetylcholine (cf. p. 97).

5.5. Conclusion

The terminals of most noradrenergic neurones appear to be endowed with both nicotine and muscarine receptors. Activation of presynaptic nicotine receptors probably leads to membrane depolarization, which in turn initiates action potentials and calcium-dependent noradrenaline release. The electrical discharges contribute to release and, under certain conditions, may be the major cause. However, release can be elicited even when action potentials are prevented by tetrodotoxin; the local depolarization and perhaps a specific enhancement of calcium conductance seem to allow sufficient calcium entry. The effect of nicotine and congeners is reduced by muscarinic agonists, prostaglandins, and a large number of apparently unrelated drugs which possibly interfere with depolarization or calcium entry in an unknown "nonspecific" manner.

The activation of presynaptic muscarine receptors does not change the basal outflow of noradrenaline. However, the release of transmitter per nerve impulse is depressed. Presynaptic muscarine receptors resemble postsynaptic ones in their sensitivity to drugs. Present evidence is compatible with the view that muscarinic drugs selectively inhibit calcium-dependent release processes, perhaps by making less calcium available for stimulus-secretion coupling.

Nicotinic agents fail to release catecholamines from the chicken heart. Muscarinic inhibition has not been found in guinea-pig atria and the rat cerebral cortex. It is not known whether the catecholaminergic fibers of these tissues are devoid of presynaptic nicotine and muscarine receptors, respectively, or whether the operation of existing receptors was concealed by the experimental procedure. – There are indications that cholinergic drugs can also *enhance* release by nerve impulses. Some results suggest that presynaptic nicotine receptors are involved. On the other hand, picomolar concentrations of acetylcholine are reported to augment release by an action resistant to both antinicotinic and antimuscarinic drugs. The occurrence and nature of these facilitation phenomena require further investigation.

In several peripheral tissues cholinergic and noradrenergic nerve terminals lie in close apposition, so that mutual influences seem possible. It is unlikely that endogenous acetylcholine ever attains the high concentrations required for the nicotinic releasing effect. Much lower concentrations are sufficient for muscarinic inhibition. Released acetylcholine might depress transmitter release from neighboring noradrenergic fibers (and vice versa). If so, the two divisions of the autonomic nervous system would act antagonistically not only at the effector cells, but also by mutual presynaptic inhibition. It must be realized, however, that experimental evidence for this physiologic function of muscarinic inhibition is restricted to only a few tissues.

6. Prostaglandins

Soon after chemically pure prostaglandins became available it was noticed that they modified biochemical as well as physical responses to catecholamines, other agonists, and sympathetic nerve impulses. There is ample evidence to show that prostaglandins can inhibit or enhance responses to noradrenaline by a postsynaptic effect (e.g., interaction on smooth muscle: SJÖSTRAND, 1972; on lipolysis: FREDHOLM and HEDQVIST, 1973b). A presynaptic component was first demonstrated by HEDQVIST and BRUNDIN (1969) and HEDQVIST (1969a). In the isolated perfused spleen of the cat, both PGE_2 and, less regularly, PGE_1 reduced the overflow of noradrenaline evoked by sympathetic nerve impulses. Since it was known that these impulses lead to formation of prostaglandins, HEDQVIST (1969a) suggested "that PGE_2, locally mobilized by sympathetic nerve stimulation, may counteract further release of noradrenaline by a negative feed back mechanism, thus exerting a braking effect on the sympathetic neuro-effector system". Evidence for a role of prostaglandins in the regulation of the per pulse release of noradrenaline has now been obtained in many organs. Their influence on autonomic neurotransmission has been reviewed by HEDQVIST (1973a and b), HORTON (1973), BRODY and KADOWITZ (1974), BALDESSARINI (1975) and STJÄRNE (1975a).

6.1. Inhibition of Release

PGE_1 and/or PGE_2 reduce the stimulation-evoked overflow of noradrenaline from cat mesenteric artery and spleen, rat iris and cerebral cortex, guinea-pig vas deferens, rabbit heart, ear artery, pulmonary artery, kidney, oviduct, and superior cervical ganglion, and human blood vessels and oviducts (Table 4). PGE_1 increases overflow per stimulus from the in vitro blood-perfused cat spleen, but not from the spleen *in situ* (BLAKELEY et al., 1969a); the increase is attributed to inhibition of platelet aggregation and maintenance of a uniform blood flow through the organ. In canine white adipose tissue, the effect of PGE_2 is erratic. No effect was found in the nictitating membrane of the cat. Few studies have been performed with other prostaglandins. $PGF_{2\alpha}$ and PGB_2 also reduce the stimulation-evoked overflow, but the concentrations required are high.

From detailed analyses (HEDQVIST, 1970c, 1973b) it has been concluded that the decrease of the stimulation-evoked overflow reflects a decrease of the release of transmitter per nerve impulse. Briefly, the evidence is as follows (cf. Table 4): (1) PGE_1, PGE_2, and $PGF_{2\alpha}$ do not change the neuronal (BLAKELEY et al., 1969a; HEDQVIST, 1970a; CIOFALO, 1973) or extraneuronal (SALT, 1972) uptake of noradrenaline, nor the activities of MAO or COMT in tissue homogenates (BHAGAT et al., 1972). (2) After treatment with labeled noradrenaline, prostaglandins reduce the stimulation-evoked overflow of total radioactivity, i.e., of labeled noradrenaline plus labeled metabolites; they do not decrease the overflow of noradrenaline by promoting its biotransformation.

Table 4. Effect of prostaglandins on transmitter overflow evoked by electrical stimulation of noradrenergic nerves

Species	Tissue	Prosta-glandin	Concentration (M)	Effect	Comment	References
Dog	Subcutaneous adipose tissue	PGE_2	3×10^{-8}–6×10^{-7}	0 or –	Blood-perfused tissue. Preinfusion of ^3H-noradrenaline. No effect on overflow of ^3H-noradrenaline and total ^3H evoked by sympathetic nerve stimulation at 4 Hz before, but decrease after injection of phenoxybenzamine. Interpretation difficult because of changes in blood flow	Fredholm and Hedqvist, 1973b
Cat	Mesenteric artery	PGE_1	3×10^{-8}	0	Strips preincubated with ^3H-noradrenaline. Effect on overflow of total ^3H evoked by field stimulation at 5–10 Hz	Hedqvist, 1974e
		PGE_2	3×10^{-7}	–		
	Spleen	PGE_1	10^{-7}–3×10^{-7}	+	Blood-perfused spleen. Effect on overflow of endogenous noradrenaline evoked by splenic nerve stimulation at 10 or 30 Hz	Blakeley et al., 1969a
	Spleen	PGE_1	2×10^{-6}–4×10^{-6}	0 or –	Perfused spleen or spleen preperfused with ^3H-noradrenaline. Effect on overflow of endogenous noradrenaline or total ^3H evoked by splenic nerve stimulation at 5–30 Hz. PGE_1 decreased overflow in 4 out of 10 experiments. Effect of PGE_2 was not prevented by phenoxybenzamine and antagonized by an increase in calcium concentration	Hedqvist, 1969a and b, 1970a and b; Hedqvist and Brundin, 1969; Dubocovich and Langer, 1975; Langer et al., 1975b
		PGE_2	6×10^{-8}–6×10^{-6}	–		
	Nictitating membrane	PGE_1	8×10^{-7}–8×10^{-6}	0	Tissue preincubated with ^3H-noradrenaline. Effect on overflow of total ^3H evoked by infratrochlear nerve stimulation. No details	Langer et al., 1975b
		PGE_2	8×10^{-7}–8×10^{-6}	0		
Rat	Iris	PGE_2	3×10^{-6}	–	Tissue preincubated with ^3H-noradrenaline. Effect on overflow of total ^3H evoked by field stimulation at 10 Hz	Bergström et al., 1973
	Cerebral cortex	PGE_1	10^{-8}–10^{-6}	–	Slices preincubated with ^3H-noradrenaline. Effect on overflow of total ^3H evoked by field stimulation at 3 or 10 Hz	Bergström et al., 1973; Starke et al., 1975c
		PGE_2	3×10^{-6}	–		
Guinea pig	Vas deferens	PGE_1	2×10^{-9}–2×10^{-6}	–	Incubated tissue or tissue preincubated with ^3H-noradrenaline. Effect on overflow of endogenous noradrenaline or total ^3H evoked by hypogastric nerve stimulation (Johnson et al., 1971) or field stimulation at 1–30 Hz was tested either in absence or presence of desipramine + normetanephrine. Effect was	Johnson et al., 1971; Stjärne, 1972b, 1973a–d, f, h, k, and l, 1976; Hedqvist, 1973a, c–e, 1974a and b
		PGE_2	3×10^{-10}–2×10^{-6}	–		

Species	Tissue	Prostaglandin	Concentration	Effect	Comment	Reference
Rabbit	Heart	PGE_1	3×10^{-9}	0	Perfused heart. Effect on overflow of endogenous noradrenaline evoked by accelerans nerve stimulation at 5 or 10 Hz, prevented by phenoxybenzamine or phentolamine and was antagonized by an increase in calcium concentration	Hedqvist et al., 1970; Wennmalm and Hedqvist, 1970; Hedqvist and Wennmalm, 1971
		PGE_2	$3 \times 10^{-8}-2 \times 10^{-6}$	—		
			3×10^{-9}	0		
		$PGF_{2\alpha}$	$3 \times 10^{-8}-2 \times 10^{-6}$	0		
	Ear artery	PGE_1	10^{-7}	—	Arteries preincubated with ^{3}H-noradrenaline. Effect on overflow of total ^{3}H evoked by field stimulation at 3 Hz	Hadházy et al., 1976
	Pulmonary artery	PGE_1	$10^{-8}-10^{-6}$	—	Strips preincubated with ^{3}H-noradrenaline. Effect on overflow of total ^{3}H evoked by field stimulation at 2 Hz was tested in the presence of cocaine + corticosterone	Starke et al., 1975c; Taube et al., 1976
		PGE_2	$10^{-8}-10^{-6}$	—		
		$PGF_{2\alpha}$	$10^{-8}-10^{-6}$	0		
			10^{-5}			
		PGB_2	$10^{-8}-10^{-6}$	0		
			10^{-5}	—		
	Kidney	PGE_2	$2 \times 10^{-7}-9 \times 10^{-7}$	—	Kidney preperfused with ^{3}H-noradrenaline. Effect on overflow of total ^{3}H evoked by renal nerve stimulation at 2–10 Hz. Effect was inversely related to frequency	Frame and Hedqvist, 1975
	Oviduct	PGE_2	$9 \times 10^{-9}-9 \times 10^{-8}$	—	Tissue preincubated with ^{3}H-noradrenaline. Effect on overflow of total ^{3}H evoked by field stimulation at 5 Hz	Moawad et al., 1975
	Superior cervical ganglion	PGE_1	10^{-6}	—	Ganglia preincubated with ^{3}H-noradrenaline. Effect on overflow of total ^{3}H evoked by cervical sympathetic nerve stimulation at 6 Hz. Release probably occurs from rostrally ascending postganglionic sympathetic fibres	Noon and Roth, 1975
Man	Arteries and veins	PGE_2	$9 \times 10^{-9}-5 \times 10^{-8}$	—	Small vessels preincubated with ^{3}H-noradrenaline. Effect on overflow of total ^{3}H evoked by field stimulation at 1 Hz was tested in presence of desipramine, normetanephrine + phentolamine	Stjärne and Gripe, 1973
	Oviduct	PGE_2	$9 \times 10^{-9}-9 \times 10^{-8}$	—	Tissue preincubated with ^{3}H-noradrenaline. Effect on overflow of total ^{3}H evoked by field stimulation at 5 Hz, in some experiments in presence of phenoxybenzamine	Hedqvist and Moawad, 1975; Moawad et al., 1975

+ = increase, − = decrease, 0 = no change. In each case, the effect measured is briefly defined under "Comment".

(3) The effect is retained after both the neuronal and extraneuronal uptake have been blocked. (4) In the guinea-pig vas deferens, PGE_1 and PGE_2 reduce the stimulation-evoked overflow of dopamine-β-hydroxylase, suggesting depression of exocytotic release (JOHNSON et al., 1971).

Postsynaptic response studies suggest that PGE_1 and/or PGE_2 inhibit noradrenaline release in several organs not included in Table 4 (BAUM and SHROPSHIRE, 1971; HEDQVIST, 1972 a and b, 1974 d; HEDQVIST and EULER, 1972; CLARK et al., 1973; ILLÉS et al., 1973; ABDEL-AZIZ, 1974; GREENBERG, 1974; AMBACHE et al., 1975; HEDQVIST and PERSSON, 1975). No or minimal inhibition of effector responses to nerve stimulation was found in the spleen of the dog (DAVIES and WITHRINGTON, 1969), the nictitating membrane of the cat (BRODY and KADOWITZ, 1974; ILLÉS et al., 1974; cf. overflow experiments by LANGER et al., 1975 b), and the vas deferens of the rat (AMBACHE et al., 1972; HEDQVIST and EULER, 1972; ILLÉS et al., 1973). The reason for these differences is not known. It cannot be excluded that in some tissues the noradrenergic fibers lack presynaptic receptors for prostaglandins.

The mechanism of the depression of release is uncertain. Some information can be derived from the following observations. (1) PGE_2 does not block impulse conduction in bovine splenic nerves; a local anesthetic effect can be excluded (HEDQVIST, 1970 b). (2) PGE_1 and PGE_2 do not change the outflow of noradrenaline from isolated storage vesicles; a granular site of action is unlikely (HEDQVIST, 1970 c, 1973 b). (3) The decrease of release is probably not due to a decrease of noradrenaline biosynthesis, since release of exogenous, previously taken up noradrenaline is equally depressed. (4) Presynaptic α-adrenoceptors are not involved, since the effect persists in the presence of α-adrenolytic drugs (Table 4). (5) The effect of PGE_2 is not changed by theophylline or dibutyryl-cyclic AMP, indicating that the prostaglandin does not act by modifying the intraneuronal cyclic AMP level (STJÄRNE, 1976). (6) PGE_1 and PGE_2 reduce not only release evoked by electrical stimulation, but also that evoked by high extracellular potassium concentrations or by nicotine (STJÄRNE, 1973 h; WESTFALL and BRASTED, 1974). They do not affect basal outflow (HEDQVIST, 1970 a; HEDQVIST et al., 1970; GEORGE, 1975). The overflow evoked by tyramine is not changed in the cat spleen (HEDQVIST, 1970 b) and not changed or even enhanced in the guinea-pig heart (WESTFALL and BRASTED, 1974), but is reduced in rat mesenteric arteries (GEORGE, 1975). Most of these data suggest that prostaglandins selectively inhibit calcium-dependent release processes; however, the inconsistent results obtained with tyramine leave some doubt. (7) Inhibition of the nerve impulse-evoked release of noradrenaline is smaller, the higher the concentration of calcium in the extracellular space (HEDQVIST, 1970 b, 1973 c, 1974 b; STJÄRNE, 1973 b and f). In contrast, inhibition of the release of dopamine-β-hydroxylase has been reported to be enhanced in a high calcium medium (JOHNSON et al., 1971). The reason for the discrepancy is not known. (8) The inhibition is inversely related to stimulation frequency (HEDQVIST, 1973 d; STJÄRNE, 1973 k; DUBOCOVICH and LANGER, 1975; FRAME and HEDQVIST, 1975; cf. postsynaptic response studies by BAUM and SHROPSHIRE, 1971; ILLÉS et al., 1973; JUNSTAD and WENNMALM, 1973 b; GREENBERG, 1974; HEDQVIST and PERSSON, 1975; HADHÁZY et al., 1976).

Much of the evidence summarized here is compatible with the idea that activation of presynaptic prostaglandin receptors decreases the availability of calcium for stimulus-secretion coupling (HEDQVIST, 1970b and c, 1973b and c, 1974b; STJÄRNE, 1973b, f, and h, 1976). Both an increase of external calcium and intraneuronal accumulation of calcium during high frequency stimulation would overcome this inhibition. However, some inconsistent findings remain to be explained. Moreover, the chain of events that leads to the proposed reduction of intraneuronal calcium remains unknown.

6.2. Facilitation of Release?

Postsynaptic response studies indicate that prostaglandins may exert a facilitatory rather than depressant presynaptic effect in some organs. In the hind paw of the dog, PGE_2, $PGF_{1\alpha}$, and $PGF_{2\alpha}$ (KADOWITZ et al., 1971, 1972) as well as PGA_2 and PGB_2 (GREENBERG et al., 1974) enhance the vasoconstrictor response to sympathetic nerve stimulation, but at identical doses leave the response to intraarterial noradrenaline unchanged. Preferential enhancement of the response to nerve stimulation has also been observed in the rabbit kidney ($PGF_{2\alpha}$) and the rat kidney (PGE_1, PGE_2, $PGF_{2\alpha}$, PGA_2; MALIK and McGIFF, 1975). The authors propose that in these cases the prostaglandins augment the release of noradrenaline. Different prostaglandins may have opposite effects in one and the same organ. As judged from the postsynaptic response, PGE_1, PGE_2, and PGA_2 reduce, whereas $PGF_{2\alpha}$ augments noradrenaline release in the rabbit kidney (MALIK and McGIFF, 1975).

The drawbacks of postsynaptic response methods in the study of noradrenaline secretion have already been pointed out (p. 8). Attempts to confirm a facilitatory presynaptic effect of prostaglandins with overflow methods have so far been unsuccessful (BRODY and KADOWITZ, 1974). In the rabbit pulmonary artery, $PGF_{2\alpha}$, a promising candidate for facilitation, caused pure inhibition (TAUBE et al., 1976). Facilitation would raise interesting questions. With sympathetic nerve impulses, endogenous prostaglandins are formed and depress further release of noradrenaline. Facilitation might occur when an exogenous prostaglandin with low intrinsic inhibitory activity competes with endogenous prostaglandins for their presynaptic receptors. Alternatively, prostaglandins might possess intrinsic facilitatory activity. If so, do facilitatory and inhibitory prostaglandins in a given tissue act on one and the same presynaptic receptor system, eliciting perhaps opposite conformational changes? And what is the basis for opposing effects of one prostaglandin in different tissues, e.g., depression by PGE_2 in the rabbit kidney, facilitation in the rat kidney?

6.3. Prostaglandin-Mediated Negative Feedback

Four lines of evidence converge to the hypothesis (HEDQVIST, 1969a) that prostaglandins, locally formed upon the arrival of noradrenergic nerve impulses, act on the secreting terminals and depress further release of noradrenaline. Firstly, exogenous prostaglandins inhibit release, as has been discussed previously. Secondly, prostaglandin formation during sympathetic nerve stimulation is well documented. Thirdly, noradrenaline release increases when the biosynthesis of prostaglandins is blocked. Fourthly, noradrenaline release decreases when their biosynthesis is enhanced. The second, third, and fourth arguments will now be considered.

Sympathetic nerve stimulation is followed by an outflow of prostaglandins in various organs such as the dog spleen (Davies et al., 1968; Gilmore et al., 1968; Bedwani and Millar, 1975), cat spleen (Bedwani and Millar, 1975), guinea-pig vas deferens (Swedin, 1971), and the heart (Samuelsson and Wennmalm, 1971) and kidney (Davis and Horton, 1972; Needleman et al., 1974) of the rabbit. In many experiments of this kind the prostaglandins have not been definitely identified and quantified (Horton, 1973). Since the tissue does not contain stores of prostaglandins, outflow is equivalent to *de novo* synthesis. The major source of the prostaglandins is probably not the nerve fibers, but extraneuronal cells, where released noradrenaline triggers their formation. This view is supported by several findings. (1) Not only nerve impulses, but also exogenous catecholamines promote prostaglandin biosynthesis (Gilmore et al., 1968; Needleman et al., 1974; Wennmalm, 1975). (2) α-Adrenolytic drugs prevent the formation of prostaglandins caused by either nerve stimulation or exogenous catecholamines (Davies et al., 1968; Needleman et al., 1974). In the rabbit heart, the outflow of prostaglandins during noradrenaline infusion is not changed by phenoxybenzamine or propranolol given separately (Junstad and Wennmalm, 1973a), but is abolished by a combination of α- and β-adrenolytic drugs (Wennmalm, 1975). (3) After noradrenergic nerves have been destroyed by surgical denervation (Gilmore et al., 1968) or 6-hydroxydopamine (Junstad and Wennmalm, 1973a), the release of prostaglandins evoked by exogenous noradrenaline is not reduced. — It cannot be excluded, however, that a minor, but perhaps functionally important fraction of the total prostaglandins formed upon nerve stimulation is neural in origin (Stjärne, 1972b, 1973d).

The feedback hypothesis predicts that more noradrenaline should be released after the formation of prostaglandins has been blocked. Experiments with three synthesis inhibitors (Flower, 1974), namely the substrate analogue 5,8,11,14-eicosatetraynoic acid (ETA) and the antiinflammatory agents indometacin and meclofenamic acid, are summarized in Table 5. In agreement with prediction, they enhance the stimulation-evoked overflow of noradrenaline from most tissues. The increase of overflow reflects an increase of release, for the following reasons (cf. Table 5). (1) At concentrations which increase the stimulation-evoked overflow, ETA and indometacin do not inhibit the uptake of noradrenaline into postganglionic sympathetic neurones (Hedqvist et al., 1971; Samuelsson and Wennmalm, 1971; Chanh et al., 1972; Zimmerman et al., 1973). On the other hand, similar concentrations of indometacin and niflumic acid, a drug related to meclofenamic acid, do inhibit the high affinity uptake of noradrenaline into rat brain synaptosomes (Clarenbach et al., 1974, 1976). Therefore, effects on uptake must not be neglected. (2) In tissues pretreated with labeled noradrenaline, ETA, indometacin, and meclofenamic acid increase the stimulation-evoked overflow of total radioactivity. (3) The increase persists after both the neuronal and extraneuronal uptake mechanisms have been blocked.

ETA, indometacin, and meclofenamic acid are certainly not "pure" prostaglandin synthesis inhibitors. In particular, indometacin inhibits phosphodiesterase (FLOWER, 1974), and this may contribute to the increase of noradrenaline release (9.2.1.). However, the uniform facilitatory effect of three chemically unrelated agents strongly suggests that they act by their common ability to block the biosynthesis of prostaglandins. — Indometacin augments the turnover of noradrenaline and its urinary excretion (STJÄRNE, 1972a; JUNSTAD and WENN-MALM, 1972; FREDHOLM and HEDQVIST, 1975b). This in vivo effect may also be at least partly due to interruption of the prostaglandin-mediated feedback.

Another prediction from the hypothesis is that less noradrenaline should be released when the biosynthesis of prostaglandins is enhanced. The rate of prostaglandin formation seems to be limited by the availability of their essential fatty acid precursors. Arachidonic acid is the precursor of the bisunsaturated class of prostaglandins, as for instance of PGE_2. In agreement with the prediction, arachidonic acid reduces the stimulation-evoked overflow of noradrenaline from the rabbit kidney. The effect is prevented by indometacin, indicating that newly formed prostaglandins rather than arachidonic acid itself are responsible (FRAME and HEDQVIST, 1975).

Taken together, these results support the feedback hypothesis. However, there are exceptions. Though prostaglandins are formed in canine subcutaneous adipose tissue during sympathetic nerve impulses (FREDHOLM et al., 1970), indometacin fails to enhance the stimulation-evoked overflow of noradrenaline (Table 5), possibly because the sympathetic nerves of this tissue are prostaglandin-resistant (FREDHOLM and HEDQVIST, 1973b). Indometacin and meclofenamic acid, in contrast to ETA, do not increase the stimulation-evoked overflow of noradrenaline from the cat spleen (Table 5); the formation of prostaglandins is very low in this organ (BEDWANI and MILLAR, 1975; DUBOCOVICH and LANGER, 1975) and may remain subthreshold for inhibition of transmitter release. It should also be recalled that presynaptic inhibition by prostaglandins could not be detected in the nictitating membrane of the cat, the dog spleen, and the rat vas deferens (Table 4 and p. 38). These tissue differences in the prostaglandin-mediated feedback mechanism contrast with the quite uniform operation of the prostaglandin-independent feedback mechanism mediated by presynaptic α-adrenoceptors (7.). There are also quantitative differences between the two feedbacks. Interruption of the prostaglandin mechanism by synthesis inhibitors leads only to a moderate increase of noradrenaline release (by about 10–40%). The increase following interruption of the α-receptor mechanism by α-adrenolytic drugs is larger (by up to 400%; see for instance STJÄRNE, 1972b, 1973k; STARKE and MONTEL, 1973a and c; DUBOCOVICH and LANGER, 1975). Apparently, the feedback mediated by presynaptic α-receptors is much more efficient.

Exogenous prostaglandins may *increase* the release of noradrenaline in certain tissues. There is little evidence that endogenous prostaglandins produce facilitation rather than depression of release, in other words a positive rather than negative feedback (MALIK and McGIFF, 1975).

Table 5. Effect of drugs which inhibit prostaglandin biosynthesis on transmitter overflow evoked by electrical stimulation of noradrenergic nerves

Species	Tissue	Inhibitor	Concentration (M)	Effect	Comment	References
Dog	Subcutaneous adipose tissue	Indometacin	10^{-5}	0	Blood-perfused tissue. Preinfusion of ^{3}H-noradrenaline. Effect on overflow of total ^{3}H evoked by sympathetic nerve stimulation at 4 Hz	FREDHOLM and HEDQVIST, 1975a
Cat	Mesenteric artery	ETA[a]	3×10^{-5}	+	Strips preincubated with ^{3}H-noradrenaline. Effect on overflow of total ^{3}H evoked by field stimulation at 5–10 Hz	HEDQVIST, 1974e
		Meclofenamic acid	3×10^{-5}	+		
	Spleen	ETA[a]	$2 \times 10^{-6} - 3 \times 10^{-6}$	+	Perfused spleen. Effect on overflow of endogenous noradrenaline evoked by splenic nerve stimulation at 10 Hz. ETA abolished outflow of prostaglandin-like material	HEDQVIST et al., 1971
	Spleen	Meclofenamic acid	3×10^{-6}	−	Perfused spleen or spleen preperfused with ^{3}H-noradrenaline. Effect on overflow of endogenous noradrenaline or total ^{3}H evoked by splenic nerve stimulation at 5–30 Hz. Either inhibitor abolished outflow of prostaglandin-like material	HOSZOWSKA and PANCZENKO, 1974; DUBOCOVICH and LANGER, 1975; LANGER et al., 1975b
		Indometacin	$3 \times 10^{-6} - 8 \times 10^{-5}$	0 or −		
Rat	Cerebral cortex	Indometacin		+	Slices preincubated with ^{3}H-noradrenaline. Effect of pretreatment with indometacin for 3 days, plus addition of 30 μM indometacin to the superfusion medium, on overflow of total ^{3}H evoked by field stimulation at 5 Hz	STARKE and MONTEL, 1973c

Species	Tissue	Agent	Concentration	Effect	Comment	References
Guinea pig	Vas deferens	ETA[a]	$3 \times 10^{-5} - 6 \times 10^{-5}$	+	Tissue preincubated with ³H-noradrenaline. Effect on overflow of total ³H evoked by field stimulation at 1–30 Hz was tested either in absence or presence of desipramine + normetanephrine. Effect was inversely related to frequency. ETA and indometacin abolished outflow of prostaglandin-like material	Stjärne, 1972b, 1973a, g-1; Fredholm and Hedqvist, 1973a; Hedqvist, 1973a, c, and e, 1974d
		Indometacin	6×10^{-6}	0 or +		
			$2 \times 10^{-5} - 6 \times 10^{-5}$	+		
		Meclofenamic acid	$7 \times 10^{-6} - 3 \times 10^{-5}$	+		
Rabbit	Heart	ETA[a]	$10^{-6} - 4 \times 10^{-5}$	+	Perfused heart. Effect on overflow of endogenous noradrenaline evoked by accelerans nerve stimulation at 2–5 Hz. ETA had no effect at 10 Hz. Either inhibitor abolished outflow of prostaglandin-like material	Samuelsson and Wennmalm, 1971; Chanh et al., 1972; Junstad and Wennmalm, 1973b; Starke and Montel, 1973a; Fuder and Muscholl, 1974
		Indometacin	3×10^{-6}	0		
			$10^{-5} - 7 \times 10^{-5}$	+		
	Pulmonary artery	Indometacin	3×10^{-5}	+	Strips preincubated with ³H-noradrenaline. Effect on overflow of total ³H evoked by field stimulation at 2 Hz was tested in presence of cocaine + corticosterone	Taube et al., 1976
	Kidney	Indometacin	4×10^{-5}	+	Kidney preperfused with ³H-noradrenaline. Effect on overflow of total ³H evoked by renal nerve stimulation at 5 Hz	Frame and Hedqvist, 1975
		Meclofenamic acid	5×10^{-5}	+		
Man	Arteries and veins	ETA[a]	5×10^{-5}	+	Small vessels preincubated with ³H-noradrenaline. Effect on overflow of total ³H evoked by field stimulation at 1 Hz was tested in presence of desipramine + normetanephrine	Stjärne and Gripe, 1973

+ = increase, − = decrease, 0 = no change. In each case, the effect measured is briefly defined under "Comment".

[a] ETA: 5,8,11,14-eicosatetraynoic acid.

6.4. Effect on Release of Other Substances

The influence of prostaglandins on catecholamine secretion from the adrenal medulla has been studied with variable results. Secretion evoked by splanchnic impulses is facilitated by PGE_1 and $PGF_{2\alpha}$ in the dog; PGE_1 may act on the cholinergic nerve endings (KAYAALP and TÜRKER, 1968; BRODY and KADOWITZ, 1974). On the other hand, PGE_1, PGE_2, and $PGF_{1\alpha}$ change neither basal catecholamine outflow from the cat adrenal medulla, nor secretion elicited by splanchnic nerve stimulation or various secretagogues (MIELE, 1969; HEDQVIST, 1973b, p. 112). Finally, the calcium-dependent basal outflow from slices of rat adrenal glands is *reduced* by PGE_1 and PGE_2, and not affected by $PGF_{1\alpha}$ (BOONYAVIROJ and GUTMAN, 1975; GUTMAN and BOONYAVIROJ, 1975).

Central dopaminergic nerve endings resemble noradrenergic ones in that they secrete less transmitter in the presence of PGE_1 or PGE_2 (BERGSTRÖM et al., 1973; BALDESSARINI, 1975, p. 107).

It has been suggested that prostaglandins evoke contractions of intestinal smooth muscle and salivation in part by activation of cholinergic neurones (SANNER, 1971; HAHN and PATIL, 1974). Prostaglandins have also been proposed as an essential link in electrosecretory coupling in cholinergic nerve endings (EHRENPREIS et al., 1973). These ideas are difficult to reconcile with the finding that neither prostaglandins nor indometacin have a consistent effect on the basal or stimulation-evoked overflow of acetylcholine from guinea-pig ileum (HADHÁZY et al., 1973; BOTTING and SALZMANN, 1974; ILLÉS et al., 1974; HAZRA, 1975).

WENNMALM and HEDQVIST (1971) observed that in the rabbit heart PGE_1 reduced the negative chronotropic effect of vagal nerve stimulation, but not that of infused acetylcholine. They concluded that PGE_1 might act on the terminals of postganglionic cholinergic nerves to reduce the release of acetylcholine, and that endogenous prostaglandins might control transmitter release not only from noradrenergic, but also from cholinergic fibers (cf. FENIUK and LARGE, 1975). JUNSTAD and WENNMALM (1974) in fact found that acetylcholine or vagal nerve stimulation increased the outflow of prostaglandins from the heart. However, in many other tissues, including some cardiac preparations, no evidence for a presynaptic inhibitory effect on cholinergic nerves has been detected (BAUM and SHROPSHIRE, 1971; HADHÁZY et al., 1973; PARK et al., 1973; ILLÉS et al., 1974; GUSTAFSSON et al., 1975; HEDQVIST and PERSSON, 1975). As mentioned above, neither prostaglandins nor indometacin change the outflow of acetylcholine from the guinea-pig ileum. Thus, it seems doubtful that the concept of a prostaglandin-mediated feedback inhibition can be extended to cholinergic neurones.

6.5. Conclusion

The terminals of most noradrenergic neurones appear to be endowed with prostaglandin receptors. Activation of these presynaptic receptors by PGE_1 or PGE_2 leads to depression of the release of noradrenaline per impulse. The prevailing evidence indicates that prostaglandins selectively inhibit calcium-dependent release processes, perhaps by diminishing the availability of calcium for stimulus-secretion coupling.

In a few tissues PGE_1 and PGE_2 fail to reduce release. The reason is not known. It cannot be excluded that some noradrenergic fibers lack presynaptic prostaglandin receptors. – Prostaglandins other than PGE_1 and PGE_2 have not been extensively tested. High concentrations of $PGF_{2\alpha}$ and PGB_2 inhibit noradrenaline release in the rabbit pulmonary artery. – Postsynaptic response studies suggest that some prostaglandins may facilitate rather than depress release in certain tissues. However, an increase of the stimulation-evoked overflow of noradrenaline has not yet been reported.

Noradrenergic neuroeffector transmission is accompanied by an outflow of prostaglandins from the tissue. The major fraction probably originates from extraneuronal cells, where released noradrenaline triggers their biosynthesis. In many tissues the prostaglandins so formed inhibit further release of noradrenaline. A local, prostaglandin-mediated, trans-synaptic negative feedback mechanism restricts the per pulse release of noradrenaline. Agents which block the biosynthesis of prostaglandins increase, whereas agents which promote prostaglandin biosynthesis reduce the release of noradrenaline. In contrast to the feedback inhibition mediated by presynaptic α-adrenoceptors, that mediated by presynaptic prostaglandin receptors does not operate in all tissues, either because some noradrenergic fibers are insensitive to prostaglandins, or because the amount formed is too low. Moreover, the prostaglandin mechanism is less effective than the α-receptor mechanism, since the enhancement of release that follows suppression of the biosynthesis of prostaglandins is much smaller than the enhancement after presynaptic α-receptor blockade.

7. Adrenergic Drugs: Presynaptic α-Adrenoceptors

The discovery that dibenamine and phenoxybenzamine increase the stimulation-evoked overflow of noradrenaline from the cat spleen (BROWN and GILLESPIE, 1956, 1957) triggered an impressive history of efforts in interpretation. Only a few points can be touched upon here. Early after 1956 the opinion prevailed that β-haloalkylamines as well as other α-adrenolytic drugs interfered with the inactivation of released noradrenaline within the tissue. BROWN and GILLESPIE (1957) originally supposed that combination of noradrenaline with postsynaptic α-receptors was a prelude to uptake into the effector cells and subsequent metabolic destruction, and that receptor blockade diverted the transmitter from this pathway to the venous blood (see the discussion by BROWN, 1965). Later explanations included inhibition of neuronal re-uptake, inhibition of extraneuronal uptake and degradation, and enhanced washout after abolition of neurogenic vasoconstriction. Though these mechanisms indeed can contribute to the increase of overflow, their significance is limited to certain animal preparations and high doses of certain α-adrenolytic drugs.

As early as 1959 FURCHGOTT pointed out the alternative "that the β-haloalkylamines in some manner cause a much greater release of transmitter from the sympathetic nerve endings". Facilitation of release by phenoxybenzamine was also considered, but rejected, by KIRPEKAR and CERVONI (1963). Since 1969, however, the idea has gained much support (HEDQVIST, 1969b, 1970c; FARNEBO and HAMBERGER, 1970, 1971a and b; HÄGGENDAL, 1970; KAUMANN, 1970; LANGER, 1970; JOHNSON et al., 1971; LANGER et al., 1971; POTTER et al., 1971; STARKE et al., 1971a; STJÄRNE and WENNMALM, 1971; WENNMALM, 1971). Not only phenoxybenzamine, but also phentolamine appears to increase the release of noradrenaline per impulse (FARNEBO and HAMBERGER, 1970, 1971a and b; FARNEBO and MALMFORS, 1971; LANGER et al., 1971; POTTER et al., 1971; STARKE et al., 1971b). In fact, this ability seems to be common to α-adrenolytic drugs[4]. If so, what is the connexion between α-receptors and release?

Having excluded inhibition of enzymatic degradation, of re-uptake and of vasoconstriction as explanations for the effect of phenoxybenzamine, HÄGGENDAL (1970) proposed that the drug might increase release per impulse either by a direct action on the nerve terminals (which he did not causally relate to α-receptor blockade), or by some feedback mechanism connected to the effector cell response. He favored the latter idea (cf. HÄGGENDAL, 1969, and the analogous view of HEDQVIST, 1969b, 1970c). The hypothesis predicts that more noradrenaline is released when the effector cell is depressed or difficult to excite, as after α-adrenolytic drugs, than when the effector cell is in a normal or easily excitable state. This kind of trans-synaptic feedback was later ruled out; it was found that α-adrenolytic drugs also facilitate release in the heart and simultaneously do not depress, but actually enhance the myocardial response (STARKE et al., 1971b; cf. HÄGGENDAL et al., 1972; ENERO and LANGER, 1973). However, the general idea that α-adrenolytic drugs interrupt a feedback mechanism operating within the noradrenergic synapse turned out to be fruitful.

In 1970 it was reported that two α-receptor agonists, namely xylazine and clonidine, reduced the stimulation-evoked overflow of noradrenaline from the cat spleen (HEISE and KRONEBERG, 1970; cf. HEISE et al., 1971) and the rabbit heart (WERNER et al., 1970; cf. STARKE et al., 1972), respectively. The decrease of overflow mirrored a depression of release. That the effect of these agonists was thus opposite to the effect of α-adrenolytic drugs (cf. FARNEBO and HAMBERGER, 1971a and b; FARNEBO and MALMFORS, 1971; STARKE, 1971a) was a further decisive hint that regulation of release occurs via an α-receptor system. A location at the site of release suggested itself.

FARNEBO and HAMBERGER (1971a) did not reject a postsynaptic location of release-regulating α-receptors and a trans-synaptic feedback loop (cf. FARNEBO and MALMFORS, 1971), but stated that "another possible mechanism is inhibition of noradrenaline release via α-adrenoceptors on the nerve terminal". Similarly, STARKE (1971a) proposed "that the adrenergic nerve terminals are endowed with structures related to the α-adrenoceptive sites of effector cells. On reaction with α-*stimulants,* e.g. with liberated noradrenaline, these neuronal α-receptors mediate ... inhibition of noradrenaline liberation; in the presence of α-*blockers,* this restriction is attenuated." While these proposals were partly based on results

---

[4] The view that α-adrenolytic drugs facilitate release was also emphasized by BURN and his colleagues in connexion with their "cholinergic link" hypothesis (BURN and GIBBONS, 1964a; BURN and NG, 1965; cf. p. 32). The authors suggested that the antagonists might all have an anticholinesterase action and might protect the acetylcholine involved in noradrenaline release against hydrolysis. This explanation now seems untenable. It is interesting, however, to reinterpret the findings of BURN's group in the light of the presynaptic α-receptor hypothesis.

with agonists, KIRPEKAR and PUIG (1971) drew the same conclusion from stop-flow experiments in the cat spleen, namely "that the noradrenaline released by nerve stimulation acts on ... presynaptic α sites to inhibit its own release by a negative feedback mechanism. Adrenoceptor blocking agents enhance ... noradrenaline overflow ... because they remove this autoinhibition by blocking the presynaptic α sites." From experiments on the cat nictitating membrane and guinea-pig atria LANGER et al. (1971) concluded "that the increased overflow of noradrenaline in the presence of several alpha blocking agents could be due to a presynaptic effect", and that there could be "alpha receptors in adrenergic nerve endings playing a role in the regulation of transmitter output". Evidence was subsequently presented showing that extracellular noradrenaline in fact depresses the release of intraneuronal noradrenaline by nerve impulses (KIRPEKAR et al., 1972b; McCulloch et al., 1972; STARKE, 1972b; see Fig. 2, p. 52)[5].

The hypothesis that drugs with an affinity to α-receptors modulate the per pulse release of noradrenaline, that they act on presynaptic α-receptors, and that these receptors mediate an exclusively presynaptic negative feedback mechanism, is now widely accepted (reviews by LANGER, 1974b; STJÄRNE, 1975a). It has the merit that it explains many hitherto puzzling effects of α-receptor agonists and antagonists on noradrenergic transmission. Interestingly, it agrees with the original idea of BROWN and GILLESPIE (1957) that phenoxybenzamine and congeners increase overflow by virtue of their ability to block α-adrenoceptors. The difference is that these authors thought α-receptors were involved in the inactivation of noradrenaline, while the current view contends that they are involved in regulation of its release. If α-receptors were sites of inactivation, both antagonists and agonists should enhance the stimulation-evoked overflow. That the agonists *reduce* overflow makes it unlikely that α-receptors contribute to a major degree to the retention or degradation of noradrenaline within the tissue.

7.1. α-Adrenergic Inhibition

α-Adrenoceptor agonists reduce the stimulation-evoked overflow of noradrenaline from all tissues studied so far (Table 6). Some findings require a special comment. Clonidine reduces overflow from the rat iris and cerebral cortex, mouse atria, and the rabbit heart and pulmonary artery, but increases overflow from the guinea-pig vas deferens (cf. p. 62). Dopamine is a potent inhibitor in the cat spleen and nictitating membrane, the guinea-pig hypothalamus, the rabbit ear artery, and human blood vessels, but not very effective in the rat cerebral cortex, the guinea-pig heart and vas deferens, and the rabbit pulmonary

---

[5] HOTTA (1969) discussed the inhibitory effect of exogenous noradrenaline on motor transmission in the vas deferens and proposed two different kinds of postsynaptic adrenergic receptors: junctional and extrajunctional. However, an alternative explanation might be that "noradrenaline acts on the prejunctional nerve terminal to depress the transmitter release following depolarization". To account for his observations, these prejunctional sites should not be blocked by α- and β-adrenolytic drugs, and therefore could not be classified as α-receptors. Nevertheless, and though the nature of the motor transmitter in the vas deferens is controversial (see footnote on p. 4), this appears to be the first consideration of a presynaptic release-inhibiting effect of noradrenaline on noradrenergic nerves.

Table 6. Effect of adrenergic agonists on transmitter overflow evoked by electrical stimulation of noradrenergic nerves

Species	Tissue	Drug	Concentration (M)	Effect	Comment	References
Dog	Cutaneous veins	Noradrenaline		−	Strips preincubated with ^{3}H-noradrenaline. Effect on overflow of total ^{3}H evoked by field stimulation was tested in presence of cocaine. No details	DALEMANS et al., 1976
		Dopamine		−		
		Epinine		−		
Cat	Aorta	Isoprenaline	10^{-8}	+	Strips prelabeled with ^{3}H-noradrenaline. Effect on overflow of total ^{3}H evoked by stimulation at 1–4 Hz. Effect was inversely related to frequency. No details	LANGER et al., 1975b
	Spleen	(−)-Noradrenaline	6×10^{-7}–6×10^{-6}	−	Perfused spleen or spleen preperfused with ^{3}H-noradrenaline. Effect on overflow of endogenous or ^{3}H-noradrenaline evoked by splenic nerve stimulation at 30 Hz. Noradrenaline and phenylephrine, but not methoxamine, isoprenaline or xylazine, increased basal outflow. Phentolamine antagonized effect of noradrenaline on stimulation-evoked, but not on basal outflow	HEISE et al., 1971; KIRPEKAR et al., 1972b, 1973
		(−)-Phenylephrine	10^{-6}–10^{-5}	−		
		(±)-Methoxamine	9×10^{-7}	0		
			9×10^{-6}	−		
		(±)-Isoprenaline	5×10^{-6}	0		
			5×10^{-5}	−		
		Xylazine	2×10^{-5}	−		
	Spleen	(−)-Noradrenaline	6×10^{-9}–6×10^{-6}	−	Perfused spleen prelabeled with ^{3}H-noradrenaline. Effect on overflow of ^{3}H-noradrenaline or total ^{3}H evoked by splenic nerve stimulation at 1–30 Hz. Effect of noradrenaline and dopamine was tested in presence of cocaine. Effect of noradrenaline was inversely related to frequency	LANGER, 1973a and b; LANGER et al., 1975a
		Dopamine	6×10^{-9}–6×10^{-6}	−		
		(−)-Isoprenaline	10^{-8}	+		
	Nictitating membrane	(−)-Noradrenaline	2×10^{-7}	−	Tissue preincubated with ^{3}H-noradrenaline. Effect on overflow of total ^{3}H evoked by infratrochlear nerve stimulation at 4 or 10 Hz was tested in presence of cocaine. Chlorpromazine selectively antagonized effect of dopamine. Either agonist increased basal outflow in absence, but not in presence of cocaine	LANGER, 1973a and b; ENERO and LANGER, 1975
		Dopamine	2×10^{-7}–7×10^{-7}	−		

Species	Tissue	Compound	Concentration	Effect	Comments	References
	Nictitating membrane	Isoprenaline	10^{-8}	+	Tissue prelabeled with ³H-noradrenaline. Effect on overflow of total ³H evoked by stimulation at 4 Hz. No details	LANGER et al., 1975b
Rat	Iris	Clonidine	$10^{-7}-10^{-6}$	–	Tissue preincubated with ³H-noradrenaline. Effect on overflow of total ³H evoked by field stimulation at 10 Hz	FARNEBO and HAMBERGER, 1971a, 1973a
	Vas deferens	(–)-Noradrenaline	9×10^{-6}	–	Tissue preincubated with ³H-noradrenaline. Effect on overflow of total ³H evoked by field stimulation at 2 Hz. No effect at 50 Hz. Effect was antagonized by phentolamine	VIZI et al., 1973
	Cerebral cortex	(–)-Noradrenaline	10^{-9}	0	Slices preincubated with ³H-noradrenaline. Effect on overflow of total ³H evoked by field stimulation at 1 (some experiments with isoprenaline) or 3–10 Hz; at 10 Hz, inhibitory effects were weaker. Effect of noradrenaline and dopamine was tested in presence of cocaine. Clonidine and oxymetazoline did not change basal outflow; isoprenaline increased it at 10 μM; noradrenaline and dopamine increased it in absence, but not in presence of cocaine	FARNEBO and HAMBERGER, 1971b, 1973b, 1974b; STARKE, 1973b; STARKE and MONTEL, 1973b–d; STARKE et al., 1975c; STARKE, unpublished
		Dopamine	$10^{-8}-10^{-6}$	–		
		Isoprenaline	$10^{-8}-10^{-6}$	0		
			$10^{-8}-10^{-6}$	0		
			10^{-5}	+		
		Clonidine	$10^{-8}-10^{-5}$	–		
			10^{-8}	0		
		Oxymetazoline	$10^{-7}-10^{-5}$	–		
Mouse	Heart	Isoprenaline	$10^{-7}-10^{-5}$	0	Atria preincubated with ³H-noradrenaline. Effect on overflow of total ³H evoked by field stimulation at 10 Hz	FARNEBO and HAMBERGER, 1974b
		Clonidine	$10^{-8}-10^{-7}$	0		
			10^{-6}	–		
	Vas deferens	(–)-Noradrenaline	$10^{-6}-10^{-5}$	–	Tissue preincubated with ³H-noradrenaline. Effect on overflow of total ³H evoked by field stimulation at 16 Hz. Since noradrenaline increased basal outflow, its effect on stimulation-evoked overflow was difficult to evaluate	FARNEBO and MALMFORS, 1971
		Methoxamine	$10^{-6}-10^{-4}$			
Guinea pig	Heart	(–)-Noradrenaline	$5 \times 10^{-7}-5 \times 10^{-5}$	–	Atria preincubated with ³H-noradrenaline. Effect on overflow of total ³H evoked by field stimulation at 5 Hz was tested in presence of cocaine. No details	McCULLOCH et al., 1972; RAND et al., 1973, 1975b
		(+)-Noradrenaline	5×10^{-7}	0		
		Adrenaline		–		
		Dopamine	5×10^{-7}	0		
	Heart	(±)-Isoprenaline	10^{-8}	+	Atria preincubated with ³H-noradrenaline. Effect on overflow of total ³H evoked by accelerans nerve stimulation at 4 Hz	ADLER-GRASCHINSKY and LANGER, 1975; LANGER et al., 1975b

Table 6 (continued)

Species	Tissue	Drug	Concentration (M)	Effect	Comment	References
Guinea pig	Vas deferens	(−)-Noradrenaline	6×10^{-8}–6×10^{-6}	−	Tissue preincubated with ³H-noradrenaline. Effect on overflow of total ³H evoked by field stimulation at 1–5 Hz was tested in presence of desipramine + normetanephrine +, in some experiments, phentolamine, indometacin, or ETA[a]. Under these conditions, (−)-noradrenaline and clonidine did not change basal outflow	STJÄRNE, 1973 a, h, i, and m, 1974, 1975b–d; HEDQVIST, 1974c
		(+)-Noradrenaline	2×10^{-7}–10^{-6}	0		
		(−)-Adrenaline	2×10^{-7}–10^{-6}	−		
		Dopamine	2×10^{-7}–5×10^{-6}	0		
		Methoxamine	10^{-7}–10^{-6}	0		
			10^{-5}–6×10^{-4}	−		
		Isoprenaline	2×10^{-7}	+		
			10^{-6}	0		
		Clonidine	10^{-8}	0		
			10^{-7}–10^{-6}	+		
	Hypothalamus	Noradrenaline	2×10^{-6}–5×10^{-5}	−	Slices preincubated with ³H-noradrenaline. Effect on overflow of total ³H evoked by field stimulation at 10 Hz was tested in presence of cocaine. Effect of noradrenaline was antagonized by phentolamine	BRYANT et al., 1975
		Adrenaline	5×10^{-6}	−		
		Dopamine		−		
		Isoprenaline	up to 5×10^{-5}	0		
Rabbit	Heart	(−)-Noradrenaline	6×10^{-9}	0	Perfused heart or heart preperfused with ¹⁴C-noradrenaline. Effect on overflow of endogenous or ¹⁴C-nor-adrenaline or total ¹⁴C evoked by accelerans nerve stimulation at 2.5–10 Hz. Effect of noradrenaline, phenylephrine, and orciprenaline was tested in presence of cocaine. Inhibitory effects were antagonized by phenoxybenzamine, but not propranolol; effect of oxymetazoline was not changed by indometacin. Nor-adrenaline and phenylephrine increased basal outflow in absence, but not in presence of cocaine; the other drugs did not change basal outflow	WERNER et al., 1970, 1972; STARKE, 1971a, 1972a and b, 1973a; STARKE et al., 1972; STARKE and ALTMANN, 1973; STARKE and MONTEL, 1973a; PACHA et al., 1975
		(−)-Phenylephrine	6×10^{-8}–6×10^{-7}	0		
			2×10^{-7}	0		
		Orciprenaline	2×10^{-6}–2×10^{-4}	−		
			10^{-6}	0		
			10^{-5}	−		
		Clonidine	2×10^{-9}	0		
			2×10^{-8}–2×10^{-4}	−		
		Naphazoline	5×10^{-9}–5×10^{-6}	−		
		Oxymetazoline	4×10^{-9}–4×10^{-6}	−		
	Ear artery	(−)-Noradrenaline	5×10^{-7}	−	Arteries preincubated with ³H-noradrenaline. Effect on overflow of total ³H evoked by periarterial nerve stimulation was tested in presence of cocaine. Effect of dopamine, but not that of (−)-noradrenaline, was antagonized by pimozide. No details	McCULLOCH et al., 1973; RAND et al., 1973, 1975a and b
		(+)-Noradrenaline	5×10^{-7}	0		
		Dopamine	5×10^{-8}–5×10^{-6}	−		
		Epinine		−		
		Tyramine	5×10^{-7}	0		

	Drug	Concentration	Effect	Comment	References
Pulmonary artery	(−)-Noradrenaline (−)-Adrenaline (−)-erythro-α-methylnoradrenaline (−)-Phenylephrine (±)-Methoxamine Clonidine Naphazoline Oxymetazoline Tramazoline	see Starke et al., 1974b, 1975b, or Table 8	−	Strips preincubated with ³H-noradrenaline. Effect on overflow of total ³H evoked by field stimulation at 1–8 Hz was tested in presence of cocaine, corticosterone + propranolol; isoprenaline had similar effects in absence of propranolol. Inhibitory effects of noradrenaline, clonidine, and oxymetazoline were antagonized by phentolamine. Effect of clonidine was less marked at 8 than at 2 Hz. At high concentrations, most drugs increased basal outflow; dopamine already at 1 μM	Starke et al., 1974b, 1975b and c; Taube et al., 1976; Starke, unpublished
	Dopamine	10^{-7}–3×10^{-7} 10^{-5}	0 −		
	(±)-Isoprenaline	10^{-9}–10^{-6} 10^{-5}	0 −		
Superior cervical ganglion	Methoxamine	3×10^{-5}	−	Ganglia preincubated with ³H-noradrenaline. Effect on overflow of total ³H evoked by cervical sympathetic nerve stimulation at 5 Hz. Release probably occurs from rostrally ascending postganglionic sympathetic fibers	Noon and Roth, 1975
Man Arteries and veins	(−)-Noradrenaline (−)-Adrenaline	2×10^{-7}–10^{-6} 2×10^{-9}–8×10^{-9} 2×10^{-7}–10^{-6}	− + −	Small vessels preincubated with ³H-noradrenaline. Effect on overflow of total ³H evoked by field stimulation at 1 Hz was tested in presence of desipramine + normetanephrine. Under these conditions, noradrenaline and dopamine did not change basal outflow	Stjärne and Gripe, 1973; Stjärne and Brundin, 1975a and b
	Dopamine (−)-Isoprenaline	2×10^{-7}–10^{-6} 2×10^{-9}–10^{-6}	− +		
Oviduct	Noradrenaline Isoprenaline	3×10^{-8}–3×10^{-6} 5×10^{-10}–10^{-8}	− +	Tissue preincubated with ³H-noradrenaline. Effect on overflow of total ³H evoked by field stimulation at 1 (isoprenaline) or 5 Hz (noradrenaline) was tested in absence (isoprenaline) or presence (noradrenaline) of desipramine + normetanephrine. Effect of isoprenaline was antagonized by propranolol	Hedqvist and Moawad, 1975

+ = increase, − = decrease, 0 = no change. In each case, the effect measured is briefly defined under "Comment". Experiments in which lysergic acid diethylamide (Hughes, 1973), 2-(2-methyl-6-ethyl-cyclohexylamino)-2-oxazoline (BAY a 6781; Werner et al., 1970, 1972), 2,6-dichlorobenzylidene amino-guanidine (guanabenz; Farnebo and Hamberger, 1974b), N-amidino-2-(2,6-dichlorophenyl)acetamide (BS 100–141; Pacha et al., 1975; Scholtysik et al., 1975), apomorphine (Langer, 1973a and b; Enero and Langer, 1975; Starke et al., 1975c) or 2-dimethylamino-5,6-dihydroxy tetralin (Strait and Bhatnagar, 1975) were used as agonists are not included in the Table.

ª ETA: 5,8,11,14-eicosatetraynoic acid.

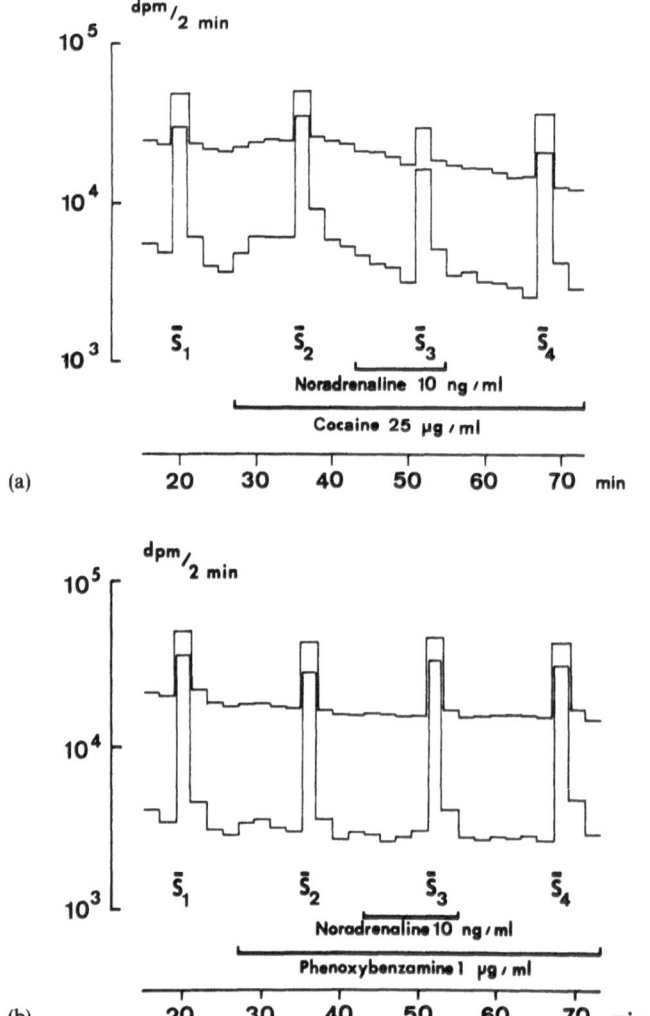

Fig. 2a and b. Effect of unlabeled (−)-noradrenaline, in the presence of cocaine (*a*) or phenoxybenzamine (*b*), on the stimulation-evoked overflow of radioactive compounds from the isolated rabbit
heart preperfused with ^{14}C-noradrenaline. *Abscissa:* min after preperfusion with ^{14}C-noradrenaline.
S_1–S_4, 1-min periods of accelerans nerve stimulation at 5 Hz (in *a* and S_1 of *b*) or 2.5 Hz (S_2–S_4
of *b*). *Ordinate:* outflow of radioactive material; *upper curve*, total ^{14}C; *lower curve*, ^{14}C-noradrenaline.
Note logarithmic ordinate scale. From STARKE (1972b)

artery. The strong effect in some tissues is probably mediated by presynaptic
dopamine receptors, which form a population distinct from the α-receptors
(8.). The effect of agonists with predominant affinity to β-adrenoceptors (isoprenaline, orciprenaline) varies. Concentrations up to 10^{-6} M cause either no change
or an increase, presumably through activation of presynaptic β-receptors (9.).
Even if low doses have no effect, higher ones may increase the stimulation-evoked
overflow, probably by inhibiting re-uptake (FARNEBO and HAMBERGER, 1974b;

STARKE et al., 1975c). However, in other cases activation of presynaptic α-receptors prevails at high doses, so that overflow is reduced (KIRPEKAR et al., 1972b, 1973; STARKE, 1973a; STARKE et al., 1975c). Adrenaline is a further compound with a dual effect. Low doses enhance overflow, probably via presynaptic β-receptors, high doses reduce overflow, probably via presynaptic α-receptors (STJÄRNE and BRUNDIN, 1975a).

The decrease of overflow reflects depression of the release per impulse rather than an increase of the degradation or retention of released noradrenaline within the tissue, for the following reasons (cf. Table 6). (1) The β-phenylethylamine derivatives listed in Table 6 do not enhance, but on the contrary inhibit the neuronal and extraneuronal uptake of noradrenaline (IVERSEN, 1975). The imidazole derivatives clonidine, naphazoline, and oxymetazoline do not change neuronal uptake at concentrations which reduce the stimulation-evoked overflow; higher concentrations inhibit uptake (FARNEBO and HAMBERGER, 1971a; STARKE, 1972a; STARKE et al., 1972). Clonidine also inhibits extraneuronal uptake (SALT, 1972). (2) In tissues treated with labeled noradrenaline, α-receptor agonists reduce the stimulation-evoked overflow of both labeled noradrenaline and total radioactivity; they do not decrease the overflow of noradrenaline by promoting its biotransformation. (3) The inhibitory effect is retained after the neuronal, or both the neuronal and extraneuronal uptake have been blocked. (4) Several agonists specified in Table 6 are taken up into noradrenergic nerves and might reduce the release of noradrenaline by being themselves released as false transmitters. In particular, unlabeled noradrenaline might reduce the release of previously stored labeled noradrenaline by diluting the specific activity of the releasable store. Experiments devised to check this possibility are shown in Figure 2a and b. In perfused rabbit hearts, the influence of infused unlabeled ($-$)-noradrenaline on the stimulation-evoked overflow of stored ^{14}C-noradrenaline was tested, firstly, in the presence of cocaine (Fig. 2a), and secondly, in the presence of phenoxybenzamine (Fig. 2b). Under the conditions used, cocaine blocked the neuronal uptake of noradrenaline more effectively than phenoxybenzamine (not shown). Thus, any decrease of ^{14}C overflow resulting from a dilution of specific activity should be more pronounced in the phenoxybenzamine group. However, the opposite was found to be the case: infused noradrenaline diminished the stimulation-evoked overflow of total ^{14}C and ^{14}C-noradrenaline in the presence of cocaine, but not in the presence of phenoxybenzamine. The results argue against a significant reduction of specific activity, and favor an action of the unlabeled noradrenaline on phenoxybenzamine-sensitive receptors.

7.2. α-Adrenolytic Facilitation

α-Adrenolytic drugs enhance the stimulation-evoked overflow of noradrenaline from all tissues studied (Table 7). In guinea-pig hypothalamic slices, phenoxybenzamine and phentolamine, in contrast to piperoxan, have no effect, though phentolamine antagonizes the inhibition caused by noradrenaline. Dihydroergotamine considerably increases the stimulation-evoked overflow from several tissues, but its effect in the rabbit heart is small. Similarly, ergotamine increases

Table 7. Increase by α-adrenolytic drugs of transmitter overflow evoked by electrical stimulation of noradrenergic nerves

Species	Tissue	Drug	References
Dog	Hind paw	Phenoxybenzamine	Zimmerman and Whitmore, 1967
	Gracilis muscle	Phenoxybenzamine	Rosell et al., 1963;
			Zimmerman and Whitmore, 1967
	Subcutaneous	Phenoxybenzamine	Fredholm and Hedqvist, 1973b
	adipose tissue	Dihydroergotamine	Fredholm and Rosell, 1970
	Spleen	Phenoxybenzamine	Potter et al., 1971
		Phentolamine	Potter et al., 1971
	Kidney	Phenoxybenzamine	Zimmerman et al., 1971
Cat	Heart	Phenoxybenzamine	Farah and Langer, 1974
	Aorta	Phentolamine	Langer et al., 1975b
	Hind limb	Phenoxybenzamine	Häggendal, 1970[a]
	Spleen	Phenoxybenzamine	Brown and Gillespie, 1957;
			Brown et al., 1961;
			Kirpekar and Cervoni, 1963[a];
			Thoenen et al., 1964a and b, 1966a and b;
			Brown, 1965;
			Geffen, 1965[a];
			Gillespie and Kirpekar, 1966;
			Kirpekar and Misu, 1967;
			Salzmann et al., 1968;
			Blakeley et al., 1969a and b;
			Hedqvist, 1969b;
			Kirpekar and Wakade, 1970[c];
			Pacha and Salzmann, 1970;
			Kirpekar and Puig, 1971;
			Cripps and Dearnaley, 1972;
			Kao and McCullough, 1973;
			Cubeddu et al., 1974a;
			Dubocovich and Langer, 1974, 1975;
			Cubeddu and Weiner, 1975a;
			Langer et al., 1975a and b;
			Summers and Blakeley, 1975
		Dibenamine	Brown and Gillespie, 1956, 1957
		Phentolamine	Brown et al., 1961;
			Hertting and Schiefthaler, 1963;
			Thoenen et al., 1964b;
			Gillespie and Kirpekar, 1966;
			Kirpekar and Wakade, 1970;
			Kirpekar and Puig, 1971;
			Cubeddu et al., 1974a, 1975;
			Dubocovich and Langer, 1975
		Azapetine	Thoenen et al., 1964b
		Piperoxan	Summers and Blakeley, 1975
		Ergotamine	Salzmann et al., 1968;
			Pacha and Salzmann, 1970
		1-Methyl-ergotamine	Pacha and Salzmann, 1970
		Dihydroergotamine	Pacha and Salzmann, 1970
		Hydergine[d]	Brown et al., 1961;
			Blakeley et al., 1963[a], 1969b;
			Gillespie and Kirpekar, 1966;

Table 7 (continued)

Species	Tissue	Drug	References
	Nictitating membrane	Phenoxybenzamine	PACHA and SALZMANN, 1970; CRIPPS and DEARNALEY, 1972[a] LANGER, 1968, 1970[e]; LANGER et al., 1970, 1971, 1972; LANGER and VOGT, 1971; ENERO et al., 1972; HENDERSON et al., 1972b, 1975; ENERO and LANGER, 1973, 1975
		Phentolamine	LANGER et al., 1970, 1971; ENERO and LANGER, 1975
		Chlorpromazine	ENERO and LANGER, 1975
	Intestine	Phenoxybenzamine	BROWN et al., 1958; BOULLIN et al., 1967
Rat	Mesenteric arteries	Phentolamine	LANGER et al., 1975b and c
	Portal vein	Phenoxybenzamine	HÄGGENDAL et al., 1972
	Iris	Phenoxybenzamine	FARNEBO and HAMBERGER, 1970[a], 1971a[a, c]
		Phentolamine	FARNEBO and HAMBERGER, 1970[a], 1971a[a], 1973a
	Vas deferens	Phentolamine	LANGER, 1970; VIZI et al., 1973
	Superior cervical ganglion	Phenoxybenzamine	VOGEL et al., 1972
	Cerebral cortex	Phenoxybenzamine	FARNEBO and HAMBERGER, 1971b, 1973b; STARKE, 1973b; STARKE and MONTEL, 1973b and d; HOWD and HORITA, 1975
		Phentolamine	FARNEBO and HAMBERGER, 1971b, 1973b, 1974b; STARKE, 1973b; STARKE and MONTEL, 1973b–d
		Chlorpromazine	FARNEBO and HAMBERGER, 1971b, 1973b
Mouse	Heart	Phenoxybenzamine	FARNEBO and HAMBERGER, 1974b
	Vas deferens	Phenoxybenzamine	FARNEBO and MALMFORS, 1971; JENKINS et al., 1975
		Phentolamine	FARNEBO and MALMFORS, 1971; HUGHES et al., 1975a
Guinea pig	Heart	Phenoxybenzamine	LANGER et al., 1970, 1971; McCULLOCH et al., 1972; RAND et al., 1973
		Phentolamine	LANGER et al., 1971; McCULLOCH et al., 1972; ADLER-GRASCHINSKY and LANGER, 1975
		Tolazoline	LANGER et al., 1971
		Azapetine	McCULLOCH et al., 1972
		Dihydroergotamine	McCULLOCH et al., 1972
	Aorta	Phenoxybenzamine	BELL, 1974
	Uterine artery	Phenoxybenzamine	BELL and VOGT, 1971[a]
	Intestine	Phenoxybenzamine	HENDERSON et al., 1972b, 1975
	Vas deferens	Phenoxybenzamine	JOHNSON et al., 1971; HEDQVIST, 1973a, 1974a and c[a, b]; STJÄRNE, 1973c[a], 1[a, b], 1976[a]

Table 7 (continued)

Species	Tissue	Drug	References
Guinea pig	Vas deferens	Phentolamine	Stjärne, 1972b[a], 1973a[a, b], d–l[a]; Hedqvist, 1973e[a, b], 1974c[a, b], d; Hughes, 1973
		Azapetine	Hedqvist, 1974c[a, b]
		Hydergine[d]	Hedqvist, 1974c[a, b]
		Chlorpromazine	Hedqvist, 1973c[a]
	Hypothalamus	Piperoxan	Bryant et al., 1975[f]
Rabbit	Heart	Phenoxybenzamine	Starke et al., 1971a; Stjärne and Wennmalm, 1971; Wennmalm, 1971; Starke, 1972a and b; Starke and Schümann, 1972
		Phentolamine	Starke et al., 1971b[a]; Starke, 1972a, 1973b[a]; Starke and Altmann, 1973; Starke and Montel, 1973a[b]
		Dihydroergotamine	Starke, 1972a
		Hydergine[d]	Stjärne and Wennmalm, 1971[a]; Wennmalm, 1971[a]
	Ear artery	Phenoxybenzamine	Rand et al., 1973
	Pulmonary artery	Phenoxybenzamine	Su and Bevan, 1970a; Borowski et al., 1976[a]
		Phentolamine	Starke et al., 1974b[a], 1975b[a]; Borowski et al., 1976[a]
		Tolazoline	Borowski et al., 1976[a]
		Azapetine	Borowski et al., 1976[a]
		Piperoxan	Borowski et al., 1976[a]
		Yohimbine	Starke et al., 1975a[a]; Borowski et al., 1976[a]
		Ergotamine	Starke, unpublished[a]
		Dihydroergotamine	Borowski et al., 1976[a]
	Portal vein	Phenoxybenzamine	Hughes, 1972[a]
	Vas deferens	Phenoxybenzamine	Hughes, 1972[a]
	Superior cervical ganglion	Phenoxybenzamine	Noon and Roth, 1975
Calf	Spleen	Phenoxybenzamine	Smith et al., 1970
Man	Arteries and veins	Phentolamine	Stjärne and Gripe, 1973[a]
	Oviduct	Phenoxybenzamine	Hedqvist and Moawad, 1975[a]

[a] Experiments performed in the presence of inhibitors of neuronal and/or extraneuronal noradrenaline uptake.

[b] Experiments performed in the presence of drugs which inhibit prostaglandin biosynthesis.

[c] The weakly α-adrenolytic β-haloalkylamine N-cyclohexylmethyl-N-ethyl-β-chloroethylamine (GD131) had no effect.

[d] Hydergine is a mixture of equal parts (weight) of dihydroergocornine, dihydroergocristine, and dihydroergokryptine.

[e] Phentolamine had no effect (25 Hz).

[f] Phenoxybenzamine and phentolamine had no effect.

the stimulation-evoked overflow from the cat spleen only within a narrow concentration range. In the rabbit pulmonary artery, ergotamine causes a minimal increase at 3×10^{-8} M, but a decrease at 10^{-7} M and higher concentrations. The slight and sometimes even inhibitory effect of some ergot alkaloids may be related to their partial agonist character; for instance, in the rabbit pulmonary artery the presynaptic agonist effect of ergotamine appears to prevail.

The increase of the stimulation-evoked overflow caused by low concentrations of α-adrenolytic drugs reflects an increase of the per pulse release of noradrenaline. Only at high concentrations, a decrease of its inactivation within the tissue may contribute (p. 58). The evidence for facilitation of release is as follows.

1. Phenoxybenzamine and phentolamine do not inhibit MAO and COMT (EISENFELD et al., 1967). — Several α-adrenolytic drugs inhibit the neuronal uptake of noradrenaline (IVERSEN, 1975). However, in general higher concentrations are required for uptake blockade than for enhancement of the stimulation-evoked overflow. Dissociation of the two effects early became apparent (KIRPEKAR and CERVONI, 1963; BOULLIN et al., 1967). Detailed dose–response studies have been performed in the cat spleen (phenoxybenzamine and phentolamine: CUBEDDU et al., 1974a and c) and nictitating membrane (phenoxybenzamine: LANGER et al., 1970, 1971, 1972; ENERO et al., 1972), the iris of the rat (phenoxybenzamine and phentolamine: FARNEBO and HAMBERGER, 1971a) and the rabbit heart (phenoxybenzamine: STARKE et al., 1971a; phentolamine: STARKE et al., 1971b; dihydroergotamine: STARKE, 1972a). The studies agree that low concentrations of phentolamine and dihydroergotamine enhance the stimulation-evoked overflow without any effect on uptake; for phenoxybenzamine, the dissociation is evident in the cat nictitating membrane and the rabbit heart, but is perhaps less clear in the cat spleen and the rat iris. — In the nictitating membrane of the cat, a 100-fold higher concentration of phenoxybenzamine is necessary to reduce extraneuronal uptake than to increase the stimulation-evoked overflow (ENERO et al., 1972; LANGER et al., 1972). — These results argue against blockade of the classical pathways of noradrenaline inactivation by low doses of phenoxybenzamine, phentolamine, and dihydroergotamine. However, a further possible pathway has recently been described by BLAKELEY et al. (1973, 1974). In the spleen and heart of cats, rapidly injected noradrenaline is in part taken up by a mechanism that cannot be blocked by combined administration of desipramine and estradiol and thus is distinct from known neuronal and extraneuronal uptake. In the spleen, the desipramine- and estradiol-resistant uptake is reduced by 9×10^{-5} M phenoxybenzamine. However, it is unlikely that this blockade significantly contributes to α-adrenolytic enhancement of the stimulation-evoked overflow. The concentration of phenoxybenzamine is 3000 times that required to augment the stimulation-evoked overflow from the spleen (3×10^{-8} M; DUBOCOVICH and LANGER, 1974). Moreover, phenoxybenzamine does not block uptake from pulses in the cat heart (BLAKELEY et al., 1973), yet increases the stimulation-evoked overflow (FARAH and LANGER, 1974). In the spleen, piperoxan enhances the stimulation-evoked overflow without affecting uptake from pulses (SUMMERS and BLAKELEY, 1975).

2. In tissues treated with labeled noradrenaline, α-adrenolytic drugs enhance the stimulation-evoked overflow of both labeled noradrenaline and total radioactivity; a block of biotransformation does not account for the increase of noradrenaline overflow (e.g., FARNEBO and HAMBERGER, 1971 a and b, 1974 b; LANGER et al., 1972; STARKE, 1973 b; STJÄRNE, 1973 e; STJÄRNE and GRIPE, 1973; CUBEDDU et al., 1974 a; STARKE et al., 1974 b, 1975 a; ADLER-GRASCHINSKY and LANGER, 1975).

3. The increase persists, after the neuronal and/or extraneuronal uptake mechanisms have been blocked (references indicated by a superscript "a" in Table 7).

4. α-Adrenolytic drugs enhance the stimulation-evoked overflow of dopamine-β-hydroxylase, indicating facilitation of exocytotic release (JOHNSON et al., 1971; POTTER et al., 1971; CUBEDDU et al., 1974 a; CUBEDDU and WEINER, 1975 a).

5. Finally, high doses of α-receptor agonists prevent the increase caused by low doses of α-adrenolytic drugs (p. 59). The interaction is difficult to explain if one assumes that the adrenolytic agents interfere with the retention or degradation of released noradrenaline; it points to an antagonism at release-modulating sites.

Taken together, these results indicate that low doses of α-adrenolytic drugs increase the stimulation-evoked overflow solely by facilitation of release. However, at high concentrations other effects may contribute. In the rabbit heart, 10^{-6} M phentolamine (sub-threshold for uptake inhibition) and cocaine have additive effects on the stimulation-evoked overflow. In contrast, in the presence of 10^{-5} M phentolamine cocaine does not cause any further increase, presumably because at this concentration phentolamine itself inhibits neuronal uptake (STARKE et al., 1971 b; cf. for phenoxybenzamine CRIPPS and DEARNALEY, 1972; HUGHES, 1972). Thus, a second component in the large effects of high doses of phentolamine and phenoxybenzamine is a block of re-uptake.

Another factor was disclosed when it was found that a high concentration of phenoxybenzamine prevents the degradation of released noradrenaline, probably by denying the transmitter access to the metabolizing enzymes (LANGER, 1968, 1970). A detailed analysis of the effect of phenoxybenzamine on the metabolic pattern of released ^{3}H-noradrenaline is shown in Figure 3. The most marked changes are, (1) a decrease of the percentage recovered as ^{3}H-DOPEG caused by 3×10^{-7} and even more by 3×10^{-6} M phenoxybenzamine; and (2) inhibition of the formation of ^{3}H-normetanephrine and ^{3}H-O-methylated deaminated products by 3×10^{-6} M phenoxybenzamine. The latter concentration almost completely prevents metabolism. It should be noted that 3×10^{-8} M phenoxybenzamine and higher concentrations cause a dose-dependent increase of the stimulation-evoked overflow of total tritium. The results indicate that 3×10^{-8} M phenoxybenzamine selectively facilitates release; owing to higher substrate concentration, the metabolites are proportionately increased, so that the percentage composition is unchanged. At 3×10^{-7} and 3×10^{-6} M, phenoxybenzamine inhibits neuronal re-uptake of noradrenaline and formation of the neuronal metabolite ^{3}H-DOPEG. At 3×10^{-6} M, it inhibits extraneuronal uptake and formation of the extraneuronal metabolite ^{3}H-normetanephrine as well (cf. p. 6). All factors combine to produce a large increase of the stimulation-evoked overflow of ^{3}H-noradrenaline (15-fold at 3×10^{-6} M).

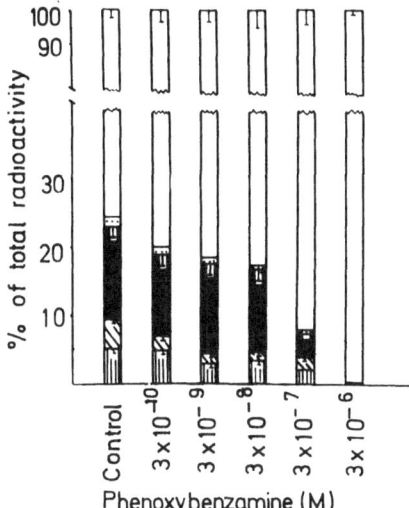

Fig. 3. Effect of phenoxybenzamine on the metabolic pattern of ^{3}H-noradrenaline released by nerve stimulation in the isolated cat spleen. Spleens were preperfused with ^{3}H-noradrenaline. The splenic nerves were stimulated at 5 Hz for 1 min. The overflow of radioactive compounds during and after stimulation (in nCi) was corrected for the spontaneous outflow. The stimulation-evoked overflow of individual metabolite fractions was then expressed as per cent of the stimulation-evoked overflow of total tritium. □, ^{3}H-noradrenaline; ▨, ^{3}H-3,4-dihydroxymandelic acid; ■, ^{3}H-DOPEG; ▨, ^{3}H-O-methylated deaminated metabolites; ▥, ^{3}H-normetanephrine. From CUBEDDU et al. (1974a)

7.3. The Presynaptic α-Receptor

α-Receptor agonists reduce and α-adrenolytic drugs enhance the release of nor-adrenaline by interaction with one and the same set of receptors. This was borne out by the demonstration that the blocking agents antagonize the inhib-itory effect of the agonists (Table 6). Phentolamine shifts dose–response curves for the presynaptic effects of clonidine, oxymetazoline, and noradrenaline to the right; the degree of the shift is similar for either agonist, indicating an action on one receptor system (STARKE et al., 1974b, 1975b).

Support for a common site comes from the observation that high doses of the agonists abolish α-adrenolytic facilitation, presumably by keeping the receptors activated even in the presence of the antagonists. This type of experi-ment is illustrated in Figure 4. Oxymetazoline, at a concentration supramaximal with respect to depression of release, prevents the effect of 10^{-6} M phentolamine (cf. MCCULLOCH et al., 1972; STARKE and ALTMANN, 1973; STARKE and MONTEL, 1973b–d; STARKE et al., 1975a).

Facilitation of release by phenoxybenzamine persists even after prolonged washout (STARKE et al., 1971a; FARAH and LANGER, 1974). When a high concen-tration of phentolamine is added before phenoxybenzamine, this long-lasting facilitation is prevented. Apparently phentolamine protects the α-receptors from irreversible blockade by phenoxybenzamine, indicating that the two drugs act on the same presynaptic receptors (FARAH and LANGER, 1974).

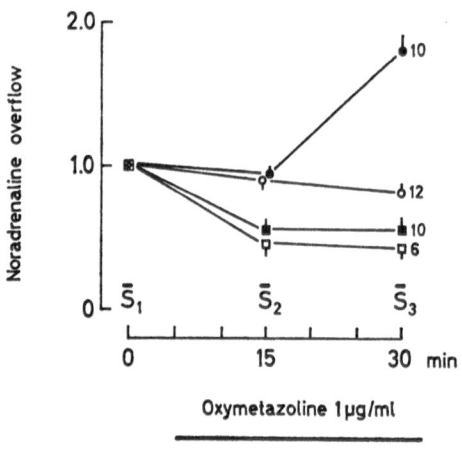

Fig. 4. Interaction of oxymetazoline and phentolamine on the stimulation-evoked overflow of nor-adrenaline from the isolated rabbit heart. S_1–S_3, 1-min periods of accelerans nerve stimulation at 5 Hz. *Ordinate:* ratio between the overflow of noradrenaline evoked by S_2 or S_3 and that evoked by S_1. ○, no drugs; □, oxymetazoline; ●, phentolamine; ■, oxymetazoline + phentolamine. Figures on the right indicate numbers of experiments. From STARKE (1972a)

The presynaptic receptors resemble classical α-adrenoceptors, since the order of the presynaptic potency of several β-phenylethylamines agrees with their known order of postsynaptic potency (KIRPEKAR et al., 1972b, 1973; STARKE, 1972b, 1973a; HEDQVIST, 1974c; BRYANT et al., 1975; RAND et al., 1975b; STARKE et al., 1975b and Table 8; STJÄRNE, 1975b). In contrast to (−)-noradren-aline, the (+)-enantiomer fails to depress release (McCULLOCH et al., 1973; RAND et al., 1973; STJÄRNE, 1974).

Yet, the agreement is not perfect. The following studies on the relative potencies of *agonists* indicate that, in a given tissue, pre- and postsynaptic α-receptors may differ in their structure. (1) Not only β-, but also α-adrenocep-tors mediate positive inotropic effects in a variety of heart preparations (e.g., GOVIER, 1968) including the perfused rabbit heart (STARKE, 1972a; WAGNER et al., 1974). In the rabbit heart, low concentrations of phenylephrine cause α-adrenergic increase of contractile force, but fail to inhibit noradrenaline release. In contrast, low concentrations of oxymetazoline or naphazoline are devoid of inotropic effects, but strongly reduce release. The different relative pre- and postsynaptic potencies suggest that the postsynaptic "myocardial phenylephrine receptors are probably not identical with those mediating the inhibition of noradrenaline secretion" (STARKE, 1972a). (2) Lysergic acid diethylamide does not elicit contraction of the guinea-pig vas deferens, but reduces the release of noradrenaline probably via presynaptic α-receptors. Again a difference be-tween pre- and postsynaptic sites seems likely (HUGHES, 1973). (3) The relative pre- and postsynaptic potencies of nine agonists were investigated in detail in the rabbit pulmonary artery (STARKE et al., 1974b, 1975b). Concentrations which caused 20% of the maximal contractions, concentrations which reduced

Table 8. Pre- and postsynaptic potencies of α-adrenoceptor agonists in the rabbit pulmonary artery

Drug	$EC_{20\,post}$ (M)	$EC_{20\,pre}$ (M)	$\dfrac{EC_{20\,pre}}{EC_{20\,post}}$
($\pm$)-Methoxamine	7.4×10^{-7}	2.4×10^{-5}	32.5
($-$)-Phenylephrine	5.4×10^{-8}	1.7×10^{-6}	30.9
($-$)-Noradrenaline	7.3×10^{-9}	1.2×10^{-8}	1.6
($-$)-Adrenaline	3.2×10^{-9}	1.9×10^{-9}	0.58
Naphazoline	3.8×10^{-8}	1.6×10^{-8}	0.41
Oxymetazoline	1.8×10^{-8}	3.1×10^{-9}	0.17
Clonidine	6.5×10^{-8}	1.0×10^{-8}	0.15
($-$)-$erythro$-α-methylnoradrenaline	6.1×10^{-8}	8.1×10^{-9}	0.13
Tramazoline	5.6×10^{-8}	3.7×10^{-9}	0.07

Artery strips were superfused with medium containing propranolol, cocaine, and corticosterone in order to block β-receptors and the neuronal and extraneuronal uptake of noradrenaline, respectively. For determination of presynaptic effects, they were preincubated with ^{3}H-noradrenaline. $EC_{20\,post}$, concentrations which caused 20% of the maximal contraction obtainable with the respective drug. $EC_{20\,pre}$, concentrations which reduced the stimulation-evoked overflow of tritium by 20%. From STARKE et al. (1975b).

transmitter release by 20%, and a potency ratio are compiled in Table 8. There is similarity between pre- and postsynaptic receptors in that the rank order of potency of β-phenylethylamines on either system is ($-$)-adrenaline> ($-$)-noradrenaline> ($-$)-phenylephrine> ($\pm$)-methoxamine. However, the analysis also reveals great differences. For instance, clonidine, α-methylnoradrenaline and tramazoline are about as potent as phenylephrine at the postsynaptic site; however, they are 170–460 times more potent at the presynaptic site. According to the $EC_{20\,pre}/EC_{20\,post}$ ratios, the agonists can be arbitrarily classified into three groups. Group 1 (ratio about 30; preferentially postsynaptic agonists) comprises methoxamine and phenylephrine. Group 2 (ratio near 1; similar pre- and postsynaptic potencies) comprises noradrenaline, adrenaline, and naphazoline. Group 3 (ratio below 0.2; preferentially presynaptic agonists) comprises oxymetazoline, clonidine, α-methylnoradrenaline, and tramazoline. The large variation of the relative potencies is compatible with the view that there are differences between pre- and postsynaptic α-adrenoceptors which make them selectively sensitive to certain agonists.

Studies with *antagonists* lead to the same conclusion. It has been known for some time that in several tissues higher doses of phenoxybenzamine or phentolamine are required to enhance the stimulation-evoked overflow of noradrenaline than to reduce the postsynaptic response (dog kidney: ZIMMERMAN et al., 1971; cat spleen: THOENEN et al., 1964a and b; SALZMANN et al., 1968; LANGER, 1973a and b; CUBEDDU et al., 1974a; DUBOCOVICH and LANGER, 1974; rat portal vein: HÄGGENDAL et al., 1972). These observations can now be explained by the existence of presynaptic α-receptors which differ from postsynaptic ones, though a greater number of spare receptors at the presynaptic site cannot be excluded (LANGER, 1973a and b; CUBEDDU et al., 1974a; DUBOCOVICH and LANGER, 1974). Pre- and postsynaptic effects of nine antagonists have been compared in detail in the rabbit pulmonary artery (BOROWSKI et al., 1976).

In analogy to the findings for agonists, some antagonists, such as phenoxybenza-mine and azapetine, preferentially block postsynaptic α-receptors. At the other extreme, yohimbine preferentially blocks presynaptic α-receptors (cf. STARKE et al., 1975a).

All these data are compatible with the view that in a given tissue pre- and postsynaptic α-receptors may differ in their structure. However, it would certainly be premature to consider postsynaptic α-receptors as one homogeneous group which can be confronted with presynaptic α-receptors as the second homogeneous group. There are indications that postsynaptic α-receptors are not all of a single type (MUJIĆ and VAN ROSSUM, 1965; SHEYS and GREEN, 1972). The same may hold good for presynaptic α-receptors. More evidence is needed before, perhaps, some generalizations can be justified.

Little is known about the presynaptic intrinsic activity of drugs with affinity to α-receptors. It has been pointed out that a presynaptic partial agonist should *facilitate* release when the concentration of noradrenaline in the fluid surround-ing the nerve endings is high (STARKE et al., 1974b). STJÄRNE (1975d) recently reported that clonidine enhances release, and simultaneously the postsynaptic response to stimulation, in the guinea-pig vas deferens. The result was unexpected in view of the numerous demonstrations of clonidine's inhibitory effect (Table 6). Moreover, there are reports that clonidine reduces rather than enhances the contractile response to stimulation in the vas deferens of the rat (VIZI et al., 1973) and guinea pig (BENTLEY and LI, 1968; EULER and HEDQVIST, 1975). Perhaps the use of uptake blocking agents in STJÄRNE's experiments led to a high biophase concentration of noradrenaline despite the low stimulation frequency of 1 Hz, thus disclosing a partial agonist character of clonidine.

Pre- and postsynaptic α-receptor mechanisms also appear to differ in their behavior after decentralization. The postsynaptic cells develop supersensitivity (TRENDELENBURG, 1963). In contrast, the presynaptic α-adrenergic mechanism in the rat iris becomes subsensitive to noradrenaline, so that release per pulse rises (FARNEBO and HAMBERGER, 1973a). No change in presynaptic sensitivity was observed in the nictitating membrane of the cat (LANGER et al., 1975b).

7.4. α-Adrenergic Negative Feedback

Facilitation by α-adrenolytic drugs is the basis for the hypothesis that previously released noradrenaline acts on the presynaptic α-receptors and inhibits release by subsequent impulses. Evidence in support of this hypothesis is, (1) an inverse relation between the biophase concentration of released noradrenaline in the vicinity of the presynaptic receptors and the quantum released per pulse; (2) an inverse relation between the biophase concentration of released noradrenaline and the degree of inhibition by exogenous agonists; (3) a positive correlation between the biophase concentration of noradrenaline and the degree of facilitation by α-adrenolytic drugs. (4) Part of the depression occurring during trains of pulses can probably be explained by α-receptor-mediated feedback inhibition.

1. If noradrenaline reduces its own further release, then the secretion per impulse should *ceteris paribus* be higher, the lower the biophase concentration of the transmitter. Experimental trial requires stimulation at constant frequency, since changes in frequency *per se* can lead to changes in release (2.2.). The biophase concentration is lowered despite a constant rate of impulse flow when the transmitter stores are partly depleted by reserpine or α-methyltyrosine. In agreement with prediction, the fraction of the total tissue noradrenaline released per pulse ("fractional overflow") is greater after depletion than in control experiments (ENERO and LANGER, 1973; CUBEDDU and WEINER, 1975a).

2. According to the hypothesis the inhibitory effect of exogenous agonists should be smaller, the higher the biophase concentration of noradrenaline. Experiments on perfused rabbit hearts are illustrated in Figure 5. The accelerans nerves were stimulated at 5 Hz. As predicted, a plot of per cent inhibition caused by oxymetazoline against the stimulation-evoked overflow of noradrenaline (thought to reflect the biophase concentration) before oxymetazoline yielded a significant negative correlation.

The effect of presynaptic modulators declines at high frequencies, perhaps because at high frequency the nerve endings are flooded with calcium (p. 5). The same reason may partly account for the frequency-dependence of α-adrenergic inhibition (cf. Table 6). However, experiments such as the one shown in Figure 5 clearly indicate that in the case of α-receptor agonists a second mechanism contributes: as the frequency rises, the presynaptic α-receptors become more and more activated by released noradrenaline, and any additional inhibition by exogenous agonists is minimized (STARKE, 1972a; STARKE and ALTMANN, 1973; VIZI et al., 1973).

3. If α-adrenolytic drugs facilitate release by abolishing the normal braking effect of noradrenaline, then facilitation should be greater, the higher the biophase concentration of the transmitter. However, the prediction holds good only for high antagonist concentrations or irreversible block; the effect of a

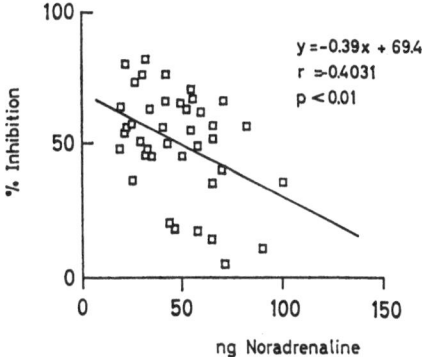

Fig. 5. Relation between pre-oxymetazoline noradrenaline overflow and degree of inhibition caused by oxymetazoline. Each point represents one experiment in an isolated rabbit heart. The accelerans nerves were stimulated twice for 1 min each at 5 Hz (S_1, S_2). Oxymetazoline, final concentration 4×10^{-6} M, was added 10 min before S_2. *Abscissa:* noradrenaline overflow evoked by S_1. *Ordinate:* per cent inhibition caused by oxymetazoline, with allowance for the decline of overflow in control experiments. From STARKE (1972a)

small dose of a competitive antagonist may be surmounted by an increase in noradrenaline concentration. Moreover, a constant frequency of stimulation is again desirable. The prediction has been borne out in experiments on tissues from cats pretreated with reserpine or α-methyltyrosine. Phenoxybenzamine increased the fractional overflow 8–13-fold in controls, but only 3-fold after noradrenaline depletion (ENERO and LANGER, 1973; CUBEDDU and WEINER, 1975a).

4. Biochemical and electrophysiologic evidence indicates that preceding noradrenergic nerve impulses can inhibit release by subsequent impulses (2.2.). It now seems likely that this phenomenon is at least partly due to α-adrenergic feedback inhibition. For instance, in guinea-pig atria the overflow per pulse decreases with successive pulses; the depression is not changed by cocaine, but attenuated by phenoxybenzamine, which interrupts the feedback loop (RAND et al., 1973). – When the sympathetic nerves of the mouse vas deferens are stimulated at more than 1 Hz, facilitation of the e.j.p.s is soon followed by depression (Fig. 6a; cf. p. 4). Under three different experimental conditions the depression is reversed. Firstly at high magnesium concentrations (BENNETT, 1973a); secondly after pretreatment with reserpine (BENNETT and MIDDLETON, 1975a); and thirdly in the presence of α-adrenolytic drugs (Fig. 6; BENNETT, 1973b; BENNETT and MIDDLETON, 1975b). Though other interpretations cannot be excluded (BENNETT, 1973a; BENNETT and MIDDLETON, 1975a) it is tempting to speculate that the common denominator is attenuation of α-adrenergic feedback inhibition. Magnesium and reserpine diminish the amount of noradrenaline released by the first pulses in the train; thus, the normal feedback is reduced, and facilitation continues unimpeded. On the other hand, and as discussed by BENNETT and MIDDLETON (1975b), α-adrenolytic drugs block the presynaptic receptors and interrupt the feedback in spite of a normal or even increased amount of noradrenaline released by the first pulses. – It would be of interest to study the effect of high magnesium on α-adrenergic inhibition and α-adrenolytic facilitation with overflow methods.

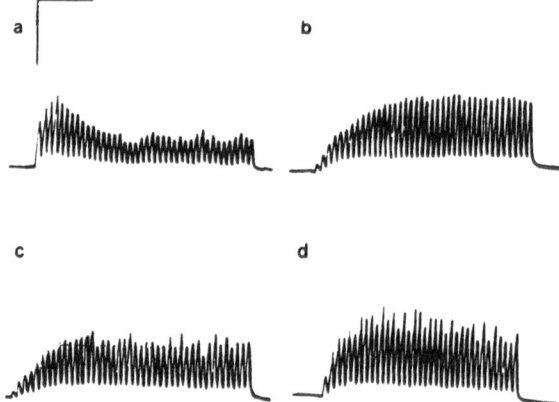

Fig. 6a–d. Effect of α-adrenolytic drugs on e.j.p.s in smooth muscle cells of the mouse vas deferens during short trains of sympathetic nerve impulses at 10 Hz. Intracellular potentials were recorded with glass microelectrodes. Calibrations: vertical, 10 mV; horizontal, 1 sec. a, control; b–d, 30 min after the addition of 3×10^{-5} M of dihydroergocornine (b), dibenamine (c), or phentolamine (d). From BENNETT and MIDDLETON (1975b)

7.5. Mechanism of Inhibition

The mechanism of α-adrenergic inhibition is not known. Some information can be derived from the following observations. (1) The effect of α-receptor agonists has been differentiated from local anesthesia and a guanethidine-like adrenergic neurone blockade. One basic difference is that α-adrenergic inhibition is never complete; a small release persists even in the presence of supramaximal agonist concentrations (STARKE, 1972a; STARKE et al., 1972). – Several α-adrenolytic drugs were found to have no effect on the electrophysiologic properties of preterminal noradrenergic nerves (KAO and McCULLOUGH, 1973; BENNETT and MIDDLETON, 1975b). (2) Agonists as well as antagonists affect noradrenaline storage vesicles (see EULER, 1972). However, it is unlikely that granular actions are important for modulation of release, since the agonists inhibit release even when their neuronal uptake, and thereby access to the granules, is blocked. (3) Inhibition is not secondary to diminished biosynthesis, since the release of exogenous, previously taken up noradrenaline is equally depressed (Table 6). (4) The effects of agonists and antagonists are not changed after the formation of prostaglandins has been blocked. Inhibition is not mediated by prostaglandins (HEDQVIST, 1973e, 1974c; STARKE and MONTEL, 1973a and c; STJÄRNE, 1973a and k). (5) α-Adrenolytic drugs probably do not act by increasing intraneuronal cyclic AMP. Phentolamine inhibits phosphodiesterase, but only at concentrations exceeding those that facilitate release. Moreover, the facilitatory effect of phentolamine is much greater than that of phosphodiesterase inhibitors which do not block α-receptors, and is not changed by the latter (CUBEDDU et al., 1975).

(6) The agonists reduce and antagonist enhance not only release evoked by electrical stimulation, but also that evoked by high potassium concentrations (KIRPEKAR and WAKADE, 1968; STARKE and MONTEL, 1973e, 1974; STJÄRNE, 1973h and k). The effect on release evoked by DMPP is inconclusive (STARKE and MONTEL, 1974); release is depressed by oxymetazoline, but also by phentolamine, and an "unspecific" interference with the chain of events triggered by presynaptic nicotine receptor activation cannot be excluded (cf. p. 25). On the other hand, postsynaptic response studies indicate that α-adrenolytic drugs are indeed able to facilitate release by nicotinic agents (FURCHGOTT et al., 1975). In contrast, the basal outflow of noradrenaline is not affected by concentrations which markedly modify the secretory response to nerve impulses (see Table 6 for agonists). Moreover, release evoked by tyramine (STARKE and MONTEL, 1973e, 1974) and the calcium-independent potassium-evoked release of *cytoplasmic* noradrenaline (PATON, 1975) are not changed. Thus, α-receptor agonists selectively inhibit, and antagonists enhance, calcium-dependent release processes.

(7) STJÄRNE (1973e) investigated the calcium-dependence of the stimulation-evoked overflow of tritium from guinea-pig vas deferens preincubated with ^{3}H-noradrenaline. In experiments performed in the presence of phentolamine or phenoxybenzamine, a double reciprocal plot of the fractional overflow against the calcium concentration yielded a straight line, indicating "that the disinhibited secretory mechanism is basically a simple function of the calcium concentration in the medium". In contrast, in control experiments in the absence of α-adreno-

lytic drugs the plot had a curved shape, suggesting that the feedback inhibition "may be in part due to restriction of calcium entry during depolarization". (8) The inhibitory effect of α-receptor agonists is inversely related to stimulation frequency (Table 6). One reason is increasing feedback inhibition with increasing biophase concentrations of noradrenaline around presynaptic receptors, so that little room is left for additional inhibition by exogenous agonists (p. 63). However, a second reason may correspond to that discussed for other presynaptic modulators, namely the intraneuronal accumulation of calcium at high frequencies, so that effects of drugs which modify calcium fluxes are minimized. – Taken together, these data are compatible with the view that the activation of presynaptic α-receptors leads to a decrease, and their blockade to an increase, of the availability of calcium for stimulus-secretion coupling (STARKE and MONTEL, 1973e, 1974; STJÄRNE, 1973e; LANGER et al., 1975a). The chain of events that leads to the proposed reduction of intraneuronal calcium remains unknown.

An interesting alternative proceeds from the assumption that calcium induces release by inhibiting the Na^+–K^+-activated ATPase of the nerve terminal membrane (see VIZI, 1975). In contrast to calcium, noradrenaline increases synaptosomal ATPase activity, and the increase is prevented by phentolamine (GILBERT et al., 1975). The authors suggest that stimulation of the enzyme may be the basis for the release-inhibiting effect of α-receptor agonists (cf. for cholinergic neurones VIZI, 1975).

7.6. Effect of Drugs with α-Receptor Affinity on Noradrenergic Synaptic Transmission

Drugs with affinity for α-receptors have two primary sites of action in noradrenergic synapses: the post- and the presynaptic α-adrenoceptors. Their overall postsynaptic effect results from two components: direct interaction with the postsynaptic receptors, and modulation of release, the modulation leading to an indirect effect on the postsynaptic receptors. The presynaptic component can be neglected when a nerve ending does not secrete, as for instance in the absence of impulses or after noradrenaline depletion. However, the presynaptic site gains functional importance as soon as the nerve ending actively secretes the transmitter. Prior to the development of the presynaptic α-receptor hypothesis, only the direct postsynaptic component was considered. The following discussion deals with the role of the presynaptic component in the overall postsynaptic effect of α-receptor agonists and antagonists.

7.6.1. Synapses with Postsynaptic β-Receptors

When drugs with α-receptor affinity are administered to actively transmitting synapses with postsynaptic receptors predominantly of the β-type, their overall postsynaptic effect should be governed by the presynaptic component. In agreement with this prediction it has been shown that α-receptor agonists such as clonidine, xylazine, and oxymetazoline reduce the positive chronotropic and inotropic effect of sympathetic nerve stimulation on various cardiac preparations

(dog: SCRIABINE et al., 1970; SCRIABINE and STAVORSKI, 1973; ROBSON and AN-
TONACCIO, 1974; see, however, ANTONACCIO et al., 1974; cat: KOBINGER, 1967;
PACHA et al., 1975; SCHOLTYSIK et al., 1975; ILHAN et al., 1976; rat: ARMSTRONG
and BOURA, 1973; DREW, 1976; guinea pig: VIZI et al., 1973; rabbit: STARKE,
1972a; WERNER et al., 1972). In analogy to the overflow studies, the inhibition
is smaller, the higher the frequency of stimulation.

The inhibition of sympathetic cardiac effects caused by the agonists is antagonized by α-adrenolytic drugs (VIZI et al., 1973; ROBSON and ANTONACCIO,
1974; PACHA et al., 1975; SCHOLTYSIK et al., 1975; DREW, 1976). Given alone,
the antagonists may enhance cardiac responses to sympathetic nerve impulses
(KAUMANN, 1970; STARKE et al., 1971b; WENNMALM, 1971; McCULLOCH et al.,
1972; STARKE and SCHÜMANN, 1972; FARAH and LANGER, 1974); it is likely
that in some of these experiments inhibition of noradrenaline re-uptake contrib-
uted to the effects. In other investigations, an α-adrenolytic increase of cardiac
responses was not observed (VIZI et al., 1973; ANTONACCIO et al., 1974; ROBSON
and ANTONACCIO, 1974; ADLER-GRASCHINSKY and LANGER, 1975). The reason
is not known.

Phenoxybenzamine enhances the relaxation of intestinal smooth muscle
caused by sympathetic nerve stimulation (KIRPEKAR and CERVONI, 1963; BURN
and GIBBONS, 1964a; cf. BURN and NG, 1965). Again, the presynaptic component,
namely facilitation of release, may be involved.

7.6.2. Synapses with Postsynaptic α-Receptors

In actively transmitting synapses with postsynaptic receptors predominantly
of the α-type, the overall postsynaptic effect of α-receptor agonists depends on
their relative pre- and postsynaptic potency, their concentration, and the
stimulation frequency. A detailed analysis has been attempted in the rabbit
pulmonary artery. In this tissue, oxymetazoline, clonidine, α-methylnoradrenaline, and tramazoline preferentially activate presynaptic α-receptors (Table 8).
Low doses should selectively depress the release of noradrenaline and reduce
the postsynaptic response to stimulation. This prediction has been verified.
The effect of tramazoline is shown in Figure 7. The inhibition caused by 3×10^{-9}
and 10^{-8} M tramazoline is restricted to low frequencies. Only at the high concentration of 10^{-6} M, which causes a large contraction, does the direct postsynaptic
action overshadow the inhibition of release. Similar results have been obtained
with oxymetazoline, clonidine, and α-methylnoradrenaline (STARKE et al., 1974b,
1975b). The prevailing effect of low doses is α-adrenergic inhibition of neurogenic
vasoconstriction, or in more general terms α-adrenergic inhibition of noradrenergic synaptic transmission. — The preferentially postsynaptic agonists, methoxamine
and phenylephrine, do not reduce the postsynaptic response at any concentration
(STARKE et al., 1975b). On the contrary, even low concentrations without overt
contractile effect tend to enhance neurogenic vasoconstriction.

Inhibition of noradrenergic synaptic transmission by α-receptor agonists,
and in particular by clonidine, has also been observed in other tissues (cat
hind limb: HAEUSLER, 1975a; vascular bed of the cat superior mesenteric artery:

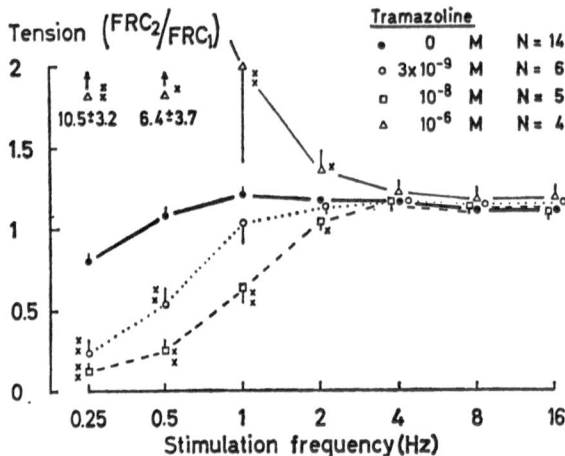

Fig. 7. Effect of tramazoline on the contractile response of rabbit pulmonary arteries to sympathetic nerve stimulation. Artery strips were superfused with propranolol, cocaine, and corticosterone in order to block β-receptors and the neuronal and extraneuronal uptake of noradrenaline, respectively. Two frequency–response curves were determined on each strip (FRC$_1$, FRC$_2$). Tramazoline was added before FRC$_2$. For each frequency, the ratio was calculated between the tension developed in FRC$_2$ (including tramazoline-induced basal tension) and that developed in FRC$_1$ (FRC$_2$/FRC$_1$). From STARKE et al. (1975 b)

SANDERS and ROSS, 1975; vasculature of the pithed rat: BENTLEY and LI, 1968; BOISSIER et al., 1968; rat vas deferens: VIZI et al., 1973; guinea-pig vas deferens see SJÖSTRAND, 1973 b; footnote p. 4; rabbit ear artery: BENTLEY and LI, 1968; BOISSIER et al., 1968; STEINSLAND and NELSON, 1975). The rabbit ear artery resembles the pulmonary artery in that clonidine is a particularly strong and methoxamine a particularly weak presynaptic agonist (STEINSLAND and NELSON, 1975). Transmission in the cat's nictitating membrane is depressed by xylazine (KRONEBERG et al., 1967) and not affected by noradrenaline (ENERO and LANGER, 1975) or clonidine (BOISSIER et al., 1968; PACHA et al., 1975), though clonidine is a preferentially presynaptic agonist in this tissue (LANGER, 1975).

The presynaptic component can contribute to the effect of antagonists, too. In the rabbit pulmonary artery, yohimbine preferentially blocks presynaptic α-receptors (p. 62). Low concentrations facilitate the stimulation-evoked release of noradrenaline and simultaneously, in contrast to what one would expect from the classical postsynaptic antagonist effect, *enhance* the contractile response: α-adrenolytic augmentation of neurogenic vasoconstriction, or in more general terms α-adrenolytic facilitation of noradrenergic synaptic transmission (STARKE et al., 1975a). Similarly, piperoxan potentiates the capsular response of the cat spleen to splenic nerve stimulation (SUMMERS and BLAKELEY, 1975). α-Adrenolytic drugs usually increase the effect of sympathetic nerve stimulation on the vas deferens (e.g., SWEDIN, 1971; footnote p. 4). According to earlier reports, α-adrenolytic drugs, in particular piperoxan and yohimbine, enhance transmission through further synapses with postsynaptic receptors of the α-type (BACQ and FREDERICQ, 1935; JANG, 1941). It seems quite likely that all these observations can be explained by preferential blockade of presynaptic α-receptors.

7.6.3. Central Synapses

Central noradrenergic neurones are endowed with presynaptic α-receptors (Tables 6 and 7). The question arises as to whether these receptors are involved in central effects of drugs with α-receptor affinity.

As first shown by ANDÉN et al. (1970), clonidine in vivo reduces the central nervous turnover of noradrenaline; the reduction is antagonized by α-adrenolytic drugs (cf. ROCHETTE et al., 1974; SCHEEL-KRÜGER et al., 1975). Several mechanisms may be responsible. (1) Clonidine may activate central postsynaptic α-receptors, presumably on non-noradrenergic neurones, and lower the impulse frequency of the noradrenergic fibers via a neurone chain (ANDÉN et al., 1970). (2) Clonidine may act on presynaptic α-receptors of the noradrenergic fibers and depress the quantum released per pulse (STARKE and ALTMANN, 1973). (3) The noradrenergic cell bodies in the locus coeruleus appear to contain adrenoceptors; their activation by clonidine slows the firing rate (SVENSSON et al., 1975). It has been proposed that these cell bodies receive an inhibitory *adrenergic* innervation and that clonidine acts as an agonist on the specific *adrenaline* receptors (BOLME et al., 1974). According to mechanisms 2 and 3, clonidine would reduce turnover by a direct action on the noradrenergic neurones. The two mechanisms might cooperate: via soma-dendritic receptors clonidine might diminish the impulse frequency toward the range optimal for presynaptic inhibition of the release per impulse.

It is at present difficult to decide which mechanism predominates. Recent observations argue against mechanism 1 as the major factor (ANDÉN et al., 1976). Clonidine increases the strength of the hind-limb flexor reflex in rats with acute upper thoracic spinal transection. The increase is thought to be mediated by spinal postsynaptic adrenoceptors, and high doses of clonidine are required. Much lower doses reduce the turnover of noradrenaline. Phenoxybenzamine markedly counteracts the effect on the flexor reflex, but only slightly counteracts the effect on turnover. Conversely, yohimbine, and less strikingly piperoxan and tolazoline, are more effective in antagonizing the effect on turnover than the effect on the reflex. These findings tend to dissociate the effect of clonidine on turnover from an action on central postsynaptic α-receptors and are more consistent with an action on a separate receptor system, either the presynaptic α-receptors on noradrenergic nerve terminals or adrenoceptors on the noradrenergic cell bodies. However, it should be stressed that all postsynaptic α-receptors in noradrenergic synapses need not be of the same kind. While the receptors mediating the decrease of turnover appear to differ from those mediating the increase of the flexor reflex, conclusions concerning a specific location are hazardous (ANDÉN et al., 1976).

Clonidine, xylazine, and related α-receptor agonists lower the arterial blood pressure by an action on the central nervous system that leads to a decrease in sympathetic and an increase in vagal tone (reviews by KOBINGER, 1974; CHALMERS, 1975; HAEUSLER, 1975b; ZWIETEN, 1975). The hypotension is due to the activation of central α-adrenoceptors (SCHMITT and SCHMITT, 1969; HEISE and KRONEBERG, 1970; HEISE et al., 1971; KOBINGER and WALLAND, 1971; SCHMITT et al., 1971). Several mechanisms may be responsible. The central nervous system appears to contain noradrenergic neurone systems which, when firing, increase the blood pressure, and others that decrease the blood pressure (CHALMERS, 1975). (1) α-Receptor agonists may reduce blood pressure by activating postsynaptic α-receptors innervated by "hypotensive" noradrenergic neurones. (2) They may also reduce blood pressure by activating presynaptic α-receptors of "hypertensive" noradrenergic neurones. (3) They may activate inhibitory adrenoceptors on the soma and dendrites of "hypertensive" noradrenergic neurones (SVENSSON et al., 1975). The two latter mechanisms would imply that the agonists act by reducing transmission through certain central noradrenergic synapses, either by decreasing release per impulse or by lowering the rate of impulse flow[6].

It is at present difficult to decide which mechanism predominates. Recent experiments argue against mechanisms 2 and 3 as major factors. When noradrenaline stores are depleted by treatment with reserpine or α-methyltyrosine, any effect of clonidine on cell bodies or terminals of noradrenergic neurones should be functionally insignificant. Yet clonidine reduces sympathetic outflow and enhances the vagally mediated cardiodepressor reflex even in this situation (HAEUSLER, 1974; KOBINGER and PICHLER, 1974). The results are compatible with the view that clonidine can lower the blood pressure via postsynaptic α-receptors on non-noradrenergic neurones. However, they probably do not exclude that under normal conditions, in the awake animal or man, inhibitory actions on the soma-dendritic or terminal part of "hypertensive" noradrenergic neurones also contribute.

Central α-adrenoceptors mediating circulatory effects appear to differ from peripheral postganglionic α-receptors (BOGAIEVSKY et al., 1974; STRUYKER BOUDIER et al., 1975; ZWIETEN, 1975; see, on the other hand, KOBINGER and PICHLER, 1975). Interestingly, two antihypertensive agents that are used clinically, namely clonidine and α-methylnoradrenaline (as metabolite of α-methyldopa), are preferentially presynaptic α-receptor agonists in the rabbit pulmonary artery and perhaps other peripheral tissues as well (Table 8 and p. 67). The central antihypertensive effect is readily antagonized by yohimbine and piperoxan, but much less by phenoxybenzamine and phentolamine (BOGAIEVSKY et al., 1974; ZWIETEN, 1975); yohimbine and piperoxan are preferentially presynaptic antagonists in the rabbit pulmonary artery and the cat spleen, respectively (STARKE et al., 1975a; SUMMERS and BLAKELEY, 1975). The results suggest some similarity between certain central receptors involved in blood pressure control and peripheral presynaptic α-receptors. However, this does not imply a presynaptic location of the central receptors.

Presynaptic α-receptors have also been proposed to contribute to the analgesic effect of clonidine (PAALZOW and PAALZOW, 1976) as well as to its influence on motor activity (STRÖMBOM, 1976) and hypothalamic self-stimulation (HUNT et al., 1976).

---

[6] The list is not meant to be exhaustive. The α-receptors responsible for the hypotensive effect of clonidine and congeners may receive an adrenergic rather than noradrenergic innervation (BOLME et al., 1974), or may not receive a catecholaminergic innervation at all (HAEUSLER, 1975b). Moreover, several brain areas may be involved, and the mechanisms of action in these areas may differ.

7.7. α-Adrenergic Inhibition of Release of Other Substances

In analogy to noradrenergic nerve endings, the chromaffin cells of the adrenal medulla have been proposed to contain inhibitory α-adrenoceptors (GUTMAN and BOONYAVIROJ, 1974; SERCK-HANSSEN, 1974; STARKE et al., 1974a) in addition to facilitatory β-receptors (SERCK-HANSSEN, 1974; cf. 9.1.). Adrenal release of catecholamines evoked by various stimuli is depressed by high concentrations of noradrenaline, adrenaline, dopamine, and isoprenaline (BÜLBRING et al., 1948; MALMÉJAC, 1955; SERCK-HANSSEN, 1974; cf. BOONYAVIROJ and GUTMAN, 1975; GUTMAN and BOONYAVIROJ, 1975). Potassium-evoked secretion from perfused bovine adrenals is not changed by oxymetazoline, possibly because the α-receptors are maximally activated by the large amounts of released catecholamines (STARKE et al., 1974a). In contrast to the agonists, α-adrenolytic drugs enhance hormone release (KIRPEKAR and CERVONI, 1963; STARKE et al., 1974a; GUTMAN and BOONYAVIROJ, 1975; LEMBECK and JUAN, 1975). The lack of effect or even inhibitory effect of phentolamine in some experiments (KIRPEKAR and CERVONI, 1963; SERCK-HANSSEN, 1974; STARKE et al., 1974a) may be explained by the competitive character of this antagonist and the high biophase concentration of catecholamines, plus perhaps an anticholinergic action if acetylcholine is the secretagogue.

In part of the work summarized here release was elicited by splanchnic nerve impulses. In these cases an effect of the drugs on the cholinergic nerve endings cannot be ruled out. Moreover, mention has been made of several experiments in which the expected effects were not found. Though these negative results are not incompatible with the idea that α-adrenoceptors on chromaffin cells mediate inhibition of release, present evidence is rather incomplete.

Effects of catecholamines on cholinergic nerve terminals have been reviewed by KOSTERLITZ and LEES (1972). As first shown by PATON and THOMPSON (1953), adrenaline reduces the stimulation-evoked release of acetylcholine from sympathetic ganglia. It seems very likely now that preganglionic cholinergic nerve endings are endowed with inhibitory α-adrenoceptors (NISHI, 1970; DAWES and VIZI, 1973).

Adrenaline and noradrenaline reduce the outflow of acetylcholine from the guinea-pig small intestine (SCHAUMANN, 1958). Part of the outflow originates from propagated electrical activity, and it seems likely that only the evoked release is depressed. Besides adrenaline and noradrenaline, other α-receptor agonists such as phenylephrine, xylazine, and clonidine diminish release, whereas isoprenaline has no effect. Moreover, the inhibition is antagonized by α- but not by β-adrenolytic drugs, indicating that it is mediated by α-adrenoceptors (KRONEBERG et al., 1967; PATON and VIZI, 1969; KOSTERLITZ et al., 1970; DECK et al., 1971). α-Adrenergic inhibition of acetylcholine release from small intestine resembles inhibition of noradrenaline release in that it is never complete, a certain minimal release per impulse persisting even in the presence of supramaximal agonist concentrations. Moreover, either inhibition is inversely related to stimulation frequency (e.g., KNOLL and VIZI, 1971). The site of the α-receptors within the complex neuronal network of the intestine is difficult to establish, though a location on the cholinergic nerve endings innervating the smooth muscle cells seems likely (Kosterlitz et al., 1970; Vizi, 1975). The inhibition appears to occur physiologically when released noradrenaline acts on neighboring cholinergic fibers (Vizi and Knoll, 1971; cf. p. 33).

The stimulation-evoked overflow of tritium from rat cerebral cortex slices preincubated with ^{3}H-serotonin is reduced by 10^{-5} M clonidine (Starke and Montel, 1973d). Lower concentrations have no effect, so that the existence of α-adrenoceptors on serotonergic nerve endings is questionable (Farnebo and Hamberger, 1974a).

The secretion of insulin (review by Porte and Robertson, 1973) and perhaps renin (review by Starke, 1972c) is inhibited by α-receptor agonists acting on the pancreatic β-cells and the renal juxtaglomerular cells, respectively. Thus, several types of cells which secrete hormones or neurotransmitters appear to possess α-adrenoceptors, the activation of which has a braking effect on secretion.

7.8. Conclusion

The terminals of all peripheral and central noradrenergic neurones tested thus far appear to be endowed with receptors which resemble the α-adrenoceptors of effector cells. Activation of these presynaptic α-receptors by exogenous agonists or previously released noradrenaline leads to depression of the release of noradrenaline per impulse. Present evidence is compatible with the view that α-receptor agonists selectively inhibit calcium-dependent release processes, perhaps by diminishing the availability of calcium for stimulus-secretion coupling.

The presynaptic α-adrenergic effect of released noradrenaline constitutes a local, presynaptic negative feedback mechanism which restricts the per pulse release of the transmitter. This α-adrenergic feedback is independent of prostaglandins; it is more effective than the prostaglandin-mediated negative feedback and, in contrast to the latter, appears to be common to all noradrenergic neurones. Several lines of evidence support its operation, the most important being that α-adrenolytic drugs increase the release of noradrenaline in all tissues studied so far. There is good evidence to show that they block the presynaptic α-receptors which are normally activated by released noradrenaline. – Biochemical and electrophysiological experiments agree that in some tissues the per pulse release of noradrenaline decreases with successive pulses. It seems likely that at least part of this depression is due to increasing α-adrenergic feedback inhibition.

High concentrations of α-adrenolytic drugs enhance the stimulation-evoked overflow of noradrenaline not only by disinhibition of release, but in addition by blocking the neuronal and extraneuronal uptake of noradrenaline and its metabolic degradation.

Pre- and postsynaptic α-receptors are basically similar. However, within a given tissue sufficient differences appear to exist to make the two systems preferentially sensitive to certain drugs. In the pulmonary artery of the rabbit low concentrations of oxymetazoline, clonidine, α-methylnoradrenaline, and tramazoline selectively activate presynaptic α-receptors, inhibit noradrenaline release, and thereby reduce the contractile response to sympathetic nerve stimulation: α-adrenergic inhibition of noradrenergic synaptic transmission. Methoxamine and phenylephrine preferentially activate postsynaptic α-receptors. Conversely, yohimbine at low concentrations selectively blocks presynaptic α-receptors, disinhibits noradrenaline release, and thereby enhances the contractile response to stimulation: α-adrenolytic facilitation of noradrenergic synaptic transmission. Phenoxybenzamine preferentially blocks postsynaptic α-receptors. Similar observations have been made in further tissues. However, in other organs, in particular in the central nervous system, the relative drug sensitivities of pre- and postsynaptic α-receptors may be quite different. It is at present not possible to consider presynaptic α-receptors as one structurally homogeneous group which can be confronted with postsynaptic α-receptors as the second structurally homogeneous group.

Clonidine-like drugs reduce the central nervous turnover of noradrenaline. Moreover, they decrease sympathetic and increase vagal tone by a central action. The effects are due to the activation of central α-adrenoceptors. The location of these receptors is not known. It cannot be ruled out that some of them are presynaptic α-receptors on central noradrenergic neurones.

8. Adrenergic Drugs: Presynaptic Dopamine Receptors

8.1. Dopaminergic Inhibition

Dopamine acts as an agonist on postsynaptic α-adrenoceptors, being $^1/_{10}$–$^1/_{100}$ as potent as noradrenaline in most tissues (SHEYS and GREEN, 1972; review by GOLDBERG, 1972). On the other hand, it is approximately equipotent with noradrenaline in reducing the stimulation-evoked overflow of noradrenaline from the spleen and nictitating membrane of the cat, the guinea-pig hypothalamus, the ear artery of the rabbit, and human blood vessels (Table 6). Arguments analogous to those for α-adrenergic inhibition (7.1.) make it likely that the decrease of overflow reflects a depression of release. LANGER (1973a) first supposed that noradrenaline and dopamine acted on one presynaptic receptor system, which differed from postsynaptic α-receptors in being particularly sensitive to dopamine. However, experiments with antagonists led to the conclusion that a discrete set of presynaptic dopamine receptors exists (LANGER, 1973b; ENERO and LANGER, 1975). In the nictitating membrane of the cat, phentolamine antagonizes the presynaptic effect of dopamine less than that of noradrenaline; the reverse holds true when chlorpromazine is used as antagonist. Moreover, the dopaminergic agonist apomorphine shares with dopamine the ability to depress noradrenaline release; its effect is blocked by chlorpromazine and the rather specific dopamine receptor antagonist pimozide, but not by phentolamine (ENERO and LANGER, 1975). Presynaptic antagonism between dopamine and pimozide has been observed in the rabbit ear artery (RAND et al., 1975a). Finally, in the cat heart the presumed dopamine receptor agonist M-7 (2-dimethylamino-5,6-dihydroxy tetralin) reduces noradrenaline release (STRAIT and BHATNAGAR, 1975).

On the other hand, in several tissues no inhibitory effect of dopamine has been obtained (Table 6; rat cerebral cortex, guinea-pig atria, guinea-pig vas deferens). In the rabbit pulmonary artery, 10^{-6} M dopamine slightly decreases stimulation-evoked overflow, but simultaneously accelerates basal outflow and elicits strong smooth muscle contraction; apomorphine causes minimal inhibition at only one concentration out of a wide range tested (STARKE et al., 1975c). Dopaminergic inhibition is clearly less common than α-adrenergic inhibition. The reason for the tissue differences is not known. It cannot be ruled out that presynaptic dopamine receptors are confined to only some noradrenergic neurones.

Postsynaptic response studies confirm the release-inhibiting effect of dopaminergic agonists. Dopamine reduces the postsynaptic response to sympathetic nerve stimulation in the heart (ILHAN and LONG, 1975; ILHAN et al., 1975; LONG et al., 1975), the superior mesenteric artery vascular bed (SANDERS and ROSS, 1975) and nictitating membrane (LANGER, 1973a; ENERO and LANGER, 1975) of the cat, the rabbit ear artery (McCULLOCH et al., 1973; STEINSLAND and NELSON, 1975), and human blood vessels (STJÄRNE and BRUNDIN, 1975b), but only at high doses with a large effect on postsynaptic α-receptors in the cat's hindlimb (HAEUSLER, 1975a). The positive chronotropic effect of accelerans nerve stimulation in cats is also reduced by N,N-dimethyldopamine, apomorphine, and M-7; the inhibition is antagonized by haloperidol, chlorpromazine, and bulbocapnine (ILHAN et al., 1974, 1975, 1976; ILHAN and LONG, 1975; LONG et al., 1975). Inhibition of the postsynaptic response is more pronounced the lower the frequency of stimulation (LONG et al., 1975; SANDERS and ROSS, 1975).

It has been suggested that especially after prolonged activity enough dopamine is released from noradrenergic nerves to cause significant activation of presynaptic dopamine receptors and a dopaminergic feedback inhibition of release (McCULLOCH et al., 1973). In support of this proposal, pimozide enhances the stimulation-evoked overflow of noradrenaline from rabbit ear arteries (RAND et al., 1975a); however, the basal outflow rises simultaneously, making interpretation difficult. Neither chlorpromazine nor pimozide increases the stimulation-evoked overflow from the cat nictitating membrane; a small increase caused by chlorpromazine in the presence of cocaine presumably reflects α-receptor blockade (LANGER, 1973b; ENERO and LANGER, 1975). The results indicate that dopamine receptors are not involved in a physiologic feedback mechanism.

Dopaminergic inhibition may help to explain some therapeutic and pharmacologic observations. ENERO and LANGER (1975) have pointed out that during treatment with dopa the plasma levels of dopamine are high; moreover, dopamine may accumulate in, and be released from, noradrenergic nerve endings. In this situation, the presynaptic dopamine receptors may gain functional importance. The ensuing impairment of postganglionic sympathetic transmission may be responsible for the orthostatic hypotension which occurs as a side effect in patients treated with dopa (ENERO and LANGER, 1975).

Dopamine dilates renal, mesenteric, and some other vascular beds in vivo. The effect cannot be completely explained by for instance histamine release, inhibition of ganglionic transmission, or a β-adrenergic action on vascular smooth muscle. It has been proposed that the smooth muscle cells contain specific dopamine receptors mediating relaxation (reviews by GOLDBERG, 1972, 1975). Unfortunately, the phenomenon has only rarely been studied in vitro. In isolated blood vessels, the relaxation has been found to be unaffected by β-adrenolytic drugs (GOLDBERG et al., 1973; TODA et al., 1975), but the opposite has also been reported (MORISHITA and FURUKAWA, 1974). The influence of dopamine antagonists could not be tested, since each of them caused relaxation. Thus, as pointed out by GOLDBERG (1975), the evidence for specific muscular dopamine receptors is incomplete.

The question arises whether a presynaptic effect on noradrenergic vasoconstrictor fibers is responsible for dopaminergic vasodilatation. If so, vasodilatation should occur only as long as an uninterrupted sympathetic impulse flow keeps the vascular smooth muscle contracted. However, dopamine produces dilatation even after the administration of ganglioplegic, adrenergic neurone blocking, or α-adrenolytic drugs (BELL et al., 1975; review by GOLDBERG, 1972). Moreover, apomorphine and N,N-dimethyldopamine appear to be as potent as or even more potent than dopamine on presynaptic receptors, but less potent or ineffective as vasodilators (cf. ENERO and LANGER, 1975; ILHAN et al., 1975; and GOLDBERG et al., 1968). Thus, presynaptic inhibition cannot be the only mechanism of dopaminergic vasodilatation, though it may contribute to it (ENERO and LANGER, 1975).

8.2. Dopaminergic Inhibition of Release of Other Substances

Apomorphine reduces, whereas chlorpromazine and pimozide enhance the stimulation-evoked overflow of tritium from rat striatal slices preincubated with ^{3}H-dopamine; the overflow probably reflects release from dopaminergic nerve endings (FARNEBO and HAMBERGER, 1971b, 1973b; see, however, BALDESSARINI, 1975, p. 106; SEEMAN and LEE, 1975). The results are analogous to those obtained with noradrenergic nerves and are compatible with the view that a negative feedback mechanism, mediated by presynaptic dopamine receptors, regulates the per pulse release of transmitter from dopaminergic neurones. Exogenous agonists as well as released dopamine activate the receptors and depress release; antagonists interrupt the feedback loop and disinhibit secretion (FARNEBO and HAMBERGER, 1971b, 1973b). The mechanism appears to be the exact counterpart of presynaptic α-adrenergic feedback inhibition of noradrenaline release. — It seems likely that the activation of presynaptic dopamine receptors of dopaminergic neurones leads to a second effect, namely inhibition of the biosynthesis of dopamine. The reduction of tyrosine hydroxylase activity is *not* a consequence of diminished release, since it takes place even when there is no release, as for instance after axotomy (KEHR et al., 1972; CARLSSON, 1975). Presynaptic, receptor-mediated inhibition of release and synthesis in dopaminergic neurones appear to be parallel, largely independent events (cf. WESTFALL et al., 1976). This second function has no equivalent in noradrenergic fibers. Neither α-receptor agonists nor antagonists affect transmitter synthesis in noradrenergic nerve terminals after the axons have been cut. Any effects that are normally observed (see, e.g., 7.6.3.)

appear to depend on an uninterrupted impulse flow and to be due to changes
in either impulse frequency or the release per impulse (GRABOWSKA and ANDÉN,
1976).

Electrophysiologic research indicates that dopamine inhibits the nerve impulse-induced release of acetylcholine from preganglionic fibers, but is less potent
than noradrenaline and adrenaline (DUN and NISHI, 1974). Dopamine is also
much less potent than noradrenaline and adrenaline in reducing the release
of acetylcholine from the guinea-pig ileum (PATON and VIZI, 1969; VIZI et al.,
1974); apomorphine has no effect (VIZI, 1975). All these cholinergic fibers probably lack specific dopamine receptors, and the inhibitory effect of high doses
of dopamine is mediated by α-adrenoceptors.

8.3. Conclusion

The terminals of many noradrenergic neurones appear to be endowed with
receptors for dopamine which are distinct from the presynaptic α-adrenoceptors.
Activation of these presynaptic receptors by dopamine, apomorphine, or related
agonists leads to depression of the release of noradrenaline per impulse. The
depression is counteracted by dopamine receptor antagonists such as pimozide.
The mechanism of dopaminergic inhibition is not known.

The inhibition has not been found in the rat cerebral cortex, guinea-pig
atria and vasa deferentia, and the rabbit pulmonary artery. It is not known
whether the noradrenergic fibers of these tissues are devoid of presynaptic dopamine receptors, or whether the operation of existing receptors was concealed
by the experimental conditions. – It seems unlikely that presynaptic dopamine
receptors normally mediate a feedback inhibition of noradrenaline release. The
relation between presynaptic dopamine receptors and the "dopamine vascular
receptor" (GOLDBERG, 1975) remains uncertain.

9. Adrenergic Drugs: Presynaptic β-Adrenoceptors.
Cyclic Nucleotides

Recent research indicates that noradrenergic nerve endings possess adrenoceptors of the β-type. Their activation increases the amount of transmitter released
per impulse. One possible mechanism is stimulation of an adenylate cyclase
of the nerve terminal membrane. There is evidence that cyclic AMP facilitates
release. β-Receptors and cyclic AMP may also modify catecholamine secretion
from chromaffin cells. Therefore, presynaptic and adrenal medullary effects
of cyclic nucleotides and drugs with affinity to β-receptors are discussed together.

9.1. β-Adrenergic Facilitation and Positive Feedback

9.1.1. Noradrenergic Nerves

As first shown by LANGER and his coworkers, low concentrations of isoprenaline (up to 10^{-6} M) increase the stimulation-evoked overflow of noradrenaline from several tissues (Table 6; cat aorta, spleen, and nictitating membrane, guinea-pig atria and vasa deferentia, human blood vessels and oviducts). Arguments analogous to those for α-adrenergic inhibition (7.1.) make it likely that the increase of overflow reflects facilitation of release. The facilitation is probably mediated by presynaptic β-receptors, since it is antagonized by propranolol (HEDQVIST and MOAWAD, 1975; LANGER et al., 1975a)[7]. It is less pronounced, the higher the frequency of stimulation (HEDQVIST and MOAWAD, 1975; LANGER et al., 1975b).

β-Receptor agonists do not uniformly enhance noradrenaline release. In several investigations no effect or even a decrease has been obtained (Table 6). Inhibition of release is probably due to the use of high concentrations, at which the α-adrenergic component of isoprenaline or orciprenaline prevails. A further possible reason for lack of facilitation is stimulation at high frequency. However, in the rat cerebral cortex and the rabbit pulmonary artery low concentrations of isoprenaline fail to enhance release even at frequencies of 1 or 2 Hz (STARKE et al., 1975c). The rise in overflow from the cortex caused by 10^{-5} M isoprenaline is probably unrelated to β-receptor activation (FARNEBO and HAMBERGER, 1974b; STARKE et al., 1975c). It cannot be ruled out that in some tissues the noradrenergic fibers are devoid of presynaptic β-receptors.

LANGER and his colleagues (ADLER-GRASCHINSKY and LANGER, 1975; LANGER et al., 1975a and b) suggested that presynaptic β-receptors play an important physiologic role. Whereas presynaptic α-receptors mediate a negative feedback, the β-receptors may mediate a positive feedback. In analogy to exogenous isoprenaline, endogenous, previously released noradrenaline may activate the receptors and promote its own further release. If so, β-adrenolytic drugs should interrupt the positive feedback loop and diminish the secretory response to stimulation. A release-inhibiting effect of propranolol has indeed been demonstrated in guinea-pig atria and the cat's hind limb (ADLER-GRASCHINSKY and LANGER, 1975; DAHLÖF et al., 1975).

Postsynaptic β-adrenoceptors are in general more sensitive than α-adrenoceptors to noradrenaline. The same may hold good for the presynaptic site. The presynaptic β-receptors may be selectively activated at low biophase concentrations of noradrenaline, for instance during low rates of impulse traffic, so that release per impulse increases. As the frequency and the biophase concentration of noradrenaline rise, the α-adrenergic negative feedback mechanism may

---

[7] (±)-, but not (+)-propranolol inhibits the vasoconstrictor response of the cat's hind limb to sympathetic nerve stimulation, even when it is given exclusively to the leg in cross-circulation experiments (ÅBLAD et al., 1970). The authors propose "that (±)-propranolol reduced the noradrenaline output from the nerve endings. The effect is probably due to β-adrenergic receptor blockade" and "to a peripheral site of attack". Though the site is not further defined, this appears to be the first indication of presynaptic β-receptors.

be triggered, so that release per impulse falls (Adler-Graschinsky and Langer, 1975; Langer et al., 1975b). It should be noted, on the other hand, that the α-adrenergic feedback operates at quite low frequencies. For instance, phentolamine increases release by 300% in the guinea-pig vas deferens and by 100% in the rat cerebral cortex during stimulation at 1 and 0.3 Hz, respectively (uptake mechanisms blocked; Stjärne, 1973k; Starke, unpublished).

In many tissues, the per pulse release of noradrenaline increases with increasing train length and stimulation frequency (2.2.). The question arises as to whether the β-adrenergic positive feedback mechanism contributes to this facilitation. If so, the facilitation should be reduced by β-adrenolytic drugs. Moreover, β-adrenergic agonists should preferentially enhance release evoked by the first pulses in a train, as long as the biophase concentration of noradrenaline and feedback facilitation are small. The possibility merits a trial; of course, it will not account for the entire facilitation observed, since closely similar phenomena are known for cholinergic fibers.

β-Adrenolytic drugs do not uniformly reduce noradrenaline release. Since β-adrenergic facilitation was not obtained in the rat cerebral cortex and the rabbit pulmonary artery, the absence of β-adrenolytic inhibition in these tissues is not surprising (Starke et al., 1975c). Another possible reason for negative results is stimulation at too high a frequency. Moreover, high concentrations of many β-adrenolytic drugs inhibit the neuronal and extraneuronal uptake of noradrenaline (Eisenfeld et al., 1967; Foo et al., 1968; Iversen, 1975). The inhibition is unrelated to β-receptor blockade. It may overshadow any depression of release and actually increase the stimulation-evoked overflow (Muscholl, 1966; Schümann et al., 1970a; Werner et al., 1971; Starke and Schümann, 1972). However, even if all those cases are left aside in which the absence of β-adrenolytic inhibition can be plausibly explained, disquieting findings remain. In the cat spleen and in human blood vessels propranolol failed to reduce release even though isoprenaline caused significant facilitation and though propranolol antagonized the facilitatory effect (Langer et al., 1975a; Stjärne and Brundin, 1975a). It might be thought that the biophase concentration of noradrenaline remained subthreshold for presynaptic β-receptor activation. However, under the same conditions, including the frequency of 1 Hz, the presynaptic α-receptors are activated, as evidenced by a strong facilitatory effect of α-adrenolytic drugs (Cubeddu and Weiner, 1975a; Stjärne and Gripe, 1973). The more sensitive β-receptor mechanism should certainly operate. Why then the lack of effect of the blocking drug? – Taken together, these results are compatible with the view that presynaptic β-receptors are less common than α-receptors. Even where they occur, they do not necessarily mediate a positive feedback reinforcing the secretion of noradrenaline.

Presynaptic β-receptors may be physiologic sites of action not only of locally released noradrenaline but also of circulating adrenaline. They may mediate enhancement of sympathetic neuroeffector transmission under conditions of increased adrenal-medullary hormone secretion (Stjärne and Brundin, 1975a).

Some β-adrenolytic drugs diminish the postsynaptic response to sympathetic nerve stimulation in tissues where the postsynaptic receptors are predominantly of the α-type; the response to exogenous noradrenaline is unchanged or even enhanced (e.g., DAY et al., 1968; ÅBLAD et al., 1970; BARRETT and NUNN, 1970; ELIASH and WEINSTOCK, 1971; MYLECHARANE and RAPER, 1973; DAHLÖF et al., 1975; see, however, ADLER-GRASCHINSKY and LANGER, 1975; DAWES and FAULKNER, 1975). DAY et al. (1968) and later authors proposed a presynaptic site of action and a local anesthetic or guanethidine-like adrenergic neurone blocking mechanism, though some observations remained difficult to reconcile with these ideas. It might be supposed that blockade of presynaptic β-receptors is involved. However, in some experiments which included various β-adrenolytic drugs and enantiomers of different blocking potency the inhibition of transmission could be dissociated from β-adrenolysis (BARRETT and NUNN, 1970; ELIASH and WEINSTOCK, 1971; MYLECHARANE and RAPER, 1973). Only a study on the cat's hind limb (ÅBLAD et al., 1970; footnote p. 77) can retrospectively be explained by presynaptic β-receptor blockade, since (+)-propranolol in contrast to (±)-propranolol had no effect (cf. DAHLÖF et al., 1975). DAHLÖF et al. (1975) have shown that in this preparation the β_1-receptor blocking agent, metoprolol, also inhibits sympathetic neuroeffector transmission, suggesting to the authors that the presynaptic β-receptors are mainly of the β_1-type.

β-Adrenolytic drugs have an important place in antihypertensive therapy. Presynaptic β-adrenolytic inhibition of transmission through peripheral or central (cf. p. 70) noradrenergic synapses may contribute to their therapeutic value (ADLER-GRASCHINSKY and LANGER, 1975; DAHLÖF et al., 1975; LANGER et al., 1975b; LJUNG et al., 1975).

9.1.2. Adrenal Medulla

It has been suggested that the chromaffin cells of the adrenal medulla possess β-receptors mediating facilitation of secretion (SERCK-HANSSEN, 1974). The evidence is that exogenous agonists enhance the release of catecholamines evoked by splanchnic discharges (BÜLBRING et al., 1948; MALMÉJAC, 1955) or injection of acetylcholine, isoprenaline being the most potent (SERCK-HANSSEN, 1974). Conversely, (±)-propranolol reduces secretion, whereas (+)-propranolol is significantly less potent (SERCK-HANSSEN, 1974; cf. GUTMAN and BOONYAVIROJ, 1975). In bovine adrenal medulla, β-receptors seem to be confined to the adrenaline-storing cells, since only the release of adrenaline is modulated, not that of noradrenaline (SERCK-HANSSEN, 1974). As mentioned above (p. 71), high doses of adrenaline, noradrenaline, dopamine, and isoprenaline exert an inhibitory effect which is possibly mediated by α-adrenoceptors on the chromaffin cells.

9.2. Cyclic Nucleotides

The mechanism of β-adrenergic enhancement of secretion from noradrenergic nerves and the adrenal medulla is not known. In either case the activation of an adenylate cyclase has been proposed, leading to increased intracellular levels of, and facilitation of release by, cyclic AMP (SERCK-HANSSEN, 1974; ADLER-GRASCHINSKY and LANGER, 1975; LANGER et al., 1975a and b).

9.2.1. Noradrenergic Nerves

Direct evidence for effects of cyclic nucleotides on transmitter release has accumulated in recent years. Both phosphodiesterase inhibitors and exogenous cyclic nucleotides rarely enhance the *basal* outflow of noradrenaline (HSU and WESTFALL, 1973; LANGER, 1973a; WOOTEN et al., 1973; WESTFALL and BRASTED, 1974; CUBEDDU et al., 1974b, 1975). Papaverine selectively increases the outflow of ^{3}H-DOPEG from tissues pretreated with ^{3}H-noradrenaline; the increase probably reflects an action on storage vesicles unrelated to phosphodiesterase inhibition (LANGER, 1973a; CUBEDDU et al., 1974b). In the guinea-pig hypogastric nerve–vas deferens preparation, dibutyryl-cyclic AMP, in the presence of phenoxybenzamine, enhances the basal outflow of both noradrenaline and dopamine-β-hydroxylase (WOOTEN et al., 1973); since the preparation contains preganglionic nerve endings and noradrenergic cell bodies, the site of action is uncertain.

Phosphodiesterase inhibitors (papaverine; dipyridamol; theophylline; aminophylline; 1-methyl-3-isobutylxanthine; 4-(3-butoxy-4-methoxybenzyl)-2-imidazolidinone) and cyclic nucleotides (monobutyryl-cyclic AMP; dibutyryl-cyclic AMP; 8-methylthio-cyclic AMP; 8-bromo-cyclic GMP; but not cyclic AMP itself) enhance the *stimulation-evoked* overflow of noradrenaline (LANGER, 1973a; WOOTEN et al., 1973; CUBEDDU et al., 1974b, 1975; LANGER et al., 1975b; see, however, STJÄRNE, 1976). Simultaneously, the overflow of dopamine-β-hydroxylase is increased, indicating facilitation of exocytotic release (WOOTEN et al., 1973; CUBEDDU et al., 1974b, 1975). The phosphodiesterase inhibitor 4-(3-butoxy-4-methoxybenzyl)-2-imidazolidinone augments the facilitation caused by monobutyryl-cyclic AMP (CUBEDDU et al., 1975). The effect of 8-bromo-cyclic GMP is analogous to that of the cyclic AMP derivatives. The compound inhibits phosphodiesterase; at least part of its effect might be mediated by accumulation of cyclic AMP. It is uncertain whether the influence of intraneuronal cyclic GMP on noradrenaline release is opposite or similar to the influence of cyclic AMP (CUBEDDU et al., 1975).

The facilitation caused by phosphodiesterase inhibitors and cyclic nucleotides is slight in comparison with that caused by α-adrenolytic drugs (CUBEDDU et al., 1975; STJÄRNE, 1976). The particularly large effect of papaverine is probably unrelated to phosphodiesterase inhibition. As mentioned above, papaverine affects noradrenaline storage granules and may share with other reserpine-like agents the ability to promote release (CUBEDDU et al., 1974b; CUBEDDU and WEINER, 1975b).

Taken together, these results are compatible with the idea that cyclic AMP can facilitate the normal process of release evoked by nerve impulses, though its presence is not a *conditio sine qua non* for release (CUBEDDU et al., 1975; STJÄRNE, 1976). As has been suggested for secretion of other substances (RASMUSSEN, 1970), cyclic AMP might cause a redistribution of intraneuronal calcium. It might also activate a protein kinase which in turn could enhance exocytosis by phosphorylation of microtubule or microfilament, or vesicle or plasma membrane protein. The results are also compatible with the view that β-adrenergic facilitation involves a rise of intraneuronal cyclic AMP (ADLER-GRASCHINSKY and LANGER, 1975; LANGER et al., 1975a and b). Of course further criteria have to be satisfied to establish this mechanism (SUTHERLAND et al., 1968). An adenylate cyclase should be demonstrated in noradrenergic nerve endings which should respond to β-receptor agonists. Moreover, the increase in cyclic AMP should precede the facilitation of release. These criteria are difficult to obtain in the case of nerves supplying a large tissue mass. Experiments on noradrenergic cell bodies would be an alternative. It should be noted, however, that the adenylate cyclase of postganglionic sympathetic cell bodies is apparently *not* activated by β-adrenergic agonists (GREENGARD and KEBABIAN, 1974).

9.2.2. Adrenal Medulla

It has been supposed for some time that caffeine and theophylline cause or facilitate hormone release from the adrenal medulla (see BERKOWITZ and SPECTOR, 1971). More direct evidence has been provided by the demonstration that methylxanthines increase *basal* catecholamine outflow from isolated adrenal glands (PEACH, 1972; POISNER, 1973a and b; RAHWAN et al., 1973; STITZEL et al., 1973; see, however, JAANUS and RUBIN, 1974). The increase appears to reflect exocytotic release (POISNER, 1973c). It occurs even in the absence of extracellular calcium, probably because intracellular calcium stores are mobilized. Not only methylxanthines but also exogenous cyclic AMP and its derivatives augment basal hormone outflow (POYART et al., 1968; IZUMI et al., 1971; PEACH, 1972; SERCK-HANSSEN, 1974; see, however, JAANUS and RUBIN, 1974), as does cyclic GMP (POISNER, 1973c). Caffeine and theophylline enhance the effect of cyclic nucleotides (IZUMI et al., 1971; PEACH, 1972). These results differ from the largely negative ones obtained in nerves. An explanation may be that the effect of cyclic AMP is confined to facilitation of an ongoing exocytotic release. Exocytosis probably contributes significantly to spontaneous release from the adrenal medulla (SMITH and WINKLER, 1972, p. 564), but minimally to spontaneous release from the nerves (the vas deferens may be an exception, since basal outflow of dopamine-β-hydroxylase is relatively high; WOOTEN et al., 1973).

Theophylline increases *acetylcholine-evoked* release of adrenaline from the bovine adrenal medulla; noradrenaline release is unchanged (SERCK-HANSSEN, 1974). In cat adrenal glands, two phosphodiesterase inhibitors as well as cyclic AMP failed to augment the secretion of total catecholamines evoked by acetylcholine, nicotine or potassium (PEACH, 1972; JAANUS and RUBIN, 1974); the results probably do not exclude a small, selective increase in adrenaline secretion.

Taken together, these data indicate that cyclic nucleotides can accelerate the basal secretion of adrenal medullary hormones. Further augmentation of release evoked by the physiologic secretagogue acetylcholine is much less certain. Consequently, the proposed role (SERCK-HANSSEN, 1974) of cyclic AMP in β-adrenergic facilitation of hormone release remains doubtful. Evidence against such a role is the failure of adrenaline and isoprenaline to increase adrenal medullary cyclic AMP levels (GUIDOTTI and COSTA, 1974; OTTEN et al., 1975).

9.3. Conclusion

The terminals of many noradrenergic neurones appear to be endowed with β-adrenoceptors. Activation of these presynaptic β-receptors by isoprenaline and presumably also by released noradrenaline leads to an increase of the release of noradrenaline per impulse. The effect of isoprenaline is antagonized by propranolol.

The presynaptic β-adrenergic effect of previously released noradrenaline constitutes a local positive feedback mechanism. After interruption of the feedback loop by β-adrenolytic agents release is depressed. It seems possible that β-adrenergic facilitation and α-adrenergic inhibition require different biophase concentrations of noradrenaline. At low concentrations, as for instance during low rates of impulse flow, facilitation prevails. At high concentrations, as for instance during high rates of impulse flow, the α-adrenergic braking mechanism is triggered.

β-Adrenergic facilitation and β-adrenolytic depression of release have not been found in all tissues studied. The reason is not known. Interestingly, propranolol fails to reduce release in some tissues where isoprenaline causes facilitation, indicating perhaps that the positive feedback may not operate even where presynaptic β-receptors exist.

One way by which β-receptor agonists could facilitate release is the activation of a neuronal adenylate cyclase and an increase of intraneuronal cyclic AMP. Both cyclic nucleotides and phosphodiesterase inhibitors enhance the stimulation-evoked release of noradrenaline. No consistent effect on basal outflow has been found. It seems unlikely that cyclic AMP is an essential link in electrosecretory coupling. Further evidence is needed to corroborate its role in β-adrenergic facilitation.

10. Narcotic Analgesics

Morphine-like analgesics have two main effects on noradrenergic nerve endings. Firstly, they inhibit the uptake of noradrenaline across the cell membrane; the effect is nonspecific. Secondly, they reduce the amount of transmitter released per impulse; the effect is specific. In this connexion specificity means that the effect is mediated by the unique morphine receptor system. Ideally, demonstration of specificity should include: that narcotic analgesics have in common the ability to exert the effect; that their rank order of potency agrees with their antinociceptive order of potency, with due allowance for the dependence of the latter on the route of administration (HERZ and TESCHEMACHER, 1971); that the relative potency of optical antipodes, as for instance levorphanol and dextrorphan, corresponds to their analgesic potency; and that the effect is counteracted by narcotic antagonists.

10.1. Effect on Neuronal Uptake

Narcotic analgesics such as morphine, levorphanol, pethidine, and methadone inhibit the neuronal uptake of noradrenaline in various tissues of the cat (DENGLER and TITUS, 1961), rat (CLOUET et al., 1973; CLOUET and WILLIAMS, 1974), mouse (CARLSSON and LINDQVIST, 1969; CARMICHAEL and ISRAEL, 1973), and rabbit (CIOFALO, 1972; MONTEL and STARKE, 1973). In vivo pretreatment of rats with morphine or heroin inhibits the subsequent in vitro uptake of noradrenaline into whole brain or hypothalamic synaptosomes (BOYKIN and MARTIN, 1973). There is general agreement that the inhibition is nonspecific. High concentrations of the drugs are needed, e.g., at least 10^{-5} M morphine. There is no correlation with analgesic potency; for instance, the weakly analgesic pethidine is a particularly potent uptake inhibitor in both the mouse brain and the rabbit heart (CARMICHAEL and ISRAEL, 1973; MONTEL and STARKE, 1973). Optical antipodes with different analgesic potency have similar effects on uptake. Finally, antagonists including naloxone, which lacks intrinsic activity on morphine receptors, also inhibit uptake and fail to antagonize the effect of the agonists.

Inhibition of re-uptake is probably the main reason for the increase of the stimulation-evoked overflow of noradrenaline caused by several analgesics in the rabbit heart, and by pethidine and a high concentration of levorphanol in the rat cerebral cortex (Table 9). In the heart, concentrations which increase the stimulation-evoked overflow are similar to those which inhibit the uptake of infused noradrenaline. Moreover, morphine does not cause a further increase after the stimulation-evoked overflow has been raised by cocaine (MONTEL and STARKE, 1973).

10.2. Effect on Release

In 1957, TRENDELENBURG reported that morphine inhibits contractions of the nictitating membrane of the cat elicited by pre- or postganglionic sympathetic nerve stimulation. Since contractions evoked by injected adrenaline or noradrenaline were not diminished, he proposed that morphine might reduce the release of the postganglionic transmitter. The detection of this effect on noradrenergic nerves lagged behind the first description, by an earlier generation, of the analogous effect on cholinergic fibers; as little as 3×10^{-8} M morphine inhibited the peristaltic reflex elicited by distension of the isolated small intestine of the guinea pig (TRENDELENBURG, 1917); the reflex is mediated by cholinergic neurones. The inhibitory effect in the nictitating membrane was shown to be specific (CAIRNIE et al., 1961). A presynaptic site of action was confirmed by the demonstration that the stimulation-evoked overflow of noradrenaline is decreased (HENDERSON et al., 1972a and b).

Narcotic analgesics reduce the stimulation-evoked overflow of noradrenaline from the nictitating membrane of the cat, the cerebral cortex, cerebellar cortex, and hypothalamus of the rat, and the vas deferens of the mouse (see, however, JENKINS et al., 1975), but not from several other organs (Table 9). In contrast

Table 9. Effect of narcotic analgesics and antagonists on transmitter overflow evoked by electrical stimulation of noradrenergic nerves

Species	Tissue	Drug	Concentration (M)	Effect	Comment	References
Cat	Nictitating membrane	Morphine	3×10^{-6}	−	Incubated tissue. Effect on overflow of endogenous noradrenaline evoked by field stimulation at 1 Hz. No effect at 15 Hz. Effect of normorphine was tested in presence of phentolamine; it was antagonized by naloxone	HENDERSON et al., 1972b, 1975
		Normorphine	5×10^{-6}	−		
Rat	Cerebral cortex Cerebellum Hypothalamus	Morphine	10^{-7}–10^{-5}	−	Slices preincubated with ^{3}H-noradrenaline. Effect on overflow of total ^{3}H evoked by field stimulation at 0.3–3 Hz. Morphine had no effect at 10 Hz. Inhibitory effects of morphine, levorphanol, and fentanyl were antagonized by naloxone. Effect of morphine was not changed by indometacin. No effect on basal outflow	MONTEL et al., 1974a, b, 1975a–c
		Levorphanol	10^{-7}–10^{-6}	−		
		Pethidine	10^{-5}	+		
			10^{-7}	0		
		Fentanyl	10^{-6}–3×10^{-6}	+		
			10^{-8}–10^{-7}	−		
		Naloxone	10^{-6}–10^{-4}	0		
Mouse	Vas deferens	Morphine	7×10^{-7}–10^{-6}	−	Incubated tissue or tissue preincubated with ^{3}H-noradrenaline or ^{3}H-tyrosine. Effect on overflow of endogenous noradrenaline or ^{3}H-catecholamines evoked by field stimulation at 0.25 or 1 Hz; in some experiments it was tested in presence of hexamethonium +phentolamine. Effect of morphine was antagonized by naloxone. No effect on basal outflow	HENDERSON et al., 1972a; HENDERSON and HUGHES, 1974; HUGHES et al., 1975a
		Normorphine	3×10^{-7}–10^{-6}	−		
		Naloxone	5×10^{-8}	0		
	Vas deferens	Morphine	10^{-6}	0	Incubated tissue. Effect on overflow of endogenous noradrenaline evoked by field stimulation at 1 Hz. No details	JENKINS et al., 1975
Guinea pig	Intestine	Morphine	10^{-6}–2×10^{-6}	0	Myenteric plexus-longitudinal muscle preparation. Effect on overflow of endogenous noradrenaline evoked by field stimulation at 2 or 16 Hz	HENDERSON et al., 1972b, 1975

Animal	Tissue	Drug	Concentration	Effect	Comment	Reference
Rabbit	Heart	Morphine	10^{-6}	0	Perfused heart. Effect on overflow of endogenous noradrenaline evoked by accelerans nerve stimulation at 2.5–10 Hz. After overflow had been raised by cocaine, morphine did not cause further increase. At high concentrations most drugs enhanced basal outflow of ^{3}H from hearts preperfused with ^{3}H-noradrenaline	MONTEL and STARKE, 1973
		Pethidine	10^{-5}–10^{-4}	+		
			10^{-7}	0		
		Fentanyl	10^{-6}–10^{-4}	+		
			10^{-7}	0		
		(−)-Methadone	10^{-6}–10^{-5}	+		
		Levallorphan	10^{-5}	+		
			10^{-4}	−		
		Naloxone	10^{-6}–10^{-5}	0		
			10^{-4}	+		
	Pulmonary artery	Morphine	10^{-7}–10^{-5}	0	Strips preincubated with ^{3}H-noradrenaline. Effect on overflow of total ^{3}H evoked by field stimulation at 2 Hz was tested in presence of cocaine+corticosterone	TAUBE et al., 1976
	Portal vein Vas deferens	Morphine		0	Incubated tissue. Effect on overflow of endogenous noradrenaline evoked by field stimulation. No details	HENDERSON et al., 1972b; HENDERSON and HUGHES, 1974

+ = increase, − = decrease, 0 = no change. In each case, the effect measured is briefly defined under "Comment".

to its congeners, pethidine increases the stimulation-evoked overflow from the rat cerebral cortex. It seems likely that its strong effect on re-uptake overshadows the inhibition of release; when neuronal uptake is blocked by cocaine throughout the experiment, pethidine in fact reduces the stimulation-evoked overflow. Similarly, the increase of overflow caused by 10^{-5} M levorphanol may reflect blockade of re-uptake (MONTEL et al., 1974b).

The decrease of the stimulation-evoked overflow results from depression of release rather than an increase of the degradation or retention of released noradrenaline within the tissue, for the following reasons (cf. Table 9). (1) The low concentrations that reduce overflow do not affect noradrenaline uptake; higher concentrations inhibit rather than enhance uptake. Moreover, morphine up to 10^{-4} M has no effect on adrenal medullary MAO activity (ANDERSON and SLOTKIN, 1975). (2) After treatment with ^{3}H-noradrenaline narcotic analgesics reduce the stimulation-evoked overflow of total tritium, i.e., of ^{3}H-noradrenaline plus ^{3}H-metabolites; they do not reduce the overflow of noradrenaline by promoting its biotransformation. (3) In rat cerebral cortex slices, narcotic analgesics diminish the stimulation-evoked overflow after re-uptake has been blocked by high concentrations of cocaine (2×10^{-5}–10^{-4} M; MONTEL et al., 1974b; STARKE, unpublished). (4) There is some evidence that morphine enhances the biosynthesis of noradrenaline in the rat brain in vivo (e.g., CICERO et al., 1973; see, however, SVENSSON and TROLIN, 1975). The possibility was considered that narcotic analgesics might reduce the stimulation-evoked overflow of tritium from brain slices preincubated with ^{3}H-noradrenaline by accelerating the formation of unlabeled noradrenaline, thereby diluting the specific activity of the neuronal store. However, the stimulation-evoked overflow was decreased even after biosynthesis had been blocked by α-methyltyrosine (MONTEL et al., 1974a and b, 1975b). (5) Naloxone prevents the effect of the agonists. The simplest explanation is antagonism at release-inhibiting receptors.

Antagonism by naloxone strongly favors the assumption that the inhibition of release is a specific effect. Further evidence is the lack of an inhibitory effect of naloxone (Table 9). The rank order of the inhibitory potency of agonists in the rat brain cortex agrees with their analgesic order of potency (fentanyl > levorphanol ≈ morphine > pethidine; MONTEL et al., 1974a and b). Finally, in contrast to levorphanol, its analgesically inactive enantiomer dextrorphan does not reduce the stimulation-evoked overflow from the rat cerebral cortex (STARKE, unpublished).

The mechanism of the inhibition is not known. Some information can be derived from the following observations. (1) Relevant concentrations of morphine are devoid of local anesthetic activity (KOSTERLITZ and WALLIS, 1964). (2) A granular site of action is unlikely, since up to 10^{-4} M morphine has no effect on the catecholamine storage vesicles of the rat adrenal medulla (ANDERSON and SLOTKIN, 1975) or on the uptake of noradrenaline into synaptic vesicles from the rat brain (BLOSSER and CATRAVAS, 1974). (3) Inhibition of release is not due to inhibition of biosynthesis, since the release of exogenous, previously stored noradrenaline is equally depressed (Table 9). (4) Morphine promotes the biosynthesis of prostaglandins (COLLIER et al., 1974). However, the inhibition of release is not mediated by prostaglandins, since it persists after blockade of their formation (MONTEL et al., 1975c). (5) Presynaptic α-adrenoceptors or muscarine receptors are not involved, since α-adrenolytic drugs (Table 9) and atropine (rat cerebral cortex; STARKE, unpublished) do not change the inhibition. (6) At concentrations which inhibit release by nerve impulses, narcotic analgesics do not affect basal outflow (Table 9). It has been briefly stated that morphine does not change the potassium-evoked overflow of noradrenaline from synaptosomes (CLOUET and WILLIAMS, 1974). More detailed analyses are not available, so that it is not possible to decide whether narcotic analgesics selectively inhibit certain, for instance calcium-dependent, modes of release. (7) The inhibition is inversely related to stimulation frequency (Table 9; cf. CAIRNIE et al., 1961). The frequency-dependence is compatible with the view that the analgesics, in analogy to other presynaptic inhibitors, diminish the availability of calcium for stimulus-secretion coupling; this diminution would be overcome when the nerve endings are flooded with calcium during high rates of impulse flow. It may be significant that morphine in vivo decreases the calcium content of various regions of the brain within 10 min (CARDENAS and ROSS, 1975). However, the hypothesis remains speculative, and more work is needed to identify the mechanism of the release-inhibiting effect.

Morphine and congeners fail to reduce the stimulation-evoked overflow of noradrenaline from the myenteric plexus-longitudinal muscle preparation of the guinea pig as well as the heart, pulmonary artery, portal vein, and vas deferens of the rabbit (Table 9). It is unlikely that depression of release was masked by inhibition of re-uptake, since the agonists were used at low concentrations. Moreover, morphine did not diminish overflow from the rabbit heart and pulmonary artery even after uptake had been blocked by cocaine. The unchanged overflow reflects an unchanged release. Postsynaptic response studies indicate the absence of a release-inhibiting effect of morphine in further tissues, namely the cat heart, spleen, and vas deferens and the vasa deferentia of the rat, guinea pig, hamster, and gerbil (CAIRNIE et al., 1961; HUGHES et al., 1975a; the sympathetic fibers innervating the guinea-pig jejunum appear to be morphine-sensitive; SZERB, 1961). The reason for these negative results is not known. An inappropriately high stimulation frequency can be excluded for most experiments (Table 9). There may be other factors which prevent expression of the effect of morphine despite the existence of presynaptic receptors (11.4.). However, the alternative that some noradrenergic neurones lack presynaptic morphine receptors cannot be ruled out.

10.3. Effect of Chronic Morphine Treatment

Uptake and release of ^{3}H-noradrenaline have recently been studied in cerebral cortex slices from rats with chronically implanted morphine pellets (MONTEL et al., 1975a). The results were interpreted in terms of withdrawal, dependence, and tolerance. When slices from dependent animals were washed free of morphine and then stimulated, the overflow of tritium and presumably the biophase concentration of noradrenaline were higher than in controls (withdrawal). Surprisingly, when morphine or levorphanol were added, they reduced the stimulation-evoked overflow at the same doses in slices from dependent rats and those from placebo-pretreated controls. Since, however, the overflow from the "dependent slices" during withdrawal was elevated, low doses of morphine or levorphanol brought it back to normal (dependence), and higher doses were required for a decrease to subnormal levels (tolerance).

These observations suggest a presynaptic type of tolerance/dependence. In untreated animals morphine reduces noradrenaline release and thereby transmission through noradrenergic synapses. Tolerance to this defect in transmission may result not from postsynaptic supersensitivity to the transmitter, as for instance an increase in the number of postsynaptic receptors (COLLIER, 1968), but from increased effectiveness of the noradrenaline release mechanism and/or decreased effectiveness of a noradrenaline inactivation mechanism. If so, the biophase concentration of noradrenaline returns toward normal despite the presence of morphine.

From their experiments MONTEL et al. (1975a) were unable to specify the nature of the presynaptic compensatory process. Since during in vitro withdrawal the anti-release effect of morphine or levorphanol was not changed, a subsensitivity of the presynaptic morphine receptors was excluded. Interestingly, the uptake of ^{3}H-noradrenaline into slices from dependent rats was slightly impaired even after morphine had been washed out (see, however, experiments in mice by CARMICHAEL and ISRAEL, 1973). Thus, one possibility of compensation might be a change in the neuronal noradrenaline uptake mechanism leading to diminished re-uptake. Chronic administration of morphine also alters the amine uptake kinetics of adrenal medullary storage vesicles (ANDERSON and SLOTKIN, 1975; SLOTKIN and ANDERSON, 1975). — It would be of obvious interest to study the transmitter economics of *peripheral* morphine-sensitive noradrenergic neurones after chronic opiate treatment.

10.4. Effect on Release of Other Substances

Morphine increases hormone secretion from the adrenal medulla (for references see ANDERSON and SLOTKIN, 1975). The increase is mainly mediated by the central nervous system, but a small direct effect may contribute (YOSHIZAKI, 1973; ANDERSON and SLOTKIN, 1976).

Conflicting results have been reported concerning the effect of morphine on the uptake of dopamine into brain slices or synaptosomes (HITZEMANN and LOH, 1973; CELSEN and KUSCHINSKY, 1974; CLOUET and WILLIAMS, 1974). Morphine reduces the potassium-evoked overflow of ^{14}C from rat striatal slices preincubated with ^{14}C-dopamine; the effect is specific (CELSEN and KUSCHINSKY, 1974). By an unspecific effect, morphine increases the stimulation-evoked overflow of previously accumulated ^{3}H-dopamine from the guinea-pig myenteric plexus-longitudinal muscle preparation (SCHULZ and CARTWRIGHT, 1974; release from the noradrenergic nerves of the tissue?). — Morphine and methadone do not change the overflow of ^{3}H-serotonin from preloaded rat striatal slices (electrical stimulation; THORNBURG and BLAKE, 1971), rabbit brain synaptosomes (high potassium in the presence of cocaine; CIOFALO, 1974) and guinea-pig myenteric plexus-longitudinal muscle preparations (electrical stimulation; SCHULZ and CARTWRIGHT, 1974).

There are parallels between the presynaptic effect of narcotic analgesics on noradrenergic neurones on the one hand, and cholinergic ones on the other hand. Morphine reduces the stimulation-evoked and basal outflow of acetylcholine from the guinea-pig ileum; part of the basal outflow results from nerve impulses (PATON, 1957; SCHAUMANN, 1957). The effect is specific (COX and WEINSTOCK, 1966) and inversely related to frequency (PATON, 1957; LEES et al., 1972). The site of action is not completely clear. However, electrophysiologic data support location of the morphine receptors on the cholinergic terminals innervating the smooth muscle (NORTH and HENDERSON, 1975).

Morphine also impairs cholinergic transmission in the hearts of the cat, rat, and rabbit, but not in the guinea-pig heart and the rabbit ileum (KOSTERLITZ and TAYLOR, 1959; KENNEDY and WEST, 1967; LEES et al., 1972). High doses of narcotic analgesics depress acetylcholine release from preganglionic autonomic, skeletal muscle motor, and invertebrate neurones; however, these effects are nonspecific (see LEES et al., 1972; BELL and REES, 1974; TREMBLAY et al., 1974). In vivo, narcotic analgesics reduce the outflow of acetylcholine from various brain regions (e.g., cats: BELESLIN and POLAK, 1965; JHAMANDAS et al., 1971; YAKSH and YAMAMURA, 1975); the effect is specific. However, it is uncertain whether the drugs act directly on cholinergic nerve endings, since in other experimental situations they may *increase* acetylcholine outflow (e.g., cat: MULLIN et al., 1973); moreover, except at very high concentrations with nonspecific effects narcotic analgesics do not diminish the overflow of acetylcholine from brain slices or synaptosomes evoked by electrical stimulation or potassium (CLOUET and WILLIAMS, 1974; SZERB, 1974). Thus, cholinergic as well as noradrenergic neurones greatly differ in their response to narcotic analgesics.

10.5. Possible Significance

It has long been an enigma why the organism should possess specific receptors for a bizarre alkaloid and related chemicals. This problem has apparently been solved by the finding that ligands for the morphine receptor occur naturally in the brain, pituitary gland, and intestine (HUGHES et al., 1975b; TERENIUS and WAHLSTRÖM, 1975; TESCHEMACHER et al., 1975; SIMANTOV and SNYDER, 1976). The compounds behave like morphine in that they diminish stimulation-evoked contractions of the guinea-pig ileum and the mouse vas deferens, and in that the inhibition is reversed by naloxone (HUGHES et al., 1975b; TESCHEMACHER et al., 1975). Thus, morphine and its congeners appear to be exogenous agonists acting on receptors meant for endogenous substances just as nicotine and muscarine are agonists related to acetylcholine.

The existence of endogenous ligands opens up the possibility that presynaptic morphine receptors have a physiologic significance. If endogenous ligands, at least under certain conditions, have access to and activate the receptors, then narcotic antagonists should displace them and enhance transmitter release. Naloxone has no effect on noradrenaline release in several tissues (Table 9). On the other hand, WATERFIELD and KOSTERLITZ (1975) have shown that naloxone enhances the stimulation-evoked overflow of acetylcholine from the guinea-pig ileum (cf. NUETEN et al., 1976). The finding is indirect evidence for the occurrence of an endogenous ligand and suggests that one of its normal functions is modulation of acetylcholine release. The lack of effect of naloxone in brain slices (Table 9) does not exclude a presynaptic inhibitory action of endogenous ligands on central noradrenergic neurones under other, perhaps more physiologic, conditions.

It is not known whether presynaptic inhibition of noradrenergic neurotransmission is responsible for therapeutic or side effects of narcotic analgesics. It seems to be the basis for some electrophysiologic phenomena (SASA et al., 1975). However, a more far-reaching importance cannot be ruled out. The analgesic effect of morphine is mimicked or enhanced by phenoxybenzamine or phentolamine (CICERO et al., 1974), a dopamine-β-hydroxylase inhibitor (CICERO et al., 1974), clonidine (PAALZOW, 1974) which at least in some noradrenergic synapses preferentially activates presynaptic α-receptors (pp. 61, 67), and lesions of noradrenergic neurone bundles (PRICE and FIBIGER, 1975). All these procedures share with morphine one consequence, namely inhibition of transmission through (some) noradrenergic synapses with postsynaptic α-receptors. The common effect, analgesia, may therefore at least partly be due to diminished activation of central postsynaptic α-receptors. Particularly suggestive of such a mechanism is the demonstration that morphine can inhibit central noradrenergic neurotransmission in a second way. It slows the spontaneous firing rate of the noradrenergic locus coeruleus neurones and counteracts the increase in firing caused by noxious stimuli; the site of action may be the noradrenergic cell bodies themselves (KORF et al., 1974). The two actions on central noradrenergic fibers may cooperate: the fall in impulse frequency may procure optimal conditions for presynaptic inhibition of the release per impulse.

The literature concerning the involvement of various neurotransmitters in the effects of narcotic analgesics is highly controversial, and the emphasis laid on a presynaptic action on noradrenergic fibers is admittedly speculative. Anyway, release-inhibiting concentrations are much lower than those required for many other actions on neurones. It seems likely that presynaptic inhibition plays an important role (cf. KOSTERLITZ and HUGHES, 1975).

10.6. Conclusion

The terminals of some noradrenergic neurones appear to be endowed with specific receptors for narcotic analgesics. Activation of these presynaptic receptors by morphine and congeners leads to depression of the release of noradrenaline per impulse. The depression is antagonized by naloxone. Its mechanism is not known. — In many tissues narcotic analgesics fail to reduce noradrenaline release. It is not known whether these noradrenergic fibers lack presynaptic morphine receptors, or whether the operation of existing receptors was concealed by the experimental conditions.

At high doses, many narcotic analgesics and their antagonists inhibit the neuronal uptake of noradrenaline, pethidine and methadone being particularly potent. The inhibition is nonspecific. Inhibition of re-uptake may overshadow the depression of release and result in enhancement of the stimulation-evoked overflow of noradrenaline and the postsynaptic response to stimulation.

During chronic treatment, central noradrenergic neurones appear to adapt themselves so that the biophase concentration of noradrenaline becomes normal despite the presence of morphine. The adaptation does not involve subsensitivity of the presynaptic morphine receptors. Rather, the basic release mechanism is enhanced and/or the neuronal re-uptake mechanism is impaired. This presynaptic type of tolerance—recovery of the biophase concentration of the transmitter—is distinct from the proposed postsynaptic tolerance which is thought to consist of an increase of the sensitivity of the postsynaptic cells to the transmitter.

Morphine and its congeners probably share a common receptor system with endogenous compounds. One function of these endogenous ligands may be presynaptic inhibition of neurotransmitter release. There is as yet no evidence for the physiologic occurrence of an "endogenous ligand" modulation of noradrenaline release.—It seems possible that inhibition of release from central noradrenergic neurones contributes to the analgesia produced by morphine-like drugs.

11. General Problems

11.1. Presynaptic Receptors—A Working Hypothesis

Many endogenous substances are able to increase or decrease the amount of noradrenaline secreted in response to nerve impulses. There is no doubt that the modulation is mediated by specific receptor systems; each can be activated or blocked only by structurally related drugs. The receptors have been called "presynaptic" under the assumption that they are located on or in the neurones. It must be admitted, however, that this location has not been demonstrated directly. It is a working hypothesis. What is its basis?

A working hypothesis should explain the observations as simply as possible. The most obvious site for a release-modulating agent to act is the releasing cell. Any view that the primary action takes place elsewhere, for example postsynaptically, requires two additional assumptions, namely that a second signal is created at the primary site, and that the neurone is endowed with receptors for the second, unknown signal. Economy clearly favors the presynaptic receptor hypothesis. For most modulators one indirect mechanism, stimulation of prostaglandin formation, has been experimentally excluded.

Neurones with different morphological and biochemical environments respond to presynaptic modulators in the same way. For instance, morphine inhibits release from noradrenergic neurones supplying two or three smooth muscle organs (cat nictitating membrane, mouse vas deferens, and perhaps guinea-pig jejunum) and three brain areas (rat cerebral cortex, cerebellar cortex, and hypothalamus). It cannot be excluded that in all these tissues the modulators primarily act upon some element outside the noradrenergic fibers which then elaborates the second signal. However, a direct action on the one ingredient common to the tissues, the neurones themselves, appears less strained.

The soma and dendrites of postganglionic sympathetic neurones possess a host of receptor systems (Trendelenburg, 1967; Haefely, 1972). Maybe the receptors are restricted to this part. However, if a nerve cell has the ability to synthesize receptors and construct post-receptor reaction chains for its soma-dendritic region, it can easily do the same for its axon terminals.

From a more general point of view the assumption of presynaptic receptors seems inevitable. It is necessitated, on the one hand, by the morphological demonstration of axo-axonic synapses: a presynaptic terminal impinges not on an effector cell or the soma or dendrites of another neurone, but on a second nerve ending. Clearly the transmitter of the first terminal acts on receptors of the second one, i.e., on receptors which are presynaptic with respect to the second terminal. Catecholaminergic neurones form axo-axonic synapses where they are the secreting component (Chiba and Doba, 1976); axo-axonic synapses where noradrenergic terminals are the receptive component have not yet been found. – The assumption of presynaptic receptors is also necessitated by the electrophysiologic demonstration of presynaptic inhibition (Eccles, 1964, p. 220). The term describes an inhibition of synaptic transmission which results from a decrease of transmitter release rather than a decrease of postsynaptic sensitivity. Again, the inhibitory modulator probably acts directly on the releasing nerve ending.

Taken together, these arguments indicate that nerve endings indeed possess receptors, and that presynaptic receptors are the most plausible way to account for the effect of release-modulating agents. Responses of effector, for instance smooth muscle, cells are assumed to be mediated by receptors located in or on these cells themselves, and not for instance on fibrocytes which then send out a second signal; direct evidence for this location is also often lacking.

A working hypothesis should be amenable to trial. One type of experiment would be to search for presynaptic receptors on noradrenergic neurones grown in organ culture. Vogel et al. (1972) have shown that phenoxybenzamine increases the stimulation-evoked overflow of tritium from cultured rat sympathetic ganglia preincubated with ^{3}H-noradrenaline. Because the preparation contains no postsynaptic element, the site of action must be presynaptic. Unfortunately, the concentration of phenoxybenzamine was very high (3×10^{-5} M). It is thus not possible to decide whether the presynaptic site was the α-adrenoceptor or the noradrenaline uptake mechanism. Anyway, the model might help to clarify whether the attribute "presynaptic" describes the location of release-modulating receptors correctly.

11.2. Where within the Neurone?

Which part of the neurone, if any, carries the receptors? The majority of experiments discussed in this review were performed on isolated preparations devoid of nerve cell bodies and dendrites. Therefore, a neuronal location would imply a place on either the preterminal, smooth part of the axon, or the thin segments connecting the varicosities, or the varicosities themselves, or several of these structures. Unequivocal distinction is not possible. However, it seems likely that at least part of the receptors are constituents of the varicosities.

One reason again concerns economy of construction. The varicosities are assumed to secrete the transmitter. They are the strategically optimal place for a release-modulating receptor.

A second argument comes from experiments on potassium-evoked release. Many modulators have been shown to exert similar effects on release by electrical stimulation and release by high potassium concentrations. The latter is independent of the propagation of action potentials along the axons and not changed by tetrodotoxin (BLAUSTEIN et al., 1972; STJÄRNE, 1973h). High potassium leads to parallel and independent depolarization of, and release from, many or all individual varicosities. Therefore, drugs that modulate this release must also exert parallel and independent effects on many or all varicosities. Interaction with receptors on the preterminal axon alone would be insufficient and can be ruled out. It cannot be excluded that the primary site of action is the intervaricose, conductive sections of the nerve terminals, from which a chain of reactions would then spread into the varicosities. However, a direct effect on the varicosities themselves seems more plausible.

Noradrenergic varicosities appear to be constructed not only to synthesize, store, secrete, and inactivate the transmitter, but also to adjust the secretion per impulse according to humoral signals. It seems possible, though by no means certain, that the varicosities themselves, perhaps their plasma membranes, carry the receptors that perceive these signals.

11.3. Mechanism of Modulation

The sequence of events between interaction with the receptor and change of release is not known for any presynaptic modulator. Current hypotheses are centered upon alterations of the availability of calcium for the release process, since calcium is thought to be rate-limiting in electro-secretory coupling. Drugs might modify the stimulus-evoked influx of calcium from the extracellular space, or its efflux, or its distribution between intraneuronal compartments. However, the evidence for the calcium availability hypothesis of presynaptic modulation is suggestive rather than conclusive.

Several presynaptic inhibitors have been shown to affect only calcium-dependent modes of release. However, this does not necessarily indicate that the inhibitors reduce the amount of calcium available for stimulus-secretion coupling. They may just as well slow down another step in calcium-mediated release which then becomes rate-limiting.

It has also been shown that some inhibitors cause a more pronounced depression, the lower the external calcium concentration. An increase in calcium shifts presynaptic dose–inhibition curves to the right. The results are compatible with the view that the inhibitors reduce the availability of calcium for electro-secretory coupling; this reduction would be overcome by an increase in extracellular calcium. However, there are alternatives. Changes in external calcium strongly affect release even in the absence of modulating drugs. When a presynaptic inhibitor is tested at different calcium levels, the cation is not the only independent variable, but the entire basic conditions are altered. A change in the degree of inhibition may depend not on the change in calcium, but the alteration of another basic condition. The interaction between presynaptic inhibitors and calcium may be a functional antagonism (BRINK, 1973), which does not allow any conclusions about the mode of action to be drawn.

All modulators appear to be less effective, the higher the frequency of stimulation. The observation is in accordance with the calcium availability hypothesis if one assumes that the relation between intraneuronal calcium and release obeys saturation kinetics and that with rising frequency calcium approaches saturation levels. If so, a drug-induced change in calcium which strongly affects release at low frequency will become negligible under calcium saturation conditions at high frequency (cf. p. 5). However, at present the relations between frequency, intraneuronal calcium, and release are hypothetical. Many unknown factors make conclusions from the frequency-dependence of presynaptic modulation highly tentative.

11.4. Tissue Differences

Noradrenergic neurones are similar in morphology and biochemistry. All the more puzzling is the finding that they greatly differ in their response to presynaptic modulators. Results obtained in some noradrenergically innervated tissues are summarized in Table 10. Only one group of drugs, namely α-receptor agonists, are effective in all tissues studied.

In principle, two reasons may account for negative results. Firstly, the terminals of resistant fibers may lack receptors. Secondly, receptors and post-receptor mechanisms may exist, but may be silent under the experimental conditions chosen. Circumstances unsuited for the demonstration of presynaptic receptors include stimulation at high frequency and excessive calcium concentrations. However, the negative results listed in Table 10 were obtained at low frequencies and moderate calcium concentrations (no details given in three cases).

An interesting possibility was suggested by HENDERSON and HUGHES (1974). They investigated the relation between frequency and noradrenaline overflow per impulse in different tissues and at different calcium concentrations. In the rabbit portal vein and vas deferens, when the calcium concentration was 2.5 mM, the overflow per pulse increased with increasing stimulation frequency, whereas in the cat nictitating membrane and the mouse vas deferens it remained constant. Increases or decreases in external calcium selectively enhanced or reduced, respectively, release at low frequency, so that the slopes of the frequency–overflow curves changed (cf. p. 5). Morphine reduced release by low frequency stimulation in the cat nictitating membrane (cf. HENDERSON et al., 1972b, 1975) and the mouse vas deferens, tissues in which the slope of the frequency–overflow curve approached zero, but not in the two other tissues. Conversely, angiotensin enhanced release by low frequency stimulation in the rabbit portal vein and

Table 10. Tissue differences in effects of presynaptic modulators of noradrenaline release

	Cat spleen	Cat nictitating membrane	Rat cerebral cortex	Guinea-pig vas deferens	Rabbit heart	Rabbit pulmonary artery
Angiotensin	0	0	0		+	+
Muscarinic agonists	−	−	0	−	−	−
PGE_1, PGE_2	−	0	−	−	−	−
Adrenergic agonists; effect on α-receptors	−	−	−	−	−	−
Dopamine; effect on dopamine receptors	−	−	0	0		0
Adrenergic agonists; effect on β-receptors	+	+	0	+		0
Narcotic analgesics	0[a]	−	−	0[b]	0	0
Slope of frequency–overflow curve at near normal calcium	Positive	Zero	Negative	Positive		Positive[c]

Effects on noradrenaline release: + increase, − decrease, 0 no change (for details, see Tables 1, 3, 4, 6, and 9). Positive, zero, and negative slopes of frequency-overflow curves indicate that the overflow of noradrenaline evoked per pulse increases, remains constant, and decreases, respectively, with increasing stimulation frequency (for references, see p. 5). [a] Postsynaptic response study (CAIRNIE et al., 1961). [b] Postsynaptic response study (HUGHES et al., 1975a). [c] STARKE, unpublished.

vas deferens, tissues with a positive frequency–overflow slope, but not in the mouse vas deferens. The authors suggest that a positive slope and a slope near zero of the frequency–overflow curve are generally associated with sensitivity to angiotensin and morphine, respectively (cf. HENDERSON et al., 1972a and b, 1975). The basis for this correlation is not known.

Slopes of frequency–overflow curves at moderate calcium concentrations are included in Table 10. In general, a comparison confirms the suggestion of HENDERSON and coworkers, with two exceptions. Firstly, angiotensin fails to enhance release in the cat spleen, an organ with a positive slope of the frequency–overflow curve; angiotensin-induced formation of prostaglandins may account for its lack of effect (p. 16). Secondly, in the rat cerebral cortex the frequency–overflow slope is *negative*, indicating perhaps that sensitivity to morphine may be associated with both negative and zero slopes. – Clearly, the suggestion merits further investigation. One way to test its validity would be to vary the slope of frequency–overflow curves in a given tissue by means of different calcium concentrations. This procedure should change the sensitivity to morphine on the one hand, and to angiotensin on the other hand, in opposite directions. For instance, a decrease of slope from a positive value to near zero should transform an angiotensin-sensitive, morphine-resistant neurone to a morphine-sensitive, angiotensin-resistant one. This would impressively illustrate that experimental conditions can determine whether existing presynaptic receptors are demonstrable.

Further research may reveal that other negative results compiled in Table 10 also reflect the choice of unsuited experimental conditions, as for instance the surprising lack of an inhibitory effect of muscarinic agonists in the rat cerebral cortex, and of prostaglandins in the cat nictitating membrane. However, it cannot be ruled out that in some tissues the nerve endings indeed lack certain presynaptic receptor systems. Unfortunately, more direct ways to establish the *absence* of a receptor are not available.

11.5. General Significance

Physiologic implications of presynaptic receptors have been discussed under the individual agents. From a general point of view, three groups of possible functions may be distinguished.

Firstly, presynaptic receptors may be the site of action of modulators originating from a remote part of the organism and transported by the blood stream. Known candidates for such a modulatory role are renin/angiotensin and the adrenal medullary hormones. Both angiotensin and adrenaline can occur in the blood at concentrations which significantly facilitate release via presynaptic angiotensin and β-receptors, respectively.

Secondly, presynaptic receptors may be the site of action of neurotransmitters or neuromodulators secreted from adjacent neurones or other cells. As mentioned above, central axo-axonic synapses with noradrenergic terminals as the receptive component have not yet been detected. Thus, the occurrence of receptors on noradrenergic terminals for presynaptic inhibition or perhaps facilitation by other neurones remains speculative. It should be noted that even non-neuronal elements of the central nervous system might influence adjacent nerve terminals by the secretion of modulators or their precursors, e.g., prostaglandins, renin, or endogenous ligands for morphine receptors. – This kind of modulation may also occur in the periphery. Morphologically specialized axo-axonic synapses between peripheral cholinergic and noradrenergic fibers are not known. Nevertheless it has been shown that acetylcholine released from postganglionic parasympathetic nerve endings can inhibit noradrenaline release from neighboring sympathetic nerve endings, and vice versa. This mutual presynaptic antagonism may supplement the postsynaptic antagonism between the two divisions of the autonomic nervous system.

Thirdly, presynaptic receptors are a site of action of the transmitter itself and of prostaglandins which are locally formed during noradrenergic synaptic transmission. Presynaptic prostaglandin receptors and adrenoceptors are constituents of feedback loops controlling the release of transmitter per impulse (Fig. 8). The prostaglandin mechanism appears to be trans-synaptic, involving the postsynaptic cell where at least a large fraction of the prostaglandins is synthesized. In contrast, the feedbacks mediated by presynaptic α- and β-receptors are entirely presynaptic, independent of the postsynaptic cell. – Structures and functions developed in the course of evolution should be of some advantage to the organism. It must be admitted that the advantage provided by the pre- and trans-synaptic feedbacks is far from being clear. Anyway, even if the sense

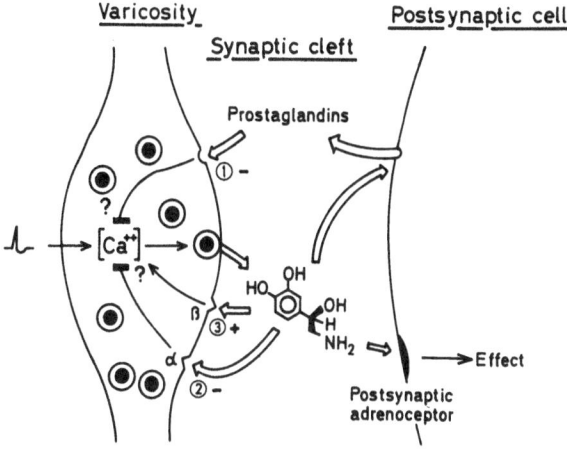

Fig. 8. Synaptic feedback mechanisms controlling the per pulse release of noradrenaline. ①, prosta-glandin-mediated negative feedback; ②, α-adrenergic negative feedback; ③, β-adrenergic positive feedback. For this figure it has been assumed that the prostaglandins originate from postsynaptic cells, and that inhibitory and facilitatory mechanisms involve a decrease and increase of intraneuronal calcium, respectively

is not clear we have to accept the fact that these mechanisms work. It is not the frequency of impulse flow alone that determines the noradrenergic input to postsynaptic cells; prostaglandin-, α- and/or β-receptor-mediated feedback mechanisms contribute to the regulation of noradrenergic transmission.

Not only noradrenergic, but also dopaminergic and serotonergic nerve terminals appear to secrete less, the higher the concentration of the respective transmitter in the biophase (FARNEBO and HAMBERGER, 1971 b), suggesting "that monoamine nerve terminals have presynaptic receptors for release regulation which are sensitive to the transmitter of the nerve terminal" (FARNEBO and HAMBERGER, 1973 b). The same may hold good for cholinergic fibers (p. 33). On the other hand, presynaptic receptors for a particular neurotransmitter are not restricted to the nerve endings which contain this transmitter. For instance, presynaptic muscarine receptors occur on noradrenergic and dopaminergic as well as on cholinergic fibers. There may be nothing specific in the existence of neurotransmitter receptors on the nerve endings releasing the transmitter. These "autoreceptors" may occur incidentally among a variety of other receptor systems. It should also be noted that the term "autoreceptor" is misleading since it suggests that neurones possess only one receptor system for their own transmitter. Reality is more complex. Noradrenergic terminals potentially possess no less than three discrete adrenoceptor populations, namely α-, β-, and dopamine receptors, with quite different structures and functions. Both muscarine and nicotine receptors may occur on cholinergic nerve endings. Unfortunately, no attractively simple general principle has emerged from the recent research on presynaptic receptor systems.

More practical implications have also been discussed under the individual agents. In general terms presynaptic receptors may be, firstly, sites of dysregulation as well as of regulation, "one more level where things may go wrong" (STJÄRNE, 1975a, p. 223). For instance, an excessive release of noradrenaline per impulse may play a role in hypertensive disease (BEVAN et al., 1975); one possible reason is a defect in presynaptic receptor function (cf. LANGER et al., 1975c). Secondly, presynaptic receptors may be sites of action of drugs. Certain presynaptic α-receptors can be activated or blocked independently of postsynaptic α-receptors. The same may hold true for presynaptic dopamine receptors on dopaminergic neurones and for other receptor systems as well. Such specific agonists and antagonists might be powerful pharmacologic tools as well as therapeutic agents (STARKE and MONTEL, 1973f; CARLSSON, 1975).

12. Summary

The release of transmitter from noradrenergic nerves per unit time depends on the rate of impulse flow on the one hand, and the release per impulse on the other hand. Several substances naturally occurring in the body are able to increase or decrease the quantum secreted per impulse. They act on receptors which are probably located on the noradrenergic nerve terminals. Presynaptic receptor systems for angiotensin, acetylcholine (nicotine and muscarine receptors), prostaglandins, catecholamines (α-, β-, and dopamine receptors) and narcotic analgesics are discussed in this review. Chapters on individual modulators cover their influence on the noradrenergic fibers of different tissues, their mode of action, physiologic and pharmacologic implications, and their action on the release of other neurotransmitters and adrenal medullary hormones. Most agents have effects on noradrenergic nerve endings apart from modulation of release; these side effects are also considered.

Some problems concern presynaptic receptor systems in general. The morphologic place of the receptors is not known with certainty. Their location on or in the nerve endings is a working hypothesis which is now widely accepted. It seems likely that the varicosities themselves carry the receptors for the humoral signals which modulate their secretory function. – Most presynaptic modulators have been shown to affect only calcium-dependent release processes. However, it is not certain whether enhancement or reduction of the stimulus-evoked increase of intraneuronal calcium are indeed the essential factors in the facilitation or depression of release. It cannot be ruled out that effects on other steps of stimulus-secretion coupling are of primary importance so that release per pulse rises or falls even though the availability of calcium for coupling is unchanged. – Presynaptic α-adrenoceptors have been found on all noradrenergic neurones studied so far. Other modulators affect release in some tissues only. Negative results may reflect the choice of experimental conditions which obscure the function of existing receptors. However, it cannot be excluded that neurones of different tissues differ in their presynaptic receptors. – In general terms, presynaptic receptors on noradrenergic fibers may serve three purposes: (1) They allow modulation of release by blood-borne agents such as angiotensin, or adrenaline originating from the adrenal medulla; (2) they also allow modulation by neurotransmitters or neuromodulators secreted from neighboring nerve terminals or non-neuronal cells; and (3) presynaptic prostaglandin, α-, and β-receptors are constituents of feedback mechanisms which operate within the noradrenergic synapse and control the release per nerve impulse. In addition to their physiologic functions, presynaptic receptor systems may also have a pathophysiologic and therapeutic significance.

Acknowledgements. I owe a debt of gratitude to Professor H.J. SCHÜMANN for his continuous interest and encouragement, and to Dr. H. MONTEL, who took part in a great deal of my work. I also wish to thank Professor U. TRENDE-LENBURG for his pertinent critical comments on the manuscript and many useful suggestions. The conscientious technical assistance of Mrs. B. GAMISCH and Mr. E. HAGELSKAMP has been an invaluable help throughout several years. Studies in the author's laboratory were supported by the Deutsche Forschungsgemeinschaft.

References

Abdel-Aziz, A.: Blockade by prostaglandins E_2 and $F_{1\alpha}$ of the response of the rabbit ileum to stimulation of sympathetic nerve and its reversal by some antihistamines, dexamphetamine and methylphenidate. Europ. J. Pharmacol. **25**, 226–230 (1974)

Åblad, B., Ek, L., Johansson, B., Waldeck, B.: Inhibitory effect of propranolol on the vasoconstrictor response to sympathetic nerve stimulation. J. Pharm. Pharmacol. **22**, 627–628 (1970)

Ackerly, J., Blumberg, A., Peach, M.: Angiotensin interactions with myocardial sympathetic neurons: Enhanced release of dopamine-β-hydroxylase during nerve stimulation. Proc. Soc. exp. Biol. (N.Y.) **151**, 650–653 (1976)

Adler-Graschinsky, E., Langer, S.Z.: Possible role of a β-adrenoceptor in the regulation of noradrenaline release by nerve stimulation through a positive feed-back mechanism. Brit. J. Pharmacol. **53**, 43–50 (1975)

Allen, G.S., Glover, A.B., McCulloch, M.W., Rand, M.J., Story, D.F.: Modulation by acetylcholine of adrenergic transmission in the rabbit ear artery. Brit. J. Pharmacol. **54**, 49–53 (1975)

Allen, G.S., Glover, A.B., Rand, M.J., Story, D.F.: Effects of acetylcholine on vasoconstriction and release of ^{3}H-noradrenaline in response to sympathetic nerve stimulation in the isolated artery of the rabbit ear. Brit. J. Pharmacol. **46**, 527–528 P (1972a)

Allen, G.S., Rand, M.J., Story, D.F.: Effects of McN-A-343 and DMPP on the uptake and release of ^{3}H-noradrenaline by guinea-pig atria. Brit. J. Pharmacol. **45**, 480–489 (1972b)

Allen, G.S., Rand, M.J., Story, D.F.: Comparison of effects of six cholinomimetic drugs on inhibition of uptake of ^{3}H-($\pm$)-noradrenaline by guinea-pig atria. Brit. J. Pharmacol. **47**, 179–180 (1973)

Allen, G.S., Rand, M.J., Story, D.F.: Effects of the muscarinic agonist McN-A-343 on the release by sympathetic nerve stimulation of (^{3}H)-noradrenaline from rabbit isolated ear arteries and guinea-pig atria. Brit. J. Pharmacol. **51**, 29–34 (1974)

Ambache, N., Dunk, L.P., Verney, J., Zar, M.A.: Inhibition of post-ganglionic motor transmission in the vas deferens by indirectly acting sympathomimetic drugs. J. Physiol. (Lond.) **227**, 433–456 (1972)

Ambache, N., Killick, S.W., Srinivasan, V., Zar, M.A.: Effects of lysergic acid diethylamide on autonomic post-ganglionic transmission. J. Physiol. (Lond.) **246**, 571–593 (1975)

Ambache, N., Zar, M.A.: Evidence against adrenergic motor transmission in the guinea-pig vas deferens. J. Physiol. (Lond.) **216**, 359–389 (1971)

Andén, N.E., Corrodi, H., Fuxe, K., Hökfelt, B., Hökfelt, T., Rydin, C., Svensson, T.: Evidence for a central noradrenaline receptor stimulation by clonidine. Life Sci. **9**, I, 513–523 (1970)

Andén, N.E., Grabowska, M., Strömbom, U.: Different alpha-adrenoceptors in the central nervous system mediating biochemical and functional effects of clonidine and receptor blocking agents. Naunyn-Schmiedeberg's Arch. Pharmacol. **292**, 43–52 (1976)

Anderson, T.R., Slotkin, T.A.: Effects of morphine on the rat adrenal medulla. Biochem. Pharmacol. **24**, 671–679 (1975)

Anderson, T.R., Slotkin, T.A.: The role of neural input in the effects of morphine on the rat adrenal medulla. Biochem. Pharmacol. **25**, 1071–1074 (1976)

Antonaccio, M.J., Halley, J., Kerwin, L.: Functional significance of α-stimulation and α-blockade on responses to cardiac nerve stimulation in anesthetized dogs. Life Sci. **15**, 765–777 (1974)

Armstrong, J.M., Boura, A.L.A.: Effects of clonidine and guanethidine on peripheral sympathetic nerve function in the pithed rat. Brit. J. Pharmacol. **47**, 850–852 (1973)

Bacq, Z.M., Fredericq, H.: Recherches sur la physiologie et la pharmacologie du système nerveux autonome. XIV. — Modifications apportées par deux dérivés de l'aminométhylbenzodioxane (883F. et 933F.) aux effets de l'adrénaline et de l'excitation sympathique sur la membrane nictitante du chat. Arch. int. Physiol. **40**, 454–466 (1935)

Baldessarini, R.J.: Release of catecholamines. In: Iversen, L.L., Iversen, S.D., Snyder, S.H. (eds.), Handbook of Psychopharmacology, Vol. 3, pp. 37–137. New York-London: Plenum Press 1975

Balfour, D.J.K.: Effects of nicotine on the uptake and retention of ^{14}C-noradrenaline and ^{14}C-5-hydroxytryptamine by rat brain homogenates. Europ. J. Pharmacol. **23**, 19–26 (1973)

Barrett, A.M., Nunn, B.: Adrenergic neuron blocking properties of ($\pm$)-propranolol and (+)-propranolol. J. Pharm. Pharmacol. **22**, 806–810 (1970)

Barth, C.: Aktivität der Catechol-ortho-methyltransferase unter Angiotensineinfluß und beim DOCA-Hochdruck der Ratte. Z. klin. Chem. **8**, 379–382 (1970)

Baum, T., Shropshire, A.T.: Influence of prostaglandins on autonomic responses. Amer. J. Physiol. **221**, 1470–1475 (1971)

Bedwani, J.R., Millar, G.C.: Prostaglandin release from cat and dog spleen. Brit. J. Pharmacol. **54**, 499–505 (1975)

Beleslin, D., Polak, R.L.: Depression by morphine and chloralose of acetylcholine release from the cat's brain. J. Physiol. (Lond.) **177**, 411–419 (1965)

Bell, C.: Differential effects of tetrodotoxin on sympathomimetic actions of nicotine and tyramine. Brit. J. Pharmacol. **32**, 96–103 (1968)

Bell, C.: Mechanism of enhancement by angiotensin II of sympathetic adrenergic transmission in the guinea pig. Circulat. Res. **31**, 348–355 (1972)

Bell, C.: Release of endogenous noradrenaline from an isolated elastic artery. J. Physiol. (Lond.) **236**, 473–482 (1974)

Bell, C., Conway, E.L., Lang, W.J., Padanyi, R.: Vascular dopamine receptors in the canine hindlimb. Brit. J. Pharmacol. **55**, 167–172 (1975)

Bell, C., Vogt, M.: Release of endogenous noradrenaline from an isolated muscular artery. J. Physiol. (Lond.) **215**, 509–520 (1971)

Bell, K.M., Rees, J.M.H.: The depressant action of morphine on transmission at a skeletal neuromuscular junction is non-specific. J. Pharm. Pharmacol. **26**, 686–691 (1974)

Benelli, G., Bella, D.D., Gandini, A.: Angiotensin and peripheral sympathetic nerve activity. Brit. J. Pharmacol. **22**, 211–219 (1964)

Bennett, M.R.: An electrophysiological analysis of the storage and release of noradrenaline at sympathetic nerve terminals. J. Physiol. (Lond.) **229**, 515–531 (1973a)

Bennett, M.R.: An electrophysiological analysis of the uptake of noradrenaline at sympathetic nerve terminals. J. Physiol. (Lond.) **229**, 533–546 (1973b)

Bennett, M.R., Florin, T.: An electrophysiological analysis of the effect of Ca ions on neuromuscular transmission in the mouse vas deferens. Brit. J. Pharmacol. **55**, 97–104 (1975)

Bennett, M.R., Middleton, J.: An electrophysiological analysis of the effects of reserpine on adrenergic neuromuscular transmission. Brit. J. Pharmacol. **55**, 79–85 (1975a)

Bennett, M.R., Middleton, J.: An electrophysiological analysis of the effects of amine-uptake blockers and α-adrenoceptor blockers on adrenergic neuromuscular transmission. Brit. J. Pharmacol. **55**, 87–95 (1975b)

Bentley, G.A., Li, D.M.F.: Studies of the new hypotensive drug ST 155. Europ. J. Pharmacol. **4**, 124–134 (1968)

Bergström, S., Farnebo, L.O., Fuxe, K.: Effect of prostaglandin E_2 on central and peripheral catecholamine neurons. Europ. J. Pharmacol. **21**, 362–368 (1973)

Berkowitz, B.A., Spector, S.: Effect of caffeine and theophylline on peripheral catecholamines. Europ. J. Pharmacol. **13**, 193–196 (1971)

Bevan, J.A., Haeusler, G.: Electrical events associated with the action of nicotine at the adrenergic nerve terminal. Arch. int. Pharmacodyn. **218**, 84–95 (1975)

Bevan, J.A., Su, C.: Distribution theory of resistance of neurogenic vasoconstriction to alpha-receptor blockade in the rabbit. Circulat. Res. **28**, 179–187 (1971)

Bevan, J.A., Su, C.: Uptake of nicotine by the sympathetic nerve terminals in the blood vessel. J. Pharmacol. exp. Ther. **182**, 419–426 (1972)

Bevan, R.D., Purdy, R.E., Su, C., Bevan, J.A.: Evidence for an increase in adrenergic nerve function in blood vessels from experimental hypertensive rabbits. Circulat. Res. **37**, 503–508 (1975)

Bhagat, B., Dhalla, N.S., Ginn, D., Montagne, A.E., Montier, A.D.: Modification by prostaglandin E_2 (PGE_2) of the response of guinea-pig isolated vasa deferentia and atria to adrenergic stimuli. Brit. J. Pharmacol. **44**, 689–698 (1972)

Bhagat, B., Robinson, I.M., West, W.L.: Mechanism of sympathomimetic responses of isolated guinea-pig atria to nicotine and dimethylphenylpiperazinium iodide. Brit. J. Pharmacol. **30**, 470–477 (1967)

Blakeley, A.G.H., Brown, L., Dearnaley, D.P., Woods, R.I.: Perfusion of the spleen with blood containing prostaglandin E_1: transmitter liberation and uptake. Proc. roy. Soc. B **174**, 281–292 (1969a)

Blakeley, A.G.H., Brown, G.L., Ferry, C.B.: Pharmacological experiments on the release of the sympathetic transmitter. J. Physiol. (Lond.) **167**, 505–514 (1963)

Blakeley, A.G.H., Brown, L., Geffen, L.B.: Uptake and re-use of sympathetic transmitter in the cat's spleen. Proc. roy. Soc. B **174**, 51–68 (1969b)

Blakeley, A.G.H., Powis, G., Summers, R.J.: Uptake of (L)-noradrenaline in the isolated cat heart perfused with blood containing desmethylimipramine (DMI) and 17-β-oestradiol (17βO). J. Physiol. (Lond.) **234**, 108–109P (1973)

Blakeley, A.G.H., Powis, G., Summers, R.J.: An uptake mechanism for L-noradrenaline in the cat spleen, associated with the nerves but distinct from uptake$_1$. J. Physiol. (Lond.) **238**, 193–206 (1974)

Blaustein, M.P., Johnson, E.M., Needleman, P.: Calcium-dependent norepinephrine release from presynaptic nerve endings in vitro. Proc. nat. Acad. Sci. (Wash.) **69**, 2237–2240 (1972)

Bloom, F.E.: Amine receptors in CNS. I. Norepinephrine. In: Iversen, L.L., Iversen, S.D., Snyder, S.H. (eds.), Handbook of Psychopharmacology, Vol. 6, pp. 1–22. New York-London: Plenum Press 1975

Blosser, J.C., Catravas, G.N.: Action of reserpine in morphine-tolerant rats: Absence of an antagonism of catecholamine depletion. J. Pharmacol. exp. Ther. **191**, 284–289 (1974)

Blumberg, A.L., Ackerly, J.A., Peach, M.J.: Differentiation of neurogenic and myocardial angiotensin II receptors in isolated rabbit atria. Circulat. Res. **36**, 719–726 (1975a)

Blumberg, A., Brooker, G., Peach, M., Westfall, T.C.: Angiotensin interactions with myocardial sympathetic neurons: Increased release of dopamine-β-hydroxylase and elevated atrial cyclic AMP concentrations. Fed. Proc. **34**, 770 Abs (1975b)

Boadle, M.C., Hughes, J., Roth, R.H.: Angiotensin accelerates catecholamine biosynthesis in sympathetically innervated tissues. Nature (Lond.) **222**, 987–988 (1969)

Boadle-Biber, M.C., Hughes, J., Roth, R.H.: Acceleration of catecholamine biosynthesis in sympathetically innervated tissues by angiotensin-II-amide. Brit. J. Pharmacol. **46**, 289–299 (1972)

Bogaievsky, D., Bogaievsky, Y., Tsoucaris-Kupfer, D., Schmitt, H.: Blockade of the central hypotensive effect of clonidine by α-adrenoreceptor antagonists in rats, rabbits and dogs. Clin. exp. Pharmacol. Physiol. **1**, 527–534 (1974)

Boissier, J.R., Giudicelli, J.F., Fichelle, J., Schmitt, H., Schmitt, H.: Cardiovascular effects of 2-(2,6-dichlorophenylamino)-2-imidazoline hydrochloride (ST 155). Europ. J. Pharmacol. **2**, 333–339 (1968)

Bolme, P., Corrodi, H., Fuxe, K., Hökfelt, T., Lidbrink, P., Goldstein, M.: Possible involvement of central adrenaline neurons in vasomotor and respiratory control. Studies with clonidine and its interactions with piperoxane and yohimbine. Europ. J. Pharmacol. **28**, 89–94 (1974)

Boonyaviroj, P., Gutman, Y.: α-Adrenergic stimulants, prostaglandins and catecholamine release from the adrenal gland in vitro. Prostaglandins **10**, 109–116 (1975)

Borowski, E., Ehrl, H., Starke, K.: Relative pre- and postsynaptic potencies of α-adrenolytic drugs. Naunyn-Schmiedeberg's Arch. Pharmacol. **293**, R2 (1976)

Botting, J.H., Salzmann, R.: The effect of indomethacin on the release of prostaglandin E_2 and acetylcholine from guinea-pig isolated ileum at rest and during field stimulation. Brit. J. Pharmacol. **50**, 119–124 (1974)

Boullin, D.J., Costa, E., Brodie, B.B.: Evidence that blockade of adrenergic receptors causes overflow of norepinephrine in cat's colon after nerve stimulation. J. Pharmacol. exp. Ther. **157**, 125–134 (1967)

Boykin, M.E., Martin, R.H.: Preliminary observations on the uptake of catecholamines in synaptosomes of opiate-treated animals. J. Pharm. Pharmacol. **25**, 484–485 (1973)

Braestrup, C., Nielsen, M.: Intra- and extraneuronal formation of the two major noradrenaline metabolites in the cns of rats. J. Pharm. Pharmacol. **27**, 413–419 (1975)

Brandão, F., Guimarães, S.: Inactivation of endogenous noradrenaline released by electrical stimulation in vitro of dog saphenous vein. Blood Vessels **11**, 45–54 (1974)

Brandon, K.W., Boyd, H.: Release of noradrenaline from the spleen of the cat by acetylcholine. Nature (Lond.) **192**, 880–881 (1961)

Brink, F.G. van den: The model of functional interaction. I. Development and first check of a new model of functional synergism and antagonism. Europ. J. Pharmacol. **22**, 270–278 (1973)

Brody, M.J., Kadowitz, P.J.: Prostaglandins as modulators of the autonomic nervous system. Fed. Proc. **33**, 48–60 (1974)

Brown, G.L.: The release and fate of the transmitter liberated by adrenergic nerves. Proc. roy. Soc. B **162**, 1–19 (1965)

Brown, G.L., Davies, B.N., Ferry, C.B.: The effect of neuronal rest on the output of sympathetic transmitter from the spleen. J. Physiol. (Lond.) **159**, 365–380 (1961)

Brown, G.L., Davies, B.N., Gillespie, J.S.: The release of chemical transmitter from the sympathetic nerves of the intestine of the cat. J. Physiol. (Lond.) **143**, 41–54 (1958)

Brown, G.L., Gillespie, J.S.: Output of sympathin from the spleen. Nature (Lond.) **178**, 980 (1956)

Brown, G.L., Gillespie, J.S.: The output of sympathetic transmitter from the spleen of the cat. J. Physiol. (Lond.) **138**, 81–102 (1957)

Brücke, F.T.: Über die Wirkung von Acetylcholin auf die Pilomotoren. Klin. Wschr. **14**, 7–9 (1935)

Bryant, B.J., McCulloch, M.W., Rand, M.J., Story, D.F.: Release of ^{3}H-(−)-noradrenaline from guinea-pig hypothalamic slices: effects of adrenoceptor agonists and antagonists. Brit. J. Pharmacol. **53**, 454P (1975)

Bülbring, E., Burn, J.H., Elio, F.J. de: The secretion of adrenaline from the perfused suprarenal gland. J. Physiol. (Lond.) **107**, 222–232 (1948)

Burn, J.H., Gibbons, W.R.: The effect of phenoxybenzamine and of tolazoline on the response to sympathetic stimulation. Brit. J. Pharmacol. **22**, 527–539 (1964a)

Burn, J.H., Gibbons, W.R.: The part played by calcium in determining the response to stimulation of sympathetic postganglionic fibres. Brit. J. Pharmacol. **22**, 540–548 (1964b)

Burn, J.H., Ng, K.K.F.: The action of pempidine and antiadrenaline substances at the sympathetic postganglionic termination. Brit. J. Pharmacol. **24**, 675–688 (1965)

Burn, J.H., Rand, M.J.: Sympathetic postganglionic mechanism. Nature (Lond.) **184**, 163–165 (1959)

Burnstock, G., Holman, M.E., Kuriyama, H.: Facilitation of transmission from autonomic nerve to smooth muscle of guinea-pig vas deferens. J. Physiol. (Lond.) **172**, 31–49 (1964)

Cabrera, R., Torrance, R.W., Viveros, H.: The action of acetyl choline and other drugs upon the terminal parts of the postganglionic sympathetic fibre. Brit. J. Pharmacol. **27**, 51–63 (1966)

Cairnie, A.B., Kosterlitz, H.W., Taylor, D.W.: Effect of morphine on some sympathetically innervated effectors. Brit. J. Pharmacol. **17**, 539–551 (1961)

Cardenas, H.L., Ross, D.H.: Morphine induced calcium depletion in discrete regions of rat brain. J. Neurochem. **24**, 487–493 (1975)

Carlsson, A.: Dopaminergic autoreceptors. In: Almgren, O., Carlsson, A., Engel, J. (eds.), Chemical Tools in Catecholamine Research, Vol. 2, pp. 219–225. Amsterdam-Oxford: North-Holland 1975

Carlsson, A., Lindqvist, M.: Central and peripheral monoaminergic membrane-pump blockade by some addictive analgesics and antihistamines. J. Pharm. Pharmacol. **21**, 460–464 (1969)

Carmichael, F.J., Israel, Y.: In vitro inhibitory effects of narcotic analgesics and other psychotropic drugs on the active uptake of norepinephrine in mouse brain tissue. J. Pharmacol. exp. Ther. **186**, 253–260 (1973)

Celsen, B., Kuschinsky, K.: Effects of morphine on kinetics of ^{14}C-dopamine in rat striatal slices. Naunyn-Schmiedeberg's Arch. Pharmacol. **284**, 159–165 (1974)

Cervoni, P., Reit, E.: Interaction of angiotensin with exogenous and neurally released norepinephrine on the cat nictitating membrane in vitro. J. Pharmacol. exp. Ther. **193**, 1–8 (1975)

Chalmers, J.P.: Brain amines and models of experimental hypertension. Circulat. Res. **36**, 469–480 (1975)

Chanh, P.H., Junstad, M., Wennmalm, Å.: Augmented noradrenaline release following nerve stimulation after inhibition of prostaglandin synthesis with indomethacin. Acta physiol. scand. **86**, 563–567 (1972)

Chevillard, C., Alexandre, J.: Action, in vitro et in vivo, de l'angiotensine II sur le captage cardiaque de la noradrénaline. Experientia (Basel) **26**, 1334–1336 (1970)

Chevillard, C., Alexandre, J.M.: In vitro effects on cardiac norepinephrine of angiotensin II and of two indirectly acting sympathomimetic amines (amphetamine and tyramine). A study of their combinations. Europ. J. Pharmacol. **19**, 223–230 (1972)

Chevillard, C., Duchène, N., Alexandre, J.M.: Selective release of newly synthesized cardiac norepinephrine induced by angiotensin II. Europ. J. Pharmacol. **15**, 8–14 (1971)

Chevillard, C., Duchene, N., Alexandre, J.M.: How does angiotensin II increase cardiac dopamine-β-hydroxylation? J. Pharm. Pharmacol. **27**, 193–196 (1975)

Chiba, T., Doba, N.: Catecholaminergic axo-axonic synapses in the nucleus of the tractus solitarius (pars commissuralis) of the cat: Possible relation to presynaptic regulation of baroreceptor reflexes. Brain Res. **102**, 255–265 (1976)

Chubb, I.W., Potter, W.P. de, Schaepdryver, A.F. de: Tyramine does not release noradrenaline from splenic nerve by exocytosis. Naunyn-Schmiedeberg's Arch. Pharmacol. **274**, 281–286 (1972)

Cicero, T.J., Meyer, E.R., Smithloff, B.R.: Alpha adrenergic blocking agents: Antinociceptive activity and enhancement of morphine-induced analgesia. J. Pharmacol. exp. Ther. **189**, 72–82 (1974)

Cicero, T.J., Wilcox, C.E., Smithloff, B.R., Meyer, E.R., Sharpe, L.G.: Effects of morphine, in vitro and in vivo, on tyrosine hydroxylase activity in rat brain. Biochem. Pharmacol. **22**, 3237–3246 (1973)

Ciofalo, F.R.: Effects of some narcotics and antagonists on synaptosomal ^{3}H-norepinephrine uptake. Life Sci. **11**, I, 573–580 (1972)

Ciofalo, F.R.: Prostaglandins and synaptosomal transport of ^{3}H-norepinephrine and ^{3}H-5-hydroxytryptamine. Res. Commun. chem. Pathol. Pharmacol. **5**, 551–554 (1973)

Ciofalo, F.R.: Methadone inhibition of ^{3}H-5-hydroxytryptamine uptake by synaptosomes. J. Pharmacol. exp. Ther. **189**, 83–89 (1974)

Clarenbach, P., Raffel, G., Meyer, D., Hertting, G.: Inhibition of uptake of catecholamines into rat brain synaptosomes by inhibitors of prostaglandin synthesis. Naunyn-Schmiedeberg's Arch. Pharmacol. **285**, R9 (1974)

Clarenbach, P., Raffel, G., Meyer, D.K., Hertting, G.: Inhibition by indomethacin and niflumic acid of catecholamine-uptake into rat hypothalamic and striatal synaptosomes. Arch. int. Pharmacodyn. **219**, 79–86 (1976)

Clark, K.E., Ryan, M.J., Brody, M.J.: Effects of prostaglandins E_1 and $F_{2\alpha}$ on uterine hemodynamics and motility. Advanc. Biosci. **9**, 779–782 (1973)

Clouet, D.H., Johnson, J.C., Ratner, M., Williams, N., Gold, G.J.: The effect of morphine on rat brain catecholamines: Turnover in vivo and uptake in isolated synaptosomes. In: Usdin, E., Snyder, S.H. (eds.), Frontiers in Catecholamine Research, pp. 1039–1042. New York: Pergamon Press 1973

Clouet, D.H., Williams, N.: The effect of narcotic analgesic drugs on the uptake and release of neurotransmitters in isolated synaptosomes. J. Pharmacol. exp. Ther. **188**, 419–428 (1974)

Collier, H.O.J.: Supersensitivity and dependence. Nature (Lond.) **220**, 228–231 (1968)

Collier, H.O.J., McDonald-Gibson, W.J., Saeed, S.A.: Morphine and apomorphine stimulate prostaglandin production by rabbit brain homogenate. Brit. J. Pharmacol. **52**, 116P (1974)

Coon, J.M., Rothman, S.: The nature of the pilomotor response to acetyl choline; some observations on the pharmacodynamics of the skin. J. Pharmacol. exp. Ther. **68**, 301–311 (1940)

Cox, B.M., Weinstock, M.: The effect of analgesic drugs on the release of acetylcholine from electrically stimulated guinea-pig ileum. Brit. J. Pharmacol. **27**, 81–92 (1966)

Cripps, H., Dearnaley, D.P.: Vascular responses and noradrenaline overflows in the isolated blood-perfused cat spleen: Some effects of cocaine, normetanephrine and α-blocking agents. J. Physiol. (Lond.) **227**, 647–664 (1972)

Cubeddu, L.X., Barnes, E.M., Langer, S.Z., Weiner, N.: Release of norepinephrine and dopamine-β-hydroxylase by nerve stimulation. I. Role of neuronal and extraneuronal uptake and of alpha presynaptic receptors. J. Pharmacol. exp. Ther. **190**, 431–450 (1974a)

Cubeddu, L.X., Barnes, E., Weiner, N.: Release of norepinephrine and dopamine-β-hydroxylase by nerve stimulation. II. Effects of papaverine. J. Pharmacol. exp. Ther. **191**, 444–457 (1974b)

Cubeddu, L.X., Barnes, E., Weiner, N.: Release of norepinephrine and dopamine-β-hydroxylase by nerve stimulation. IV. An evaluation of a role for cyclic adenosine monophosphate. J. Pharmacol. exp. Ther. **193**, 105–127 (1975)

Cubeddu, L.X., Langer, S.Z., Weiner, N.: The relationships between alpha receptor block, inhibition of norepinephrine uptake and the release and metabolism of ^{3}H-norepinephrine. J. Pharmacol. exp. Ther. **188**, 368–385 (1974c)

Cubeddu, L.X., Weiner, N.: Nerve stimulation-mediated overflow of norepinephrine and dopamine-β-hydroxylase. III. Effects of norepinephrine depletion on the alpha presynaptic regulation of release. J. Pharmacol. exp. Ther. **192**, 1–14 (1975a)

Cubeddu, L.X., Weiner, N.: Release of norepinephrine and dopamine-β-hydroxylase by nerve stimulation. V. Enhanced release associated with a granular effect of a benzoquinolizine derivative with reserpine-like properties. J. Pharmacol. exp. Ther. **193**, 757–774 (1975b)

Dahlöf, C., Åblad, B., Borg, K.O., Ek, L., Waldeck, B.: Prejunctional inhibition of adrenergic nervous vasomotor control due to β receptor blockade. In: Almgren, O., Carlsson, A., Engel, J. (eds.), Chemical Tools in Catecholamine Research, Vol. 2, pp. 201–210. Amsterdam-Oxford: North-Holland 1975

Dalemans, P., Janssens, W., Verbeuren, T., Vanhoutte, P.M.: Effects of naturally occurring catecholamines on adrenergic neuro-effector interaction in isolated cutaneous veins. Arch. int. Pharmacodyn. **220**, 333–334 (1976)

Davey, M.J., Hayden, M.L., Scholfield, P.C.: The effects of bretylium on C fibre excitation and noradrenaline release by acetylcholine and electrical stimulation. Brit. J. Pharmacol. **34**, 377–387 (1968)

Davies, B.N., Horton, E.W., Withrington, P.G.: The occurrence of prostaglandin E$_2$ in splenic venous blood of the dog following splenic nerve stimulation. Brit. J. Pharmacol. **32**, 127–135 (1968)

Davies, B.N., Withrington, P.G.: Actions of prostaglandins A$_1$, A$_2$, E$_1$, E$_2$, F$_{1\alpha}$, and F$_{2\alpha}$ on splenic vascular and capsular smooth muscle and their interactions with sympathetic nerve stimulation, catecholamines and angiotensin. In: Mantegazza, P., Horton, E.W. (eds.), Prostaglandins, Peptides and Amines, pp. 53–56. London-New York: Academic Press 1969

Davila, D., Khairallah, P.A.: Effect of ions on inhibition of norepinephrine uptake by angiotensin. Arch. int. Pharmacodyn. **185**, 357–364 (1970)

Davila, D., Khairallah, P.A.: Angiotensin and biosynthesis of norepinephrine. Arch. int. Pharmacodyn. **193**, 307–314 (1971)

Davis, H.A., Horton, E.W.: Output of prostaglandins from the rabbit kidney, its increase on renal nerve stimulation and its inhibition by indomethacin. Brit. J. Pharmacol. **46**, 658–675 (1972)

Dawes, P.M., Faulkner, D.C.: The effect of propranolol on vascular responses to sympathetic nerve stimulation. Brit. J. Pharmacol. **53**, 517–524 (1975)

Dawes, P.M., Vizi, E.S.: Acetylcholine release from the rabbit isolated superior cervical ganglion preparation. Brit. J. Pharmacol. **48**, 225–232 (1973)

Day, M.D., Moore, A.F.: Interaction of angiotensin II with noradrenaline and other spasmogens on rabbit isolated aortic strips. Arch. int. Pharmacodyn. **219**, 29–44 (1976)

Day, M.D., Owen, D.A.A., Warren, P.R.: An adrenergic neuron blocking action of propranolol in isolated tissues. J. Pharm. Pharmacol. **20**, 130–134S (1968)

Deck, R., Oberdorf, A., Kroneberg, G.: Die Wirkung von 2-(2,6-Dichlorphenylamino)-2-imidazolin-Hydrochlorid (Clonidin) auf die Kontraktion und die Acetylcholin-Freisetzung am isolierten, koaxial elektrisch gereizten Meerschweinchenileum. Arzneimittel-Forsch. **21**, 1580–1584 (1971)

Dengler, H.J., Titus, E.O.: Die Aufnahme von H^3-Noradrenalin in Gewebeschnitte und deren Beeinflussung durch Pharmaka. Naunyn-Schmiedeberg's Arch. exp. Path. Pharmakol. **241**, 523 (1961)

Douglas, W.W., Kanno, T., Sampson, S.R.: Effects of acetylcholine and other medullary secreta-gogues and antagonists on the membrane potential of adrenal chromaffin cells: An analysis employing techniques of tissue culture. J. Physiol. (Lond.) **188**, 107–120 (1967)

Drew, G.M.: Effects of α-adrenoceptor agonists and antagonists on pre- and postsynaptically located α-adrenoceptors. Europ. J. Pharmacol. **36**, 313–320 (1976)

Dubey, M.P., Muscholl, E., Pfeiffer, A.: Muscarinic inhibition of potassium-induced noradrenaline release and its dependence on the calcium concentration. Naunyn-Schmiedeberg's Arch. Pharmacol. **291**, 1–15 (1975)

Dubocovich, M., Langer, S.Z.: Effects of flow-stop on the metabolism of noradrenaline released by nerve stimulation in the perfused spleen. Naunyn-Schmiedeberg's Arch. Pharmacol. **278**, 179–194 (1973)

Dubocovich, M.L., Langer, S.Z.: Negative feed-back regulation of noradrenaline release by nerve stimulation in the perfused cat's spleen: Differences in potency of phenoxybenzamine in blocking the pre- and post-synaptic adrenergic receptors. J. Physiol. (Lond.) **237**, 505–519 (1974)

Dubocovich, M.L., Langer, S.Z.: Evidence against a physiological role of prostaglandins in the regulation of noradrenaline release in the cat spleen. J. Physiol. (Lond.) **251**, 737–762 (1975)

Dun, N., Nishi, S.: Effects of dopamine on the superior cervical ganglion of the rabbit. J. Physiol. (Lond.) **239**, 155–164 (1974)

Eccles, J.C.: The Physiology of Synapses. Berlin-Göttingen-Heidelberg: Springer 1964

Ehrenpreis, S., Greenberg, J., Belman, S.: Prostaglandins reverse inhibition of electrically-induced contractions of guinea pig ileum by morphine, indomethacin and acetylsalicylic acid. Nature (Lond.) **245**, 280–282 (1973)

Eisenfeld, A.J., Axelrod, J., Krakoff, L.: Inhibition of the extraneuronal accumulation and metabo-lism of norepinephrine by adrenergic blocking agents. J. Pharmacol. exp. Ther. **156**, 107–113 (1967)

Eliash, S., Weinstock, M.: Role of adrenergic neurone blockade in the hypotensive action of propranolol. Brit. J. Pharmacol. **43**, 287–294 (1971)

Elie, R., Panisset, J.C.: Effect of angiotensin and atropine on the spontaneous release of acetylcholine from cat cerebral cortex. Brain Res. **17**, 297–305 (1970)

Endoh, M., Tamura, K., Hashimoto, K.: Negative and positive inotropic responses of the blood-perfused canine papillary muscle to acetylcholine. J. Pharmacol. exp. Ther. **175**, 377–387 (1970)

Enero, M.A., Langer, S.Z.: Influence of reserpine-induced depletion of noradrenaline on the negative feed-back mechanism for transmitter release during nerve stimulation. Brit. J. Pharmacol. **49**, 214–225 (1973)

Enero, M.A., Langer, S.Z.: Inhibition by dopamine of ^{3}H-noradrenaline release elicited by nerve stimulation in the isolated cat's nictitating membrane. Naunyn-Schmiedeberg's Arch. Pharmacol. **289**, 179–203 (1975)

Enero, M.A., Langer, S.Z., Rothlin, R.P., Stefano, F.J.E.: Role of the α-adrenoceptor in regulating noradrenaline overflow by nerve stimulation. Brit. J. Pharmacol. **44**, 672–688 (1972)

Euler, U.S. von: Synthesis, uptake and storage of catecholamines in adrenergic nerves, the effect of drugs. In: Blaschko, H., Muscholl, E. (eds.), Catecholamines. Handbuch der experimentellen Pharmakologie, Vol. 33, pp. 186–230. Berlin-Heidelberg-New York: Springer 1972

Euler, U.S. von, Hedqvist, P.: Evidence for an α- and β₂-receptor mediated inhibition of the twitch response in the guinea pig vas deferens by noradrenaline. Acta physiol. scand. **93**, 572–573 (1975)

Euler, U.S. von, Lishajko, F.: Noradrenaline release from isolated nerve granules. Acta physiol. scand. **51**, 193–203 (1961)

Farah, M.B., Langer, S.Z.: Protection by phentolamine against the effects of phenoxybenzamine on transmitter release elicited by nerve stimulation in the perfused cat heart. Brit. J. Pharmacol. **52**, 549–557 (1974)

Farnebo, L.O., Hamberger, B.: Effects of desipramine, phentolamine and phenoxybenzamine on the release of noradrenaline from isolated tissues. J. Pharm. Pharmacol. **22**, 855–857 (1970)

Farnebo, L.O., Hamberger, B.: Drug-induced changes in the release of (^{3}H)-noradrenaline from field stimulated rat iris. Brit. J. Pharmacol. **43**, 97–106 (1971a)

Farnebo, L.O., Hamberger, B.: Drug-induced changes in the release of ^{3}H-monoamines from field stimulated rat brain slices. Acta physiol. scand., Suppl. **371**, 35–44 (1971b)

Farnebo, L.O., Hamberger, B.: Chronic decentralization prevents α-receptor mediated regulation of noradrenaline release in the field stimulated rat iris. Brain Res. **62**, 477–482 (1973a)

Farnebo, L.O., Hamberger, B.: Catecholamine release and receptors in brain slices. In: Usdin, E., Snyder, S.H. (eds.), Frontiers in Catecholamine Research, pp. 589–593. New York: Pergamon Press 1973b

Farnebo, L.O., Hamberger, B.: Regulation of (^{3}H)5-hydroxytryptamine release from rat brain slices. J. Pharm. Pharmacol. **26**, 642–644 (1974a)

Farnebo, L.O., Hamberger, B.: Influence of α- and β-adrenoceptors on the release of noradrenaline from field stimulated atria and cerebral cortex slices. J. Pharm. Pharmacol. **26**, 644–646 (1974b)

Farnebo, L.O., Malmfors, T.: ^{3}H-noradrenaline release and mechanical response in the field stimulated mouse vas deferens. Acta physiol. scand., Suppl. **371**, 1–18 (1971)

Feniuk, W., Large, B.J.: The effects of prostaglandins E_1, E_2 and $F_{2\alpha}$ on vagal bradycardia in the anaesthetized mouse. Brit. J. Pharmacol. **55**, 47–49 (1975)

Ferry, C.B.: The sympathomimetic effect of acetylcholine on the spleen of the cat. J. Physiol. (Lond.) **167**, 487–504 (1963)

Ferry, C.B.: Cholinergic link hypothesis in adrenergic neuroeffector transmission. Physiol. Rev. **46**, 420–456 (1966)

Fischer, J.E., Weise, V.K., Kopin, I.J.: Interactions of bretylium and acetylcholine at sympathetic nerve endings. J. Pharmacol. exp. Ther. **153**, 523–529 (1966)

Flower, R.J.: Drugs which inhibit prostaglandin biosynthesis. Pharmacol. Rev. **26**, 33–67 (1974)

Foo, J.W., Jowett, A., Stafford, A.: The effects of some β-adrenoreceptor blocking drugs on the uptake and release of noradrenaline by the heart. Brit. J. Pharmacol. **34**, 141–147 (1968)

Fozard, J.R., Muscholl, E.: Effects of several muscarinic agonists on cardiac performance and the release of noradrenaline from sympathetic nerves of the perfused rabbit heart. Brit. J. Pharmacol. **45**, 616–629 (1972)

Fozard, J.R., Muscholl, E.: Atropine-resistant effects of the muscarinic agonists McN-A-343 and AHR 602 on cardiac performance and the release of noradrenaline from sympathetic nerves of the perfused rabbit heart. Brit. J. Pharmacol. **50**, 531–541 (1974)

Fozard, J.R., Mwaluko, G.M.P.: Mechanism of the indirect sympathomimetic effect of 5-hydroxy-tryptamine on the isolated heart of the rabbit. Brit. J. Pharmacol. **57**, 115–125 (1976)

Frame, M.H., Hedqvist, P.: Evidence for prostaglandin mediated prejunctional control of renal sympathetic transmitter release and vascular tone. Brit. J. Pharmacol. **54**, 189–196 (1975)

Fredholm, B., Hedqvist, P.: Increased release of noradrenaline from stimulated guinea pig vas deferens after indomethacin treatment. Acta physiol. scand. **87**, 570–572 (1973a)

Fredholm, B.B., Hedqvist, P.: Role of pre- and postjunctional inhibition by prostaglandin E_2 of lipolysis induced by sympathetic nerve stimulation in dog subcutaneous adipose tissue in situ. Brit. J. Pharmacol. **47**, 711–718 (1973b)

Fredholm, B.B., Hedqvist, P.: Indomethacin and the role of prostaglandins in adipose tissue. Biochem. Pharmacol. **24**, 61–66 (1975a)

Fredholm, B.B., Hedqvist, P.: Indomethacin-induced increase in noradrenaline turnover in some rat organs. Brit. J. Pharmacol. **54**, 295–300 (1975b)

Fredholm, B.B., Rosell, S.: Fate of ^{3}H-noradrenaline in canine subcutaneous adipose tissue. Acta physiol. scand. **80**, 404–411 (1970)

Fredholm, B.B., Rosell, S., Strandberg, K.: Release of prostaglandin-like material from canine subcutaneous adipose tissue by nerve stimulation. Acta physiol. scand. **79**, 18–19A (1970)

Fuder, H., Muscholl, E.: The effect of methacholine on noradrenaline release from the rabbit heart perfused with indometacin. Naunyn-Schmiedeberg's Arch. Pharmacol. **285**, 127–132 (1974)

Furchgott, R.F.: The receptors for epinephrine and norepinephrine (adrenergic receptors). Pharmacol. Rev. **11**, 429–441 (1959)

Furchgott, R.F.: The classification of adrenoceptors (adrenergic receptors). An evaluation from the standpoint of receptor theory. In: Blaschko, H., Muscholl, E. (eds.), Catecholamines. Handbuch der experimentellen Pharmakologie, Vol. 33, pp. 283–335. Berlin-Heidelberg-New York: Springer 1972

Furchgott, R.F., Steinsland, O.S., Wakade, T.D.: Studies on prejunctional muscarinic and nicotinic receptors. In: Almgren, O., Carlsson, A., Engel, J. (eds.), Chemical Tools in Catecholamine Research, Vol. 2, pp. 167–174. Amsterdam-Oxford: North-Holland 1975

Gagnon, D.J., Sirois, P., Boucher, P.J.: Stimulation by angiotensin II of the release of vasopressin from incubated rat neurohypophyses – possible involvement of cyclic AMP. Clin. exp. Pharmacol. Physiol. **2**, 305–313 (1975)

Garcia, A.G., Kirpekar, S.M.: Release of noradrenaline from the cat spleen by sodium deprivation. Brit. J. Pharmacol. **47**, 729–747 (1973)

Garcia, A.G., Kirpekar, S.M.: On the mechanism of release of norepinephrine from cat spleen slices by sodium deprivation and calcium pretreatment. J. Pharmacol. exp. Ther. **192**, 343–350 (1975)

Geffen, L.B.: The effect of desmethylimipramine upon the overflow of sympathetic transmitter from the cat's spleen. J. Physiol. (Lond.) **181**, 69–70P (1965)

George, A.J.: The effect of prostaglandin E_1 and E_2 on drug-induced release of (^{3}H)-noradrenaline from rat mesenteric arteries. Brit. J. Pharmacol. **55**, 243P (1975)

Gilbert, J.C., Wyllie, M.G., Davison, D.V.: Nerve terminal ATPase as possible trigger for neuro-transmitter release. Nature (Lond.) **255**, 237–238 (1975)

Gillespie, J.S., Kirpekar, S.M.: The uptake and release of radioactive noradrenaline by the splenic nerves of cats. J. Physiol. (Lond.) **187**, 51–68 (1966)

Gilmore, N., Vane, J.R., Wyllie, J.H.: Prostaglandins released by the spleen. Nature (Lond.) **218**, 1135–1140 (1968)

Giorguieff, M.F., Le Floc'h, M.L., Westfall, T.C., Glowinski, J., Besson, M.J.: Nicotinic effect of acetylcholine on the release of newly synthesized (^{3}H)dopamine in rat striatal slices and cat caudate nucleus. Brain Res. **106**, 117–131 (1976)

Göthert, M.: Effects of halothane on the sympathetic nerve terminals of the rabbit heart. Differences in membrane actions of halothane and tetracaine. Naunyn-Schmiedeberg's Arch. Pharmacol. **286**, 125–143 (1974)

Göthert, M., Kennerknecht, E., Thielecke, G.: Inhibition of receptor-mediated noradrenaline release from the sympathetic nerves of the isolated rabbit heart by anaesthetics and alcohols in proportion to their hydrophobic property. Naunyn-Schmiedeberg's Arch. Pharmacol. **292**, 145–152 (1976)

Goldberg, L.I.: Cardiovascular and renal actions of dopamine: Potential clinical applications. Pharmacol. Rev. **24**, 1–29 (1972)

Goldberg, L.I.: The dopamine vascular receptor. Biochem. Pharmacol. **24**, 651–653 (1975)

Goldberg, L.I., Sonneville, P.F., McNay, J.L.: An investigation of the structural requirements for dopamine-like renal vasodilation: Phenylethylamines and apomorphine. J. Pharmacol. exp. Ther. **163**, 188–197 (1968)

Goldberg, L.I., Tjandramaga, T.B., Anton, A.H., Toda, N.: New investigations of the cardiovascular actions of dopamine. In: Usdin, E., Snyder, S.H. (eds.), Frontiers in Catecholamine Research, pp. 513–521. New York: Pergamon Press 1973

Gomer, S.K., Zimmerman, B.G.: Effect of angiotensin II on uptake and release of dl-^{3}H-metarami-nol. Proc. Soc. exp. Biol. (N.Y.) **142**, 787–792 (1973)

Goodman, F.R.: Effects of nicotine on distribution and release of ^{14}C-norepinephrine and ^{14}C-dopamine in rat brain striatum and hypothalamus slices. Neuropharmacol. **13**, 1025–1032 (1974)

Govier, W.C.: Myocardial alpha adrenergic receptors and their role in the production of a positive inotropic effect by sympathomimetic agents. J. Pharmacol. exp. Ther. **159**, 82–90 (1968)

Grabowska, M., Andén, N.E.: Noradrenaline synthesis and utilization: Control by nerve impulse flow under normal conditions and after treatment with alpha-adrenoreceptor blocking agents. Naunyn-Schmiedeberg's Arch. Pharmacol. **292**, 53–58 (1976)

Graefe, K.H.: Methodology of catecholamine transport studies: Definition of terms. In: Paton, D.M. (ed.), The Mechanism of Neuronal and Extraneuronal Transport of Catecholamines, pp. 7–35. New York: Raven Press 1976

Graefe, K.H., Stefano, F.J.E., Langer, S.Z.: Preferential metabolism of ($-$)-^{3}H-norepinephrine through the deaminated glycol in the rat vas deferens. Biochem. Pharmacol. **22**, 1147–1160 (1973)

Greenberg, R.: The effects of indomethacin and eicosa-5,8,11,14-tetraynoic acid on the response of the rabbit portal vein to electrical stimulation. Brit. J. Pharmacol. **52**, 61–68 (1974)

Greenberg, S., Howard, L., Wilson, W.R.: Comparative effects of prostaglandins A_2 and B_2 on vascular and airway resistances and adrenergic neurotransmission. Canad. J. Physiol. Pharmacol. **52**, 699–705 (1974)

Greengard, P., Kebabian, J.W.: Role of cyclic AMP in synaptic transmission in the mammalian peripheral nervous system. Fed. Proc. **33**, 1059–1067 (1974)

Guidotti, A., Costa, E.: A role for nicotinic receptors in the regulation of the adenylate cyclase of adrenal medulla. J. Pharmacol. exp. Ther. **189**, 665–675 (1974)

Gustafsson, L., Hedqvist, P., Lagercrantz, H.: Potentiation by prostaglandins E_1, E_2, and $F_{2\alpha}$ of the contraction response to transmural stimulation in the bovine iris sphincter muscle. Acta physiol. scand. **95**, 26–33 (1975)

Gutman, Y., Boonyaviroj, P.: Suppression by noradrenaline of catecholamine secretion from adrenal medulla. Europ. J. Pharmacol. **28**, 384–386 (1974)

Gutman, Y., Boonyaviroj, P.: Regulation of catecholamine release from adrenal medulla in vitro by α and β receptors and by prostaglandins. In: Abstr. 6th Int. Congr. Pharmacol., Helsinki, 1975, p. 423

Hadházy, P., Illés, P., Knoll, J.: The effects of PGE_1 on responses to cardiac vagus nerve stimulation and acetylcholine release. Europ. J. Pharmacol. **23**, 251–255 (1973)

Hadházy, P., Vizi, E.S., Magyar, K., Knoll, J.: Inhibition of adrenergic neurotransmission by prostaglandin E_1 (PGE_1) in the rabbit ear artery. Neuropharmacol. **15**, 245–250 (1976)

Haefely, W.: Electrophysiology of the adrenergic neuron. In: Blaschko, H., Muscholl, E. (eds.), Catecholamines. Handbuch der experimentellen Pharmakologie, Vol. 33, pp. 661–725. Berlin-Heidelberg-New York: Springer 1972

Häggendal, J.: On release of transmitter from adrenergic nerve terminals at nerve activity. Acta physiol. scand., Suppl. **330**, 29 (1969)

Häggendal, J.: Some further aspects on the release of the adrenergic transmitter. In: Schümann, H.J., Kroneberg, G. (eds.), New Aspects of Storage and Release Mechanisms of Catecholamines, pp. 100–109. Berlin-Heidelberg-New York: Springer 1970

Häggendal, J., Johansson, B., Jonason, J., Ljung, B.: Effects of phenoxybenzamine on transmitter release and effector response in the isolated portal vein. J. Pharm. Pharmacol. **24**, 161–164 (1972)

Haeusler, G.: Clonidine-induced inhibition of sympathetic nerve activity: No indication for a central presynaptic or an indirect sympathomimetic mode of action. Naunyn-Schmiedeberg's Arch. Pharmacol. **286**, 97–111 (1974)

Haeusler, G.: The importance of the presynaptic α-adrenergic regulation of noradrenaline release from vascular nerves in vivo. Naunyn-Schmiedeberg's Arch. Pharmacol. **287**, R19 (1975a)

Haeusler, G.: Cardiovascular regulation by central adrenergic mechanisms and its alteration by hypotensive drugs. Circulat. Res. **36–37**, I-223-232 (1975b)

Haeusler, G., Haefely, W., Huerlimann, A.: On the mechanism of the adrenergic nerve blocking action of bretylium. Naunyn-Schmiedebergs Arch. Pharmakol. **265**, 260–277 (1969a)

Haeusler, G., Thoenen, H., Haefely, W., Huerlimann, A.: Electrical events in cardiac adrenergic nerves and noradrenaline release from the heart induced by acetylcholine and KCl. Naunyn-Schmiedebergs Arch. Pharmakol. exp. Path. **261**, 389–411 (1968)

Haeusler, G., Thoenen, H., Haefely, W., Huerlimann, A.: Elektrosekretorische Koppelung bei der Noradrenalinfreisetzung aus adrenergen Nervenfasern durch nicotinartig wirkende Substanzen. Naunyn-Schmiedebergs Arch. Pharmakol. exp. Path. **263**, 217–218 (1969b)

Hahn, R.A., Patil, P.N.: Further observations on the interaction of prostaglandin $F_{2\alpha}$ with cholinergic mechanisms in canine salivary glands. Europ. J. Pharmacol. **25**, 279–286 (1974)

Hall, G.H., Turner, D.M.: Effects of nicotine on the release of ^{3}H-noradrenaline from the hypothalamus. Biochem. Pharmacol. **21**, 1829–1838 (1972)

Hazra, J.: Evidence against prostaglandin E having a physiological role in acetylcholine liberation from Auerbach's plexus of guinea-pig ileum. Experientia (Basel) **31**, 565–566 (1975)

Hedqvist, P.: Modulating effect of prostaglandin E_2 on noradrenaline release from the isolated cat spleen. Acta physiol. scand. **75**, 511–512 (1969a)

Hedqvist, P.: Antagonism between prostaglandin E_2 and phenoxybenzamine on noradrenaline release from the cat spleen. Acta physiol. scand. **76**, 383–384 (1969b)

Hedqvist, P.: Control by prostaglandin E_2 of sympathetic neurotransmission in the spleen. Life Sci. **9**, I, 269–278 (1970a)

Hedqvist, P.: Antagonism by calcium of the inhibitory action of prostaglandin E_2 on sympathetic neurotransmission in the cat spleen. Acta physiol. scand. **80**, 269–275 (1970b)

Hedqvist, P.: Studies on the effect of prostaglandins E_1 and E_2 on the sympathetic neuromuscular transmission in some animal tissues. Acta physiol. scand., Suppl. **345** (1970c)

Hedqvist, P.: Prostaglandin induced inhibition of neurotransmission in the isolated guinea pig seminal vesicle. Acta physiol. scand. **84**, 506–511 (1972a)

Hedqvist, P.: Prostaglandin-induced inhibition of vascular tone and reactivity in the cat's hindleg in vivo. Europ. J. Pharmacol. **17**, 157–162 (1972b)

Hedqvist, P.: Prostaglandin mediated control of sympathetic neuroeffector transmission. Advanc. Biosci. **9**, 461–473 (1973a)

Hedqvist, P.: Autonomic neurotransmission. In: Ramwell, P.W. (ed.), The Prostaglandins, Vol. 1, pp. 101–131. New York-London: Plenum Press 1973b

Hedqvist, P.: Prostaglandin as a tool for local control of transmitter release from sympathetic nerves. Brain Res. **62**, 483–488 (1973c)

Hedqvist, P.: Aspects on prostaglandin and α-receptor mediated control of transmitter release from adrenergic nerves. In: Usdin, E., Snyder, S.H. (eds.), Frontiers in Catecholamine Research, pp. 583–587. New York: Pergamon Press 1973d

Hedqvist, P.: Dissociation of prostaglandin and α-receptor mediated control of adrenergic transmitter release. Acta physiol. scand. **87**, 42–43A (1973e)

Hedqvist, P.: Prostaglandin action on noradrenaline release and mechanical responses in the stimulated guinea pig vas deferens. Acta physiol. scand. **90**, 86–93 (1974a)

Hedqvist, P.: Interaction between prostaglandins and calcium ions on noradrenaline release from the stimulated guinea pig vas deferens. Acta physiol. scand. **90**, 153–157 (1974b)

Hedqvist, P.: Role of the α-receptor in the control of noradrenaline release from sympathetic nerves. Acta physiol. scand. **90**, 158–165 (1974c)

Hedqvist, P.: Restriction of transmitter release from adrenergic nerves mediated by prostaglandins and α-adrenoreceptors. Pol. J. Pharmacol. Pharm. **26**, 119–125 (1974d)

Hedqvist, P.: Effect of prostaglandins and prostaglandin synthesis inhibitors on norepinephrine release from vascular tissue. In: Robinson, H.J., Vane, J.R. (eds.), Prostaglandin Synthetase Inhibitors, pp. 303–309. New York: Raven Press 1974e

Hedqvist, P., Brundin, J.: Inhibition by prostaglandin E$_1$ of noradrenaline release and of effector response to nerve stimulation in the cat spleen. Life Sci. **8**, I, 389–395 (1969)

Hedqvist, P., Euler, U.S. von: Prostaglandin-induced neurotransmission failure in the field-stimulated, isolated vas deferens. Neuropharmacol. **11**, 177–187 (1972)

Hedqvist, P., Moawad, A.: Presynaptic α- and β-adrenoceptor mediated control of noradrenaline release in human oviduct. Acta physiol. scand. **95**, 494–496 (1975)

Hedqvist, P., Persson, N.Å.: Prostaglandin action on adrenergic and cholinergic responses in the rabbit and guinea pig intestine. In: Almgren, O., Carlsson, A., Engel, J. (eds.), Chemical Tools in Catecholamine Research, Vol. 2, pp. 211–218. Amsterdam-Oxford: North-Holland 1975

Hedqvist, P., Stjärne, L., Wennmalm, Å.: Inhibition by prostaglandin E$_2$ of sympathetic neurotransmission in the rabbit heart. Acta physiol. scand. **79**, 139–141 (1970)

Hedqvist, P., Stjärne, L., Wennmalm, Å.: Facilitation of sympathetic neurotransmission in the cat spleen after inhibition of prostaglandin synthesis. Acta physiol. scand. **83**, 430–432 (1971)

Hedqvist, P., Wennmalm, Å.: Comparison of the effects of prostaglandins E$_1$, E$_2$ and F$_{2α}$ on the sympathetically stimulated rabbit heart. Acta physiol. scand. **83**, 156–162 (1971)

Heikkila, R.E., Orlansky, H., Cohen, G.: Studies on the distinction between uptake inhibition and release of (^{3}H)dopamine in rat brain tissue slices. Biochem. Pharmacol. **24**, 847–852 (1975)

Heise, A., Kroneberg, G.: Periphere und zentrale Kreislaufwirkung des α-Sympathicomimeticums 2-(2,6-Xylidino)-5,6-dihydro-4H-1,3-thiazinhydrochlorid (BAY 1470). Naunyn-Schmiedebergs Arch. Pharmakol. **266**, 350 (1970)

Heise, A., Kroneberg, G., Schlossmann, K.: α-sympathicomimetische Eigenschaften als Ursache der blutdrucksteigernden und blutdrucksenkenden Wirkung von BAY 1470 (2-(2,6-Xylidino)-5,6-dihydro-4H-1,3-thiazinhydrochlorid). Naunyn-Schmiedebergs Arch. Pharmakol. **268**, 348–360 (1971)

Henderson, G., Hughes, J.: Modulation of frequency-dependent noradrenaline release by calcium, angiotensin and morphine. Brit. J. Pharmacol. **52**, 455–456P (1974)

Henderson, G., Hughes, J., Kosterlitz, H.W.: A new example of a morphine-sensitive neuro-effector junction: adrenergic transmission in the mouse vas deferens. Brit. J. Pharmacol. **46**, 764–766 (1972a)

Henderson, G., Hughes, J., Kosterlitz, H.W.: The effects of morphine on the release of noradrenaline from the cat isolated nictitating membrane and the guinea-pig ileum myenteric plexus-longitudinal muscle preparation. Brit. J. Pharmacol. **53**, 505–512 (1975)

Henderson, G., Hughes, J., Thompson, J.W.: The variation of noradrenaline output with frequency of nerve stimulation and the effect of morphine on the cat nictitating membrane and on the guinea-pig myenteric plexus. Brit. J. Pharmacol. **46**, 524–525P (1972b)

Henseling, M., Eckert, E., Trendelenburg, U.: The effect of cocaine on the distribution of labelled noradrenaline in rabbit aortic strips and on efflux of radioactivity from the strips. Naunyn-Schmiedeberg's Arch. Pharmacol. **292**, 231–241 (1976)

Hertting, G., Axelrod, J.: Fate of tritiated noradrenaline at the sympathetic nerve-endings. Nature (Lond.) **192**, 172–173 (1961)

Hertting, G., Schiefthaler, T.: Beziehung zwischen Durchflußgröße und Noradrenalinfreisetzung bei Nervenreizung der isoliert durchströmten Katzenmilz. Naunyn-Schmiedebergs Arch. exp. Path. Pharmakol. **246**, 13–14 (1963)

Hertting, G., Suko, J.: Influence of angiotensin, vasopressin or changes in flow rate on vasoconstriction, changes in volume and (^{3}H)-noradrenaline release following postganglionic sympathetic nerve stimulation in the isolated cat spleen. Brit. J. Pharmacol. **26**, 686–696 (1966)

Hertting, G., Widhalm, S.: Über den Mechanismus der Noradrenalin-Freisetzung aus sympathischen Nervenendigungen. Naunyn-Schmiedebergs Arch. exp. Path. Pharmakol. **250**, 257–258 (1965)

Herz, A., Teschemacher, H.J.: Activities and sites of antinociceptive action of morphine-like analgesics and kinetics of distribution following intravenous, intracerebral and intraventricular application. Advanc. Drug Res. **6**, 79–119 (1971)

Hitzemann, R.J., Loh, H.H.: Effect of morphine on the transport of dopamine into mouse brain slices. Europ. J. Pharmacol. **21**, 121–129 (1973)

Hoffmann, F., Hoffmann, E.J., Middleton, S., Talesnik, J.: The stimulating effect of acetylcholine on the mammalian heart and the liberation of an epinephrine-like substance by the isolated heart. Amer. J. Physiol. **144**, 189–198 (1945)

Horton, E.W.: Prostaglandins at adrenergic nerve-endings. Brit. med. Bull. **29**, 148–151 (1973)

Hoszowska, A., Panczenko, B.: Effects of inhibition of prostaglandin biosynthesis on noradrenaline release from isolated perfused spleen of the cat. Pol. J. Pharmacol. Pharm. **26**, 137–142 (1974)

Hotta, Y.: Some properties of the junctional and extrajunctional receptors in the vas deferens of the guinea-pig. Agents Actions **1**, 69–77 (= (13)-(21)) (1969)

Howd, R.A., Horita, A.: The effects of drugs on accumulation and release of catecholamines in rat brain slices. Arch. int. Pharmacodyn. **218**, 231–238 (1975)

Hsu, C.Y., Westfall, T.C.: Mechanism of the release of ^{3}H-norepinephrine (NE) induced by aminophylline in the perfused guinea-pig heart. Fed. Proc. **32**, 784 Abs (1973)

Hubbard, J.I.: Microphysiology of vertebrate neuromuscular transmission. Physiol. Rev. **53**, 674–723 (1973)

Hubbard, J.I., Quastel, D.M.J.: Micropharmacology of vertebrate neuromuscular transmission. Ann. Rev. Pharmacol. **13**, 199–216 (1973)

Hughes, J.: Evaluation of mechanisms controlling the release and inactivation of the adrenergic transmitter in the rabbit portal vein and vas deferens. Brit. J. Pharmacol. **44**, 472–491 (1972)

Hughes, J.: Inhibition of noradrenaline release by lysergic acid diethylamide. Brit. J. Pharmacol. **49**, 706–708 (1973)

Hughes, J., Kosterlitz, H.W., Leslie, F.M.: Effect of morphine on adrenergic transmission in the mouse vas deferens. Assessment of agonist and antagonist potencies of narcotic analgesics. Brit. J. Pharmacol. **53**, 371–381 (1975a)

Hughes, J., Roth, R.H.: Enhanced release of transmitter during sympathetic nerve stimulation in the presence of angiotensin. Brit. J. Pharmacol. **37**, 516–517P (1969)

Hughes, J., Roth, R.H.: Evidence that angiotensin enhances transmitter release during sympathetic nerve stimulation. Brit. J. Pharmacol. **41**, 239–255 (1971)

Hughes, J., Roth, R.H.: Variation in noradrenaline output with changes in stimulus frequency and train length: Role of different noradrenaline pools. Brit. J. Pharmacol. **51**, 373–381 (1974)

Hughes, J., Smith, T.W., Kosterlitz, H.W., Fothergill, L.A., Morgan, B.A., Morris, H.R.: Identification of two related pentapeptides from the brain with potent opiate agonist activity. Nature (Lond.) **258**, 577–579 (1975b)

Hume, W.R., Lande, I.S. de la, Waterson, J.G.: Effect of acetylcholine on the response of the isolated rabbit ear artery to stimulation of the perivascular sympathetic nerves. Europ. J. Pharmacol. **17**, 227–233 (1972)

Hunt, G.E., Atrens, D.M., Chesher, G.B., Becker, F.T.: α-Noradrenergic modulation of hypothalamic self-stimulation: Studies employing clonidine, 1-phenylephrine and α-methyl-p-tyrosine. Europ. J. Pharmacol. **37**, 105–111 (1976)

Ilhan, M., Long, J.P.: Inhibition of the sympathetic nervous system by dopamine. Arch. int. Pharmacodyn. **216**, 4–10 (1975)

Ilhan, M., Long, J.P., Cannon, J.G.: Inhibition of responses to stimulation of the cardioaccelerator nerves of the cat by N,N-dimethyldopamine and apomorphine. Arch. int. Pharmacodyn. **212**, 247–254 (1974)

Ilhan, M., Long, J.P., Cannon, J.G.: Bulbocapnine's ability to antagonize the adrenergic inhibitory action of dopamine and analogs. Europ. J. Pharmacol. 33, 13–18 (1975)

Ilhan, M., Long, J.P., Cannon, J.G.: Effects of some dopamine analogs and haloperidol on response to stimulation of adrenergic nerves using cat atria in vitro. Arch. int. Pharmacodyn. 219, 193–204 (1976)

Illés, P., Hadházy, P., Torma, Z., Vizi, E.S., Knoll, J.: The effect of number of stimuli and rate of stimulation on the inhibition by PGE_1 of adrenergic transmission. Europ. J. Pharmacol. 24, 29–36 (1973)

Illés, P., Vizi, E.S., Knoll, J.: Adrenergic neuroeffector junctions sensitive and insensitive to the effect of PGE_1. Pol. J. Pharmacol. Pharm. 26, 127–136 (1974)

Iversen, L.L.: Uptake mechanisms for biogenic amines. In: Iversen, L.L., Iversen, S.D., Snyder, S.H. (eds.), Handbook of Psychopharmacology, Vol. 3, pp. 381–442. New York-London: Plenum Press 1975

Izumi, F., Oka, M., Kashimoto, T.: Possible role of cyclic-3′,5′-adenosine monophosphate in the release of catecholamines from the adrenal medulla. Med. J. Osaka Univ. 21, 241–249 (1971)

Jaanus, S.D., Rubin, R.P.: Analysis of the role of cyclic adenosine 3′,5′-monophosphate in catecholamine release. J. Physiol. (Lond.) 237, 465–476 (1974)

Jang, C.S.: The potentiation and paralysis of adrenergic effects by ergotoxine and other substances. J. Pharmacol. exp. Ther. 71, 87–94 (1941)

Janowsky, D.S., Davis, J.M., Fann, W.E., Freeman, J., Nixon, R., Michelakis, A.A.: Angiotensin effect on uptake of norepinephrine by synaptosomes. Life Sci. 11, I, 1–11 (1972)

Jenkins, D.A., Marshall, I., Nasmyth, P.A.: Is morphine inhibition of the twitch response of the mouse vas deferens mediated via noradrenaline? Brit. J. Pharmacol. 55, 267–268P (1975)

Jhamandas, K., Phillis, J.W., Pinsky, C.: Effects of narcotic analgesics and antagonists on the in vivo release of acetylcholine from the cerebral cortex of the cat. Brit. J. Pharmacol. 43, 53–66 (1971)

Johnson, D.G., Thoa, N.B., Weinshilboum, R., Axelrod, J., Kopin, I.J.: Enhanced release of dopamine-β-hydroxylase from sympathetic nerves by calcium and phenoxybenzamine and its reversal by prostaglandins. Proc. nat. Acad. Sci. (Wash.) 68, 2227–2230 (1971)

Junstad, M., Wennmalm, Å.: Increased renal excretion of noradrenaline in rats after treatment with prostaglandin synthesis inhibitor indomethacin. Acta physiol. scand. 85, 573–576 (1972)

Junstad, M., Wennmalm, Å.: On the release of prostaglandin E_2 from the rabbit heart following infusion of noradrenaline. Acta physiol. scand. 87, 573–574 (1973a)

Junstad, M., Wennmalm, Å.: Prostaglandin mediated inhibition of noradrenaline release at different nerve impulse frequencies. Acta physiol. scand. 89, 544–549 (1973b)

Junstad, M., Wennmalm, Å.: Release of prostaglandin from the rabbit isolated heart following vagal nerve stimulation or acetylcholine infusion. Brit. J. Pharmacol. 52, 375–379 (1974)

Kadowitz, P.J., Sweet, C.S., Brody, M.J.: Differential effects of prostaglandins E_1, E_2, $F_{1\alpha}$ and $F_{2\alpha}$ on adrenergic vasoconstriction in the dog hindpaw. J. Pharmacol. exp. Ther. 177, 641–649 (1971)

Kadowitz, P.J., Sweet, C.S., Brody, M.J.: Enhancement of sympathetic neurotransmission by prostaglandin $F_{2\alpha}$ in the cutaneous vascular bed of the dog. Europ. J. Pharmacol. 18, 189–194 (1972)

Kaneko, Y., Takeda, T., Nakajima, K., Ueda, H.: Effect of angiotensin on the pressor response to tyramine in normotensive subjects and hypertensive patients. Circulat. Res. 19, 673–680 (1966)

Kao, C.Y., McCullough, J.R.: Electrophysiological properties of splenic nerve in relation to norepinephrine overflow. J. Pharmacol. exp. Ther. 185, 49–59 (1973)

Kato, A.C., Collier, B., Ilson, D., Wright, J.M.: The effect of atropine upon acetylcholine release from cat superior cervical ganglia and rat cortical slices: Measurement by a radio-enzymic method. Canad. J. Physiol. Pharmacol. 53, 1050–1057 (1975)

Kaumann, A.J.: Adrenergic receptors in heart muscle: Relations among factors influencing the sensitivity of the cat papillary muscle to catecholamines. J. Pharmacol. exp. Ther. 173, 383–398 (1970)

Kayaalp, S.O., Türker, R.K.: Effect of hemicholinium (HC-3) on the catecholamine releasing action of prostaglandin E_1. Europ. J. Pharmacol. 3, 139–142 (1968)

Kehr, W., Carlsson, A., Lindqvist, M., Magnusson, T., Atack, C.: Evidence for a receptor-mediated feedback control of striatal tyrosine hydroxylase activity. J. Pharm. Pharmacol. 24, 744–747 (1972)

Kennedy, B.L., West, T.C.: Effect of morphine on electrically-induced release of autonomic mediators in the rabbit sinoatrial node. J. Pharmacol. exp. Ther. **157**, 149–158 (1967)

Khairallah, P.A.: Action of angiotensin on adrenergic nerve endings: inhibition of norepinephrine
uptake. Fed. Proc. **31**, 1351–1357 (1972)

Kilbinger, H., Lindmar, R., Löffelholz, K., Muscholl, E., Patil, P.N.: Storage and release of
false transmitters after infusion of (+)- and (−)-α-methyldopamine. Naunyn-Schmiedebergs
Arch. Pharmakol. **271**, 234–248 (1971)

Kilbinger, H., Wagner, P.: Inhibition by oxotremorine of acetylcholine resting release from guinea
pig-ileum longitudinal muscle strips. Naunyn-Schmiedeberg's Arch. Pharmacol. **287**, 47–60 (1975)

Kiran, B.K., Khairallah, P.A.: Angiotensin and norepinephrine efflux. Europ. J. Pharmacol. **6**,
102–108 (1969)

Kirpekar, S.M., Cervoni, P.: Effect of cocaine, phenoxybenzamine and phentolamine on the catecholamine output from spleen and adrenal medulla. J. Pharmacol. exp. Ther. **142**, 59–70 (1963)

Kirpekar, S.M., Furchgott, R.F., Wakade, A.R., Prat, J.C.: Inhibition by sympathomimetic amines
of the release of norepinephrine evoked by nerve stimulation in the cat spleen. J. Pharmacol.
exp. Ther. **187**, 529–538 (1973)

Kirpekar, S.M., Misu, Y.: Release of noradrenaline by splenic nerve stimulation and its dependence
on calcium. J. Physiol. (Lond.) **188**, 219–234 (1967)

Kirpekar, S.M., Prat, J.C., Puig, M., Wakade, A.R.: Modification of the evoked release of noradrenaline from the perfused cat spleen by various ions and agents. J. Physiol. (Lond.) **221**, 601–615
(1972a)

Kirpekar, S.M., Prat, J.C., Wakade, A.R.: Effect of calcium on the relationship between frequency
of stimulation and release of noradrenaline from the perfused spleen of the cat. Naunyn-
Schmiedeberg's Arch. Pharmacol. **287**, 205–212 (1975)

Kirpekar, S.M., Puig, M.: Effect of flow-stop on noradrenaline release from normal spleens and spleens
treated with cocaine, phentolamine or phenoxybenzamine. Brit. J. Pharmacol. **43**, 359–369 (1971)

Kirpekar, S.M., Wakade, A.R.: Release of noradrenaline from the cat spleen by potassium. J.
Physiol. (Lond.) **194**, 595–608 (1968)

Kirpekar, S.M., Wakade, A.R.: Effect of β-haloalkylamines and ephedrine on noradrenaline release
from the intact spleen of the cat. Brit. J. Pharmacol. **39**, 533–541 (1970)

Kirpekar, S.M., Wakade, A.R., Steinsland, O.S., Prat, J.C., Furchgott, R.F.: Inhibition of the
evoked release of norepinephrine (NE) by sympathomimetic amines. Fed. Proc. **31**, 566 Abs
(1972b)

Knoll, J., Vizi, E.S.: Effect of frequency of stimulation on the inhibition by noradrenaline of
the acetylcholine output from parasympathetic nerve terminals. Brit. J. Pharmacol. **42**, 263–272
(1971)

Kobinger, W.: Über den Wirkungsmechanismus einer neuen antihypertensiven Substanz mit Imid-
azolinstruktur. Naunyn-Schmiedebergs Arch. Pharmakol. exp. Path. **258**, 48–58 (1967)

Kobinger, W.: Medicinal chemistry related to the central regulation of blood pressure. II. Pharmacological part. In: Maas, J. (ed.), Medicinal Chemistry IV. Proceedings of the 4th International
Symposium on Medicinal Chemistry, pp. 107–120. Amsterdam-Oxford-New York: Elsevier 1974

Kobinger, W., Pichler, L.: Evidence for direct α-adrenoceptor stimulation of effector neurons
in cardiovascular centers by clonidine. Europ. J. Pharmacol. **27**, 151–154 (1974)

Kobinger, W., Pichler, L.: Investigation into some imidazoline compounds, with respect to peripheral
α-adrenoceptor stimulation and depression of cardiovascular centers. Naunyn-Schmiedeberg's
Arch. Pharmacol. **291**, 175–191 (1975)

Kobinger, W., Walland, A.: Involvement of adrenergic receptors in central vagus activity. Europ.
J. Pharmacol. **16**, 120–122 (1971)

Kopin, I.J.: Metabolic degradation of catecholamines. The relative importance of different pathways
under physiological conditions and after administration of drugs. In: Blaschko, H., Muscholl,
E. (eds.), Catecholamines. Handbuch der experimentellen Pharmakologie, Vol. 33, pp. 270–282.
Berlin-Heidelberg-New York: Springer 1972

Korf, J., Bunney, B.S., Aghajanian, G.K.: Noradrenergic neurons: morphine inhibition of spontaneous activity. Europ. J. Pharmacol. **25**, 165–169 (1974)

Kosterlitz, H.W., Hughes, J.: Some thoughts on the significance of enkephalin, the endogenous
ligand. Life Sci. **17**, 91–96 (1975)

Kosterlitz, H.W., Lees, G.M.: Interrelationships between adrenergic and cholinergic mechanisms. In: Blaschko, H., Muscholl, E. (eds.), Catecholamines. Handbuch der experimentellen Pharmakologie, Vol. 33, pp. 762–812. Berlin-Heidelberg-New York: Springer 1972

Kosterlitz, H.W., Lydon, R.J., Watt, A.J.: The effects of adrenaline, noradrenaline and isoprenaline on inhibitory α- and β-adrenoceptors in the longitudinal muscle of the guinea-pig ileum. Brit. J. Pharmacol. **39**, 398–413 (1970)

Kosterlitz, H.W., Taylor, D.W.: The effect of morphine on vagal inhibition of the heart. Brit. J. Pharmacol. **14**, 209–214 (1959)

Kosterlitz, H.W., Wallis, D.I.: The action of morphine-like drugs on impulse transmission in mammalian nerve fibres. Brit. J. Pharmacol. **22**, 499–510 (1964)

Krauss, K.R., Carpenter, D.O., Kopin, I.J.: Acetylcholine-induced release of norepinephrine in the presence of tetrodotoxin. J. Pharmacol. exp. Ther. **173**, 416–421 (1970)

Krnjević, K.: Chemical nature of synaptic transmission in vertebrates. Physiol. Rev. **54**, 418–540 (1974)

Kroneberg, G., Oberdorf, A., Hoffmeister, F., Wirth, W.: Zur Pharmakologie von 2-(2,6-Dimethylphenylamino)-4H-5,6-dihydro-1,3-thiazin (Bayer 1470), eines Hemmstoffes adrenergischer und cholinergischer Neurone. Naunyn-Schmiedebergs Arch. Pharmakol. exp. Path. **256**, 257–280 (1967)

Kuchii, M., Miyahara, J.T., Shibata, S.: (^{3}H)-adenine nucleotide and (^{3}H)-noradrenaline release evoked by electrical field stimulation, perivascular nerve stimulation and nicotine from the taenia of the guinea-pig caecum. Brit. J. Pharmacol. **49**, 258–267 (1973)

Langer, S.Z.: The effects of phenoxybenzamine on metabolism of ^{3}H-noradrenaline released from the isolated nictitating membrane. Brit. J. Pharmacol. **34**, 222–223P (1968)

Langer, S.Z.: The metabolism of (^{3}H)noradrenaline released by electrical stimulation from the isolated nictitating membrane of the cat and from the vas deferens of the rat. J. Physiol. (Lond.) **208**, 515–546 (1970)

Langer, S.Z.: The regulation of transmitter release elicited by nerve stimulation through a presynaptic feed-back mechanism. In: Usdin, E., Snyder, S.H. (eds.), Frontiers in Catecholamine Research, pp. 543–549. New York: Pergamon Press 1973a

Langer, S.Z.: Effects of dopamine on the presynaptic negative feedback mechanism that regulates noradrenaline release by nerve stimulation. In: Proceedings of the 2nd Meeting on Adrenergic Mechanisms, Porto, 1973b, pp. 49–50

Langer, S.Z.: Selective metabolic pathways for noradrenaline in the peripheral and in the central nervous system. Med. Biol. **52**, 372–383 (1974a)

Langer, S.Z.: Presynaptic regulation of catecholamine release. Biochem. Pharmacol. **23**, 1793–1800 (1974b)

Langer, S.Z.: Prejunctional regulatory mechanisms for noradrenaline release elicited by nerve stimulation. In: Abstracts of the symposium Chemical Tools in Catecholamine Research, Göteborg, 1975

Langer, S.Z., Adler, E., Enero, M.A., Stefano, F.J.E.: The role of the alpha receptor in regulating noradrenaline overflow by nerve stimulation. Proc. XXVth Int. Congr. Physiol. Sci., Munich, 1971, p. 335

Langer, S.Z., Dubocovich, M.L., Celuch, S.M.: Prejunctional regulatory mechanisms for noradrenaline release elicited by nerve stimulation. In: Almgren, O., Carlsson, A., Engel, J. (eds.), Chemical Tools in Catecholamine Research, Vol. 2, pp. 183–191. Amsterdam-Oxford: North-Holland 1975a

Langer, S.Z., Enero, M.A.: The potentiation of responses to adrenergic nerve stimulation in the presence of cocaine: Its relationship to the metabolic fate of released norepinephrine. J. Pharmacol. exp. Ther. **191**, 431–443 (1974)

Langer, S.Z., Enero, M.A., Adler-Graschinsky, E., Dubocovich, M.L., Celuchi, S.M.: Presynaptic regulatory mechanisms for noradrenaline release by nerve stimulation. In: Davies, D.S., Reid, J.L. (eds.), Central Action of Drugs in Blood Pressure Regulation, pp. 133–150. Tunbridge Wells: Pitman 1975b

Langer, S.Z., Enero, M.A., Stefano, F.J.E., Rothlin, R.P.: Acciones de la fenoxibenzamina sobre la liberación de noradrenalina por estimulación nerviosa en la membrana nictitante aislada de gato. Medicina (B. Aires) **30**, 557–558 (1970)

Langer, S.Z., Granata, A.R., Enero, M.A., Krieger, E.M.: Overflow of labelled transmitter elicited by nerve stimulation in the perfused mesenteric arteries of rats after the development of neurogenic hypertension. Blood Vessels 12, 368–369 (1975c)

Langer, S.Z., Rubio, M.C.: Effects of noradrenaline metabolites on the adrenergic receptors. Naunyn-Schmiedeberg's Arch. Pharmacol. 276, 71–88 (1973)

Langer, S.Z., Stefano, F.J.E., Enero, M.A.: Pre- and postsynaptic origin of the norepinephrine metabolites formed during transmitter release elicited by nerve stimulation. J. Pharmacol. exp. Ther. 183, 90–102 (1972)

Langer, S.Z., Vogt, M.: Noradrenaline release from isolated muscles of the nictitating membrane of the cat. J. Physiol. (Lond.) 214, 159–171 (1971)

Langley, A.E., Gardier, R.W.: Regulation of norepinephrine and dopamine-β-hydroxylase release during sympathetic nerve stimulation. Fed. Proc. 33, 523 Abs (1974)

Lees, G.M., Kosterlitz, H.W., Waterfield, A.A.: Characteristics of morphine-sensitive release of neuro-transmitter substances. In: Kosterlitz, H.W., Collier, H.O.J., Villareal, J.E. (eds.), Agonist and Antagonist Actions of Narcotic Analgesic Drugs, pp. 142–152. London-Basingstoke: Macmillan 1972

Lembeck, F., Juan, H.: Are there therapeutic indications of intravenous injection of calcium gluconate? Arzneimittelforsch. 25, 1570–1574 (1975)

Levin, J.A.: The uptake and metabolism of ^{3}H-*l*- and ^{3}H-*dl*-norepinephrine by intact rabbit aorta and by isolated adventitia and media. J. Pharmacol. exp. Ther. 190, 210–226 (1974)

Levy, M.N., Blattberg, B.: Effect of vagal stimulation on the overflow of norepinephrine into the coronary sinus during cardiac sympathetic nerve stimulation in the dog. Circulat. Res. 38, 81–85 (1976)

Liao, J.C., Zimmerman, B.G., Bergen, F.H. van: Adrenergic responses in canine cutaneous vasculature during acute hemorrhagic hypotension. Amer. J. Physiol. 228, 752–755 (1975)

Lindmar, R., DeSantis, V.P.: The significance of noradrenaline and adrenaline as adrenergic transmitters in the chicken. Naunyn-Schmiedeberg's Arch. Pharmacol. 282, R58 (1974)

Lindmar, R., Löffelholz, K., Muscholl, E.: Unterschiede zwischen Tyramin und Dimethylphenylpiperazin in der Ca^{++}-Abhängigkeit und im zeitlichen Verlauf der Noradrenalin-Freisetzung am isolierten Kaninchenherzen. Experientia (Basel) 23, 933–934 (1967a)

Lindmar, R., Löffelholz, K., Muscholl, E.: A muscarinic mechanism inhibiting the release of noradrenaline from peripheral adrenergic nerve fibres by nicotinic agents. Brit. J. Pharmacol. 32, 280–294 (1968)

Lindmar, R., Muscholl, E.: Die Wirkung von Cocain, Guanethidin, Reserpin, Hexamethonium, Tetracain und Psicain auf die Noradrenalin-Freisetzung aus dem Herzen. Naunyn-Schmiedeberg's Arch. exp. Path. Pharmakol. 242, 214–227 (1961)

Lindmar, R., Muscholl, E., Sprenger, E.: Funktionelle Bedeutung der Freisetzung von Dihydroxyephedrin und Dihydroxypseudoephedrin als „falschen" sympathischen Überträgerstoffen am Herzen. Naunyn-Schmiedebergs Arch. Pharmakol. exp. Path. 256, 1–25 (1967b)

Ljung, B., Åblad, B., Dahlöf, C., Henning, M., Hultberg, E.: Impaired vasoconstrictor nerve function in spontaneously hypertensive rats after long-term treatment with propranolol and metroprolol. Blood Vessels 12, 311–315 (1975)

Löffelholz, K.: Untersuchungen über die Noradrenalin-Freisetzung durch Acetylcholin am perfundierten Kaninchenherzen. Naunyn-Schmiedebergs Arch. Pharmakol. exp. Path. 258, 108–122 (1967)

Löffelholz, K.: Autoinhibition of nicotinic release of noradrenaline from postganglionic sympathetic nerves. Naunyn-Schmiedebergs Arch. Pharmakol. 267, 49–63 (1970a)

Löffelholz, K.: Nicotinic drugs and postganglionic sympathetic transmission. Naunyn-Schmiedebergs Arch. Pharmakol. 267, 64–73 (1970b)

Löffelholz, K.: Muscarinic inhibition of adrenergic neurotransmission. Gordon Conference on Catecholamines, Andover, New Hampshire (1975)

Löffelholz, K., Lindmar, R., Muscholl, E.: Der Einfluß von Atropin auf die Noradrenalin-Freisetzung durch Acetylcholin. Naunyn-Schmiedebergs Arch. Pharmakol. exp. Path. 257, 308 (1967)

Löffelholz, K., Muscholl, E.: A muscarinic inhibition of the noradrenaline release evoked by postganglionic sympathetic nerve stimulation. Naunyn-Schmiedebergs Arch. Pharmakol. 265, 1–15 (1969)

Löffelholz, K., Muscholl, E.: Inhibition by parasympathetic nerve stimulation of the release of the adrenergic transmitter. Naunyn-Schmiedebergs Arch. Pharmakol. 267, 181–184 (1970)

Long, J.P., Heintz, S., Cannon, J.G., Kim, J.: Inhibition of the sympathetic nervous system by 5,6-dihydroxy-2-dimethylamino tetralin (M-7), apomorphine and dopamine. J. Pharmacol. exp. Ther. **192**, 336–342 (1975)

Luchelli-Fortis, M.A., Langer, S.Z.: Selective inhibition by hydrocortisone of ^{3}H-normetanephrine formation during ^{3}H-transmitter release elicited by nerve stimulation in the isolated nerve–muscle preparation of the cat nictitating membrane. Naunyn-Schmiedeberg's Arch. Pharmacol. **287**, 261–275 (1975)

Malik, K.U., Ling, G.M.: The effect of 1,1-dimethyl-4-phenylpiperazinium on the response of mesenteric arteries to sympathetic nerve stimulation. J. Pharm. Pharmacol. **21**, 514–519 (1969a)

Malik, K.U., Ling, G.M.: Modification by acetylcholine of the response of rat mesenteric arteries to sympathetic stimulation. Circulat. Res. **25**, 1–9 (1969b)

Malik, K.U., McGiff, J.C.: Modulation by prostaglandins of adrenergic transmission in the isolated perfused rabbit and rat kidney. Circulat. Res. **36**, 599–609 (1975)

Malik, K.U., Nasjletti, A.: Facilitation of adrenergic transmission by locally generated angiotensin II in rat mesenteric arteries. Circulat. Res. **38**, 26–30 (1976)

Malméjac, J.: Action of adrenaline on synaptic transmission and on adrenal medullary secretion. J. Physiol. (Lond.) **130**, 497–512 (1955)

McCubbin, J.W., Page, I.H.: Renal pressor system and neurogenic control of arterial pressure. Circulat. Res. **12**, 553–559 (1963)

McCulloch, M.W., Rand, M.J., Story, D.F.: Inhibition of ^{3}H-noradrenaline release from sympathetic nerves of guinea-pig atria by a presynaptic α-adrenoceptor mechanism. Brit. J. Pharmacol. **46**, 523–524P (1972)

McCulloch, M.W., Rand, M.J., Story, D.F.: Evidence for a dopaminergic mechanism for modulation of adrenergic transmission in the rabbit ear artery. Brit. J. Pharmacol. **49**, 141–142P (1973)

Miele, E.: Lack of effect of prostaglandins E$_1$ and F$_{1\alpha}$ on adreno-medullary catecholamine secretion evoked by various agents. In: Mantegazza, P., Horton, E.W. (eds.), Prostaglandins, Peptides and Amines, pp. 85–93. London-New York: Academic Press 1969

Moawad, A., Hedqvist, P., Bygdeman, M.: Noradrenaline release following nerve stimulation and its modification by prostaglandin E$_2$ in human and rabbit oviduct. Acta physiol. scand. **95**, 142–144 (1975)

Montel, H., Starke, K.: Effects of narcotic analgesics and their antagonists on the rabbit isolated heart and its adrenergic nerves. Brit. J. Pharmacol. **49**, 628–641 (1973)

Montel, H., Starke, K., Taube, H.D.: Morphine tolerance and dependence in noradrenaline neurones of the rat cerebral cortex. Naunyn-Schmiedeberg's Arch. Pharmacol. **288**, 415–426 (1975a)

Montel, H., Starke, K., Taube, H.D.: Influence of morphine and naloxone on the release of noradrenaline from rat cerebellar cortex slices. Naunyn-Schmiedeberg's Arch. Pharmacol. **288**, 427–433 (1975b)

Montel, H., Starke, K., Taube, H.D.: Narcotic analgesics and central noradrenaline neurones. An in vitro study. In: Abstracts of the symposium Acute Effects of Narcotic Analgesics, Nokkala, Finland, 1975c, pp. 45–47

Montel, H., Starke, K., Weber, F.: Influence of morphine and naloxone on the release of noradrenaline from rat brain cortex slices. Naunyn-Schmiedeberg's Arch. Pharmacol. **283**, 357–369 (1974a)

Montel, H., Starke, K., Weber, F.: Influence of fentanyl, levorphanol and pethidine on the release of noradrenaline from rat brain cortex slices. Naunyn-Schmiedeberg's Arch. Pharmacol. **283**, 371–377 (1974b)

Morishita, H., Furukawa, T.: Possible mechanisms involved in the depressor response to dopamine. Arch. int. Pharmacodyn. **212**, 317–327 (1974)

Mujić, M., Rossum, J.M. van: Comparative pharmacodynamics of sympathomimetic imidazolines; studies on intestinal smooth muscle of the rabbit and the cardiovascular system of the cat. Arch. int. Pharmacodyn. **155**, 432–449 (1965)

Mullin, W.J., Phillis, J.W., Pinsky, C.: Morphine enhancement of acetylcholine release from the brain in unanaesthetized cats. Europ. J. Pharmacol. **22**, 117–119 (1973)

Musacchio, J.M.: Enzymes involved in the biosynthesis and degradation of catecholamines. In: Iversen, L.L., Iversen, S.D., Snyder, S.H. (eds.), Handbook of Psychopharmacology, Vol. 3, pp. 1–35. New York-London: Plenum Press 1975

Muscholl, E.: Release of catecholamines from the heart. In: Mechanisms of Release of Biogenic Amines, pp. 247–260. Oxford: Pergamon Press 1966

Muscholl, E.: Cholinomimetic drugs and release of the adrenergic transmitter. In: Schümann, H.J., Kroneberg, G. (eds.), New Aspects of Storage and Release Mechanisms of Catecholamines, pp. 168–186. Berlin-Heidelberg-New York: Springer 1970

Muscholl, E.: Regulation of catecholamine release. The muscarinic inhibitory mechanism. In: Usdin, E., Snyder, S.H. (eds.), Frontiers in Catecholamine Research, pp. 537–542. New York: Pergamon Press 1973a

Muscholl, E.: Muscarinic inhibition of the norepinephrine release from peripheral sympathetic fibres. In: Pharmacology and the Future of Man. Proc. 5th Int. Congr. Pharmacol., Vol. 4, pp. 440–457. Basel: Karger 1973b

Muscholl, E.: Introduction. In: Proceedings of the 2nd Meeting on Adrenergic Mechanisms, Porto, 1973c, pp. 33–39

Muscholl, E., Maître, L.: Release by sympathetic stimulation of α-methylnoradrenaline stored in the heart after administration of α-methyldopa. Experientia (Basel) 19, 658–659 (1963)

Muscholl, E., Ritzel, H., Rössler, K.: The time course of noradrenaline release caused by high potassium–low sodium solution. Effects of methacholine or decrease of the calcium ion concentration. Brit. J. Pharmacol. 55, 248P (1975)

Mylecharane, E.J., Raper, C.: Further studies on the adrenergic neuron blocking activity of some β-adrenoceptor antagonists and guanethidine. J. Pharm. Pharmacol. 25, 213–220 (1973)

Nedergaard, O.A., Bevan, J.A.: Nicotine and nicotine monomethiodide induced potentiation of adrenergic neuroeffector transmission in a blood vessel. In: Abstr. 4th Int. Congr. Pharmacol., Basel, 1969a, p. 144

Nedergaard, O.A., Bevan, J.A.: Effects of nicotine, dimethylphenylpiperazinium and cholinergic blocking agents at adrenergic nerve endings of the rabbit pulmonary artery. J. Pharmacol. exp. Ther. 168, 127–136 (1969b)

Nedergaard, O.A., Schrold, J.: Release of ³H-noradrenaline from incubated and superfused rabbit pulmonary artery. Acta physiol. scand. 89, 296–305 (1973)

Needleman, P., Douglas, J.R., Jakschik, B., Stoecklein, P.B., Johnson, E.M.: Release of renal prostaglandin by catecholamines: Relationship to renal endocrine function. J. Pharmacol. exp. Ther. 188, 453–460 (1974)

Nishi, S.: Cholinergic and adrenergic receptors at sympathetic preganglionic nerve terminals. Fed. Proc. 29, 1957–1965 (1970)

Noon, J.P., Roth, R.H.: Some physiological and pharmacological characteristics of the stimulus induced release of norepinephrine from the rabbit superior cervical ganglion. Naunyn-Schmiedeberg's Arch. Pharmacol. 291, 163–174 (1975)

North, R.A., Henderson, G.: Action of morphine on guinea-pig myenteric plexus and mouse vas deferens studied by intracellular recording. Life Sci. 17, 63–66 (1975)

Nueten, J.M. van, Janssen, P.A.J., Fontaine, J.: Unexpected reversal effects of naloxone on the guinea pig ileum. Life Sci. 18, 803–809 (1976)

Otten, U., Mueller, R.A., Thoenen, H.: Effect of hypophysectomy on cAMP changes in rat adrenal medulla evoked by catecholamines and carbamylcholine. Naunyn-Schmiedeberg's Arch. Pharmacol. 289, 157–170 (1975)

Paalzow, G., Paalzow, L.: Clonidine antinociceptive activity: Effects of drugs influencing central monoaminergic and cholinergic mechanisms in the rat. Naunyn-Schmiedeberg's Arch. Pharmacol. 292, 119–126 (1976)

Paalzow, L.: Analgesia produced by clonidine in mice and rats. J. Pharm. Pharmacol. 26, 361–363 (1974)

Pacha, W., Salzmann, R.: Inhibition of the re-uptake of neuronally liberated noradrenaline and α-receptor blocking action of some ergot alkaloids. Brit. J. Pharmacol. 38, 439–440P (1970)

Pacha, W., Salzmann, R., Scholtysik, G.: Inhibitory effects of clonidine and BS 100–141 on responses to sympathetic nerve stimulation in cats and rabbits. Brit. J. Pharmacol. 53, 513–516 (1975)

Panisset, J.C.: Cholinergic effects of angiotensin and bradykinin. Advanc. exp. Med. Biol. 2, 620–625 (1968)

Panisset, J.C., Bourdois, P.: Effect of angiotensin on the response to noradrenaline and sympathetic nerve stimulation, and on ³H-noradrenaline uptake in cat mesenteric blood vessels. Canad. J. Physiol. Pharmacol. 46, 125–131 (1968)

Park, M.K., Dyer, D.C., Vincenzi, F.F.: Prostaglandin E₂ and its antagonists: Effects on autonomic transmission in the isolated sino-atrial node. Prostaglandins 4, 717–730 (1973)

Paton, D.M.: Mechanism of K^+-induced release of cytoplasmic noradrenaline from adrenergic nerves. Brit. J. Pharmacol. **55**, 247P (1975)

Paton, D.M. (ed.): The Mechanism of Neuronal and Extraneuronal Transport of Catecholamines. New York: Raven Press 1976

Paton, W.D.M.: The action of morphine and related substances on contraction and on acetylcholine output of coaxially stimulated guinea-pig ileum. Brit. J. Pharmacol. **12**, 119–127 (1957)

Paton, W.D.M., Thompson, J.W.: The mechanism of action of adrenaline on the superior cervical ganglion of the cat. Proc. XIXth Int. Congr. Physiol. Sci., Montreal, 1953, pp. 664–665

Paton, W.D.M., Vizi, E.S.: The inhibitory action of noradrenaline and adrenaline on acetylcholine output by guinea-pig ileum longitudinal muscle strip. Brit. J. Pharmacol. **35**, 10–28 (1969)

Peach, M.J.: Stimulation of release of adrenal catecholamine by adenosine $3':5'$-cyclic monophosphate and theophylline in the absence of extracellular Ca^{2+}. Proc. nat. Acad. Sci. (Wash.) **69**, 834–836 (1972)

Peach, M.J., Bumpus, F.M., Khairallah, P.A.: Inhibition of norepinephrine uptake in hearts by angiotensin II and analogs. J. Pharmacol. exp. Ther. **167**, 291–299 (1969)

Peach, M.J., Ober, M.: Inhibition of angiotensin-induced adrenal catecholamine release by 8-substituted analogs of angiotensin II. J. Pharmacol. exp. Ther. **190**, 49–58 (1974)

Peskar, B., Hertting, G.: Release of prostaglandins from isolated cat spleen by angiotensin and vasopressin. Naunyn-Schmiedeberg's Arch. Pharmacol. **279**, 227–234 (1973)

Poisner, A.M.: Direct stimulant effect of aminophylline on catecholamine release from the adrenal medulla. Biochem. Pharmacol. **22**, 469–476 (1973a)

Poisner, A.M.: Caffeine-induced catecholamine secretion: Similarity to caffeine-induced muscle contraction. Proc. Soc. exp. Biol. (N.Y.) **142**, 103–105 (1973b)

Poisner, A.M.: Mechanisms of exocytosis. In: Usdin, E., Snyder, S.H. (eds.), Frontiers in Catecholamine Research, pp. 477–482. New York: Pergamon Press 1973c

Poisner, A.M., Douglas, W.W.: The need for calcium in adrenomedullary secretion evoked by biogenic amines, polypeptides, and muscarinic agents. Proc. Soc. exp. Biol. (N.Y.) **123**, 62–64 (1966)

Polak, R.L.: Stimulating action of atropine on the release of acetylcholine by rat cerebral cortex in vitro. Brit. J. Pharmacol. **41**, 600–606 (1971)

Porte, D., Robertson, R.P.: Control of insulin secretion by catecholamines, stress, and the sympathetic nervous system. Fed. Proc. **32**, 1792–1796 (1973)

Potter, W.P. de, Chubb, I.W., Put, A., Schaepdryver, A.F. de: Facilitation of the release of noradrenaline and dopamine-β-hydroxylase at low stimulation frequencies by α-blocking agents. Arch. int. Pharmacodyn. **193**, 191–197 (1971)

Poyart, C.F., Papayoanou, J., Nahas, G.G.: Effet de l'ouabaine et de l'A.M.P. cyclique sur le débit en catécholamines de la surrénale isolée. J. Physiol. (Paris) **60**, Suppl., 523 (1968)

Price, M.T.C., Fibiger, H.C.: Ascending catecholamine systems and morphine analgesia. Brain Res. **99**, 189–193 (1975)

Rahwan, R.G., Borowitz, J.L., Miya, T.S.: The role of intracellular calcium in catecholamine secretion from the bovine adrenal medulla. J. Pharmacol. exp. Ther. **184**, 106–118 (1973)

Rand, M.J., Hope, W., McCulloch, M.W., Story, D.F.: Interaction of pimozide with prejunctional dopamine receptors in the rabbit ear artery. Clin. exp. Pharmacol. Physiol. **2**, 439–440 (1975a)

Rand, M.J., Story, D.F., Allen, G.S., Glover, A.B., McCulloch, M.W.: Pulse-to-pulse modulation of noradrenaline release through a prejunctional α-receptor auto-inhibitory mechanism. In: Usdin, E., Snyder, S.H. (eds.), Frontiers in Catecholamine Research, pp. 579–581. New York: Pergamon Press 1973

Rand, M.J., Story, D.F., McCulloch, M.W.: Inhibitory feedback modulation of adrenergic transmission. Clin. exp. Pharmacol. Physiol., Suppl. **2**, 21–26 (1975b)

Rand, M.J., Varma, B.: The effects of cholinomimetic drugs on responses to sympathetic nerve stimulation and noradrenaline in the rabbit ear artery. Brit. J. Pharmacol. **38**, 758–770 (1970)

Rasmussen, H.: Cell communication, calcium ion, and cyclic adenosine monophosphate. Science **170**, 404–412 (1970)

Regoli, D., Park, W.K., Rioux, F.: Pharmacology of angiotensin. Pharmacol. Rev. **26**, 69–123 (1974)

Reit, E.: Actions of angiotensin on the adrenal medulla and autonomic ganglia. Fed. Proc. **31**, 1338–1343 (1972)

Richardson, J.A., Woods, E.F.: Release of norepinephrine from the isolated heart. Proc. Soc. exp. Biol. (N.Y.) **100**, 149–151 (1959)

Robson, R.D., Antonaccio, M.J.: Effect of clonidine on responses to cardiac nerve stimulation as a function of impulse frequency and stimulus duration in vagotomized dogs. Europ. J. Pharmacol. **29**, 182–186 (1974)

Rochette, L., Bralet, A.M., Bralet, J.: Influence de la clonidine sur la synthèse et la libération de la noradrénaline dans différentes structures cérébrales du rat. J. Pharmacol. (Paris) **5**, 209–220 (1974)

Rosell, S., Kopin, I.J., Axelrod, J.: Fate of H^3-noradrenaline in skeletal muscle before and following sympathetic stimulation. Amer. J. Physiol. **205**, 317–321 (1963)

Ross, G.: Effects of intracoronary infusions of acetylcholine and nicotine on the dog heart in vivo. Brit. J. Pharmacol. **48**, 612–619 (1973)

Roth, R.H.: Action of angiotensin on adrenergic nerve endings: Enhancement of norepinephrine biosynthesis. Fed. Proc. **31**, 1358–1364 (1972)

Roth, R.H., Hughes, J.: Acceleration of protein synthesis by angiotensin — Correlation with angiotensin's effect on catecholamine biosynthesis. Biochem. Pharmacol. **21**, 3182–3187 (1972)

Rubin, R.P.: The role of calcium in the release of neurotransmitter substances and hormones. Pharmacol. Rev. **22**, 389–428 (1970)

Salt, P.J.: Inhibition of noradrenaline uptake$_2$ in the isolated rat heart by steroids, clonidine and methoxylated phenylethylamines. Europ. J. Pharmacol. **20**, 329–340 (1972)

Salzmann, R., Pacha, W., Taeschler, M., Weidmann, H.: The effect of ergotamine on humoral and neuronal actions in the nictitating membrane and spleen of the cat. Naunyn-Schmiedebergs Arch. Pharmakol. exp. Path. **261**, 360–378 (1968)

Samuelsson, B., Wennmalm, Å.: Increased nerve stimulation induced release of noradrenaline from the rabbit heart after inhibition of prostaglandin synthesis. Acta physiol. scand. **83**, 163–168 (1971)

Sanders, K.M., Ross, G.: Inhibition of in vivo neural vasoconstriction by exogenous catecholamines. Blood Vessels **12**, 13–20 (1975)

Sanner, J.: Prostaglandin inhibition with a dibenzoxazepine hydrazide derivative and morphine. Ann. N.Y. Acad. Sci. **180**, 396–406 (1971)

Sasa, M., Munekiyo, K., Takaori, S.: Morphine interference with noradrenaline-mediated inhibition from the locus coeruleus. Life Sci. **17**, 1373–1380 (1975)

Schaumann, W.: Inhibition by morphine of the release of acetylcholine from the intestine of the guinea-pig. Brit. J. Pharmacol. **12**, 115–118 (1957)

Schaumann, W.: Zusammenhänge zwischen der Wirkung der Analgetica und Sympathicomimetica auf den Meerschweinchen-Dünndarm. Naunyn-Schmiedeberg's Arch. exp. Path. Pharmakol. **233**, 112–124 (1958)

Scheel-Krüger, J., Braestrup, C., Nielsen, M.: Feedback regulation of brain noradrenaline synthesis and release in vivo after treatment with amphetamines and tricyclic antidepressant drugs. In: Almgren, O., Carlsson, A., Engel, J. (eds.), Chemical Tools in Catecholamine Research, Vol. 2, pp. 227–234. Amsterdam-Oxford: North-Holland 1975

Schmitt, H., Schmitt, H.: α-Sympathomimetic drugs with centrally mediated hypotensive effect. In: Abstr. 4th Int. Congr. Pharmacol., Basel, 1969, pp. 315–316

Schmitt, H., Schmitt, H., Fenard, S.: Evidence for an α-sympathomimetic component in the effects of catapresan on vasomotor centres: Antagonism by piperoxane. Europ. J. Pharmacol. **14**, 98–100 (1971)

Scholtysik, G., Lauener, H., Eichenberger, E., Bürki, H., Salzmann, R., Müller-Schweinitzer, E., Waite, R.: Pharmacological actions of the antihypertensive agent N-amidino-2-(2,6-dichlorophenyl)acetamide hydrochloride (BS 100–141). Arzneimittelforsch. **25**, 1483–1491 (1975)

Schümann, H.J.: Discussion. In: Schümann, H.J., Kroneberg, G. (eds.), New Aspects of Storage and Release Mechanisms of Catecholamines, p. 210. Berlin-Heidelberg-New York: Springer 1970

Schümann, H.J., Philippu, A.: Untersuchungen zum Mechanismus der Freisetzung von Brenzcatechinaminen durch Tyramin. Naunyn-Schmiedeberg's Arch. exp. Path. Pharmakol. **241**, 273–280 (1961)

Schümann, H.J., Starke, K., Werner, U.: Interactions of inhibitors of noradrenaline uptake and angiotensin on the sympathetic nerves of the isolated rabbit heart. Brit. J. Pharmacol. **39**, 390–397 (1970a)

Schümann, H.J., Starke, K., Werner, U., Hellerforth, R.: The influence of angiotensin on the uptake of noradrenaline by the isolated heart of the rabbit. J. Pharm. Pharmacol. **22**, 441–446 (1970b)

Schulz, R., Cartwright, C.: Effect of morphine on serotonin release from myenteric plexus of the guinea pig. J. Pharmacol. exp. Ther. **190**, 420–430 (1974)

Scriabine, A., Stavorski, J.M.: Effect of clonidine on cardiac acceleration in vagotomized dogs. Europ. J. Pharmacol. **24**, 101–104 (1973)

Scriabine, A., Stavorski, J., Wenger, H.C., Torchiana, M.L., Stone, C.A.: Cardiac slowing effects of clonidine (ST-155) in dogs. J. Pharmacol. exp. Ther. **171**, 256–264 (1970)

Seeman, P., Lee, T.: Antipsychotic drugs: Direct correlation between clinical potency and presynaptic action on dopamine neurons. Science **188**, 1217–1219 (1975)

Serck-Hanssen, G.: Effects of theophylline and propranolol on acetylcholine-induced release of adrenal medullary catecholamines. Biochem. Pharmacol. **23**, 2225–2234 (1974)

Severs, W.B., Daniels-Severs, A.E.: Effects of angiotensin on the central nervous system. Pharmacol. Rev. **25**, 415–449 (1973)

Severs, W.B., Summy-Long, J., Daniels-Severs, A., Connor, J.D.: Influence of adrenergic blocking drugs on central angiotensin effects. Pharmacol. **5**, 205–214 (1971)

Sheys, E.M., Green, R.D.: A quantitative study of alpha adrenergic receptors in the spleen and aorta of the rabbit. J. Pharmacol. exp. Ther. **180**, 317–325 (1972)

Simantov, R., Snyder, S.H.: Isolation and structure identification of a morphine-like peptide "enkephalin" in bovine brain. Life Sci. **18**, 781–788 (1976)

Sjöstrand, N.O.: A note of the dual effect of prostaglandin E_1 on the responses of the guinea-pig vas deferens to nerve stimulation. Experientia (Basel) **28**, 431–432 (1972)

Sjöstrand, N.O.: Effects of acetylcholine and some other smooth muscle stimulants on the electrical and mechanical responses of the guinea-pig vas deferens to nerve stimulation. Acta physiol. scand. **89**, 1–9 (1973a)

Sjöstrand, N.O.: Effects of adrenaline, noradrenaline and isoprenaline on the electrical and mechanical responses of the guinea-pig vas deferens to nerve stimulation. Acta physiol. scand. **89**, 10–18 (1973b)

Sjöstrand, N.O., Swedin, G.: On the mechanism of the enhancement by smooth muscle stimulants of the motor responses of the guinea-pig vas deferens to nerve stimulation. Acta physiol. scand. **90**, 513–521 (1974)

Slotkin, T.A., Anderson, T.R.: Chronic morphine administration alters epinephrine uptake in rat adrenal storage vesicles. Neuropharmacol. **14**, 159–161 (1975)

Smith, A.D., Potter, W.P. de, Moerman, E.J., Schaepdryver, A.F. de: Release of dopamine-β-hydroxylase and chromogranin A upon stimulation of the splenic nerve. Tissue Cell **2**, 547–568 (1970)

Smith, A.D., Winkler, H.: Fundamental mechanisms in the release of catecholamines. In: Blaschko, H., Muscholl, E. (eds.), Catecholamines. Handbuch der experimentellen Pharmakologie, Vol. 33, pp. 538–617. Berlin-Heidelberg-New York: Springer 1972

Sorimachi, M., Oesch, F., Thoenen, H.: Effects of colchicine and cytochalasin B on the release of ^{3}H-norepinephrine from guinea-pig atria evoked by high potassium, nicotine and tyramine. Naunyn-Schmiedeberg's Arch. Pharmacol. **276**, 1–12 (1973)

Starke, K.: Interactions of angiotensin and cocaine on the output of noradrenaline from isolated rabbit hearts. Naunyn-Schmiedebergs Arch. Pharmakol. **265**, 383–386 (1970)

Starke, K.: Influence of α-receptor stimulants on noradrenaline release. Naturwissenschaften **58**, 420 (1971a)

Starke, K.: Action of angiotensin on uptake, release and metabolism of ^{14}C-noradrenaline by isolated rabbit hearts. Europ. J. Pharmacol. **14**, 112–123 (1971b)

Starke, K.: Alpha sympathomimetic inhibition of adrenergic and cholinergic transmission in the rabbit heart. Naunyn-Schmiedeberg's Arch. Pharmacol. **274**, 18–45 (1972a)

Starke, K.: Influence of extracellular noradrenaline on the stimulation-evoked secretion of noradrenaline from sympathetic nerves: Evidence for an α-receptor-mediated feed-back inhibition of noradrenaline release. Naunyn-Schmiedeberg's Arch. Pharmacol. **275**, 11–23 (1972b)

Starke, K.: Beziehungen zwischen dem Renin-Angiotensin-System und dem vegetativen Nervensystem. Klin. Wschr. **50**, 1069–1081 (1972c)

Starke, K.: Influence of phenylephrine and orciprenaline on the release of noradrenaline. Experientia (Basel) **29**, 579–580 (1973a)

Starke, K.: Regulation of catecholamine release: α-Receptor mediated feed-back control in peripheral and central neurones. In: Usdin, E., Snyder, S.H. (eds.), Frontiers in Catecholamine Research, pp. 561–565. New York: Pergamon Press 1973b

Starke, K., Altmann, K.P.: Inhibition of adrenergic neurotransmission by clonidine: An action on prejunctional α-receptors. Neuropharmacol. **12**, 339–347 (1973)

Starke, K., Borowski, E., Endo, T.: Preferential blockade of presynaptic α-adrenoceptors by yohimbine. Europ. J. Pharmacol. **34**, 385–388 (1975 a)

Starke, K., Endo, T., Taube, H.D.: Relative pre- and postsynaptic potencies of α-adrenoceptor agonists in the rabbit pulmonary artery. Naunyn-Schmiedeberg's Arch. Pharmacol. **291**, 55–78 (1975 b)

Starke, K., Endo, T., Taube, H.D., Borowski, E.: Presynaptic receptor systems on noradrenergic nerves. In: Almgren, O., Carlsson, A., Engel, J. (eds.), Chemical Tools in Catecholamine Research, Vol. 2, pp. 193–200. Amsterdam-Oxford: North-Holland 1975 c

Starke, K., Görlitz, B.D., Montel, H., Schümann, H.J.: Local α-adrenoceptor mediated feed-back inhibition of catecholamine release from the adrenal medulla? Experientia (Basel) **30**, 1170–1171 (1974 a)

Starke, K., Montel, H.: Sympathomimetic inhibition of noradrenaline release: Mediated by prostaglandins? Naunyn-Schmiedeberg's Arch. Pharmacol. **278**, 111–116 (1973 a)

Starke, K., Montel, H.: Alpha-receptor-mediated modulation of transmitter release from central noradrenergic neurones. Naunyn-Schmiedeberg's Arch. Pharmacol. **279**, 53–60 (1973 b)

Starke, K., Montel, H.: Interaction between indomethacin, oxymetazoline and phentolamine on the release of (^{3}H)noradrenaline from brain slices. J. Pharm. Pharmacol. **25**, 758–759 (1973 c)

Starke, K., Montel, H.: Involvement of α-receptors in clonidine-induced inhibition of transmitter release from central monoamine neurones. Neuropharmacol. **12**, 1073–1080 (1973 d)

Starke, K., Montel, H.: Influence of drugs with affinity for alfa-adrenoceptors on noradrenaline release by potassium and tyramine. In: Proceedings of the 2nd Meeting on Adrenergic Mechanisms, Porto, 1973 e, pp. 53–54

Starke, K., Montel, H.: Lokale Rückkopplungssteuerungen der Noradrenalinfreisetzung als mögliche Angriffspunkte antihypertensiver Pharmaka. Therapiewoche **23**, 4263–4264 (1973 f)

Starke, K., Montel, H.: Influence of drugs with affinity for α-adrenoceptors on noradrenaline release by potassium, tyramine and dimethylphenylpiperazinium. Europ. J. Pharmacol. **27**, 273–280 (1974)

Starke, K., Montel, H., Gayk, W., Merker, R.: Comparison of the effects of clonidine on pre- and postsynaptic adrenoceptors in the rabbit pulmonary artery. Naunyn-Schmiedeberg's Arch. Pharmacol. **285**, 133–150 (1974 b)

Starke, K., Montel, H., Schümann, H.J.: Influence of cocaine and phenoxybenzamine on noradrenaline uptake and release. Naunyn-Schmiedebergs Arch. Pharmakol. **270**, 210–214 (1971 a)

Starke, K., Montel, H., Wagner, J.: Effect of phentolamine on noradrenaline uptake and release. Naunyn-Schmiedebergs Arch. Pharmakol. **271**, 181–192 (1971 b)

Starke, K., Schümann, H.J.: Interactions of angiotensin, phenoxybenzamine and propranolol on noradrenaline release during sympathetic nerve stimulation. Europ. J. Pharmacol. **18**, 27–30 (1972)

Starke, K., Wagner, J., Schümann, H.J.: Adrenergic neuron blockade by clonidine: Comparison with guanethidine and local anesthetics. Arch. int. Pharmacodyn. **195**, 291–308 (1972)

Starke, K., Werner, U., Hellerforth, R., Schümann, H.J.: Influence of peptides on the output of noradrenaline from isolated rabbit hearts. Europ. J. Pharmacol. **9**, 136–140 (1970)

Starke, K., Werner, U., Schümann, H.J.: Wirkung von Angiotensin auf Funktion und Noradrenalinabgabe isolierter Kaninchenherzen in Ruhe und bei Sympathicusreizung. Naunyn-Schmiedebergs Arch. Pharmakol. **265**, 170–186 (1969)

Steinsland, O.S., Furchgott, R.F.: Desensitization of the adrenergic neurons of the isolated rabbit ear artery to nicotinic agonists. J. Pharmacol. exp. Ther. **193**, 138–148 (1975)

Steinsland, O.S., Furchgott, R.F., Kirpekar, S.M.: Inhibition of adrenergic neurotransmission by parasympathomimetics in the rabbit ear artery. J. Pharmacol. exp. Ther. **184**, 346–356 (1973)

Steinsland, O.S., Nelson, S.H.: "Alpha-adrenergic" inhibition of the response of the isolated rabbit ear artery to brief intermittent sympathetic nerve stimulation. Blood Vessels **12**, 378–379 (1975)

Stitzel, R.E., Robinson, R.L., Stevens, P.: Effects of Ba^{++} and aminophylline on the release of ^{3}H-ATP, ATP and catecholamines from bovine adrenal glands. Fed. Proc. **32**, 784 Abs (1973)

Stjärne, L.: Enhancement by indomethacin of cold-induced hypersecretion of noradrenaline in the rat in vivo—by suppression of PGE mediated feed-back control? Acta physiol. scand. **86**, 388–397 (1972 a)

Stjärne, L.: Prostaglandin E restricting noradrenaline secretion – neural in origin? Acta physiol. scand. **86**, 574–576 (1972b)

Stjärne, L.: Alpha-adrenoceptor mediated feed-back control of sympathetic neurotransmitter secretion in guinea-pig vas deferens. Nature New Biol. (Lond.) **241**, 190–191 (1973a)

Stjärne, L.: Kinetics of secretion of sympathetic neurotransmitter as a function of external calcium: Mechanism of inhibitory effect of prostaglandin E. Acta physiol. scand. **87**, 428–430 (1973b)

Stjärne, L.: Lack of correlation between profiles of transmitter efflux and of muscular contraction in response to nerve stimulation in isolated guinea-pig vas deferens. Acta physiol. scand. **88**, 137–144 (1973c)

Stjärne, L.: Uncompetitive character of inhibition by prostaglandin E_2 of the enhancing effect of α-adrenoceptor blocking drug on noradrenaline secretion in isolated guinea-pig vas deferens. Acta physiol. scand. **89**, 278–282 (1973d)

Stjärne, L.: Michaelis-Menten kinetics of secretion of sympathetic neurotransmitter as a function of external calcium: Effect of graded alpha-adrenoceptor blockade. Naunyn-Schmiedeberg's Arch. Pharmacol. **278**, 323–327 (1973e)

Stjärne, L.: Inhibitory effect of prostaglandin E_2 on noradrenaline secretion from sympathetic nerves as a function of external calcium. Prostaglandins **3**, 105–109 (1973f)

Stjärne, L.: Dual alpha-adrenoceptor mediated control of secretion of sympathetic neurotransmitter: one mechanism dependent and one independent of prostaglandin E. Prostaglandins **3**, 111–116 (1973g)

Stjärne, L.: Comparison of secretion of sympathetic neurotransmitter induced by nerve stimulation with that evoked by high potassium, as triggers of dual alpha-adrenoceptor mediated negative feed-back control of noradrenaline secretion. Prostaglandins **3**, 421–426 (1973h)

Stjärne, L.: Role of alpha-adrenoceptors in prostaglandin E mediated negative feedback control of the secretion of noradrenaline from the sympathetic nerves of isolated guinea-pig vas deferens. Prostaglandins **4**, 845–851 (1973i)

Stjärne, L.: Frequency dependence of dual negative feedback control of secretion of sympathetic neurotransmitter in guinea-pig vas deferens. Brit. J. Pharmacol. **49**, 358–360 (1973k)

Stjärne, L.: Prostaglandin- versus α-adrenoceptor-mediated control of sympathetic neurotransmitter secretion in guinea-pig isolated vas deferens. Europ. J. Pharmacol. **22**, 233–238 (1973l)

Stjärne, L.: Mechanisms of catecholamine secretion. Dual feedback control of sympathetic neurotransmitter secretion; role of calcium. In: Usdin, E., Snyder, S.H. (eds.), Frontiers in Catecholamine Research, pp. 491–496. New York: Pergamon Press 1973m

Stjärne, L.: Stereoselectivity of presynaptic α-adrenoceptors involved in feedback control of sympathetic neurotransmitter secretion. Acta physiol. scand. **90**, 286–288 (1974)

Stjärne, L.: Basic mechanisms and local feedback control of secretion of adrenergic and cholinergic neurotransmitters. In: Iversen, L.L., Iversen, S.D., Snyder, S.H. (eds.): Handbook of Psychopharmacology, Vol. 6, pp. 179–233. New York-London: Plenum Press 1975a

Stjärne, L.: Selectivity for catecholamines of presynaptic alpha-receptors involved in feedback control of sympathetic neurotransmitter secretion in guinea-pig vas deferens. Naunyn-Schmiedeberg's Arch. Pharmacol. **288**, 295–303 (1975b)

Stjärne, L.: Pre- and post-junctional receptor-mediated cholinergic interactions with adrenergic transmission in guinea-pig vas deferens. Naunyn-Schmiedeberg's Arch. Pharmacol. **288**, 305–310 (1975c)

Stjärne, L.: Clonidine enhances the secretion of sympathetic neurotransmitter from isolated guinea-pig tissues. Acta physiol. scand. **93**, 142–144 (1975d)

Stjärne, L.: Rate limiting factors in sympathetic neurotransmitter secretion. Acta physiol. scand. **93**, 220–227 (1975e)

Stjärne, L.: Relative importance of calcium and cyclic AMP for noradrenaline secretion from sympathetic nerves of guinea-pig vas deferens and for prostaglandin E-induced depression of noradrenaline secretion. Neurosci. **1**, 19–22 (1976)

Stjärne, L., Brundin, J.: Dual adrenoceptor-mediated control of noradrenaline secretion from human vasoconstrictor nerves: Facilitation by β-receptors and inhibition by α-receptors. Acta physiol. scand. **94**, 139–141 (1975a)

Stjärne, L., Brundin, J.: Affinity of noradrenaline and dopamine for neural α-receptors mediating negative feedback control of noradrenaline secretion in human vasoconstrictor nerves. Acta physiol. scand. **95**, 89–94 (1975b)

Stjärne, L., Gripe, K.: Prostaglandin-dependent and -independent feedback control of noradrenaline secretion in vasoconstrictor nerves of normotensive human subjects. Naunyn-Schmiedeberg's Arch. Pharmacol. **280**, 441–446 (1973)

Stjärne, L., Wennmalm, Å.: Quantitative estimation of secretion and reuptake of adrenergic transmitter in the rabbit heart. Acta physiol. scand. **81**, 286–288 (1971)

Story, D.F., Allen, G.S., Glover, A.B., Hope, W., McCulloch, M.W., Rand, M.J., Sarantos, C.: Modulation of adrenergic transmission by acetylcholine. Clin. exp. Pharmacol. Physiol., Suppl. **2**, 27–33 (1975)

Strait, M.R., Bhatnagar, R.K.: Inhibition of norepinephrine release elicited by nerve stimulation by an analog of apomorphine. Fed. Proc. **34**, 740 Abs (1975)

Strömbom, U.: Catecholamine receptor agonists. Effects on motor activity and rate of tyrosine hydroxylation in mouse brain. Naunyn-Schmiedeberg's Arch. Pharmacol. **292**, 167–176 (1976)

Struyker Boudier, H., Boer, J. de, Smeets, G., Lien, E.J., Rossum, J. van: Structure activity relationships for central and peripheral alpha adrenergic activities of imidazoline derivatives. Life Sci. **17**, 377–385 (1975)

Su, C., Bevan, J.A.: The release of H^3-norepinephrine in arterial strips studied by the technique of superfusion and transmural stimulation. J. Pharmacol. exp. Ther. **172**, 62–68 (1970a)

Su, C., Bevan, J.A.: Blockade of the nicotine-induced norepinephrine release by cocaine, phenoxybenzamine and desipramine. J. Pharmacol. exp. Ther. **175**, 533–540 (1970b)

Summers, R.J., Blakeley, A.G.H.: The effects of piperoxan (933F) on uptake of noradrenaline and overflow of transmitter in the isolated blood perfused cat spleen. In: Abstr. 6th Int. Congr. Pharmacol., Helsinki, 1975, p. 491

Sutherland, E.W., Robison, G.A., Butcher, R.W.: Some aspects of the biological role of adenosine 3',5'-monophosphate (cyclic AMP). Circulation **37**, 279–306 (1968)

Svensson, T.H., Bunney, B.S., Aghajanian, G.K.: Inhibition of both noradrenergic and serotonergic neurons in brain by the α-adrenergic agonist clonidine. Brain Res. **92**, 291–306 (1975)

Svensson, T.H., Trolin, G.: Brain noradrenaline neurons: Drugs affecting impulse flow, transmitter turnover and blood pressure. In: Almgren, O., Carlsson, A., Engel, J. (eds.), Chemical Tools in Catecholamine Research, Vol. 2, pp. 119–125. Amsterdam-Oxford: North-Holland 1975

Swedin, G.: Studies on neurotransmission mechanisms in the rat and guinea-pig vas deferens. Acta physiol. scand., Suppl. **369** (1971)

Sweet, C.S., Ferrario, C.M., Khosla, M.C., Bumpus, F.M.: Antagonism of peripheral and central effects of angiotensin II by (1-sarcosine, 8-isoleucine)angiotensin II. J. Pharmacol. exp. Ther. **185**, 35–41 (1973)

Szerb, J.C.: The effect of morphine on the adrenergic nerves of the isolated guinea-pig jejunum. Brit. J. Pharmacol. **16**, 23–31 (1961)

Szerb, J.C.: Lack of effect of morphine in reducing the release of labelled acetylcholine from brain slices stimulated electrically. Europ. J. Pharmacol. **29**, 192–194 (1974)

Szerb, J.C., Somogyi, G.T.: Depression of acetylcholine release from cerebral cortical slices by cholinesterase inhibition and by oxotremorine. Nature New Biol. (Lond.) **241**, 121–122 (1973)

Taube, H.D., Endo, T., Bangerter, A., Starke, K.: Presynaptic receptor systems on the noradrenergic nerves of the rabbit pulmonary artery. Naunyn-Schmiedeberg's Arch. Pharmacol. **293**, R2 (1976)

Taube, H.D., Montel, H., Hau, G., Starke, K.: Phencyclidine and ketamine: Comparison with the effect of cocaine on the noradrenergic neurones of the rat brain cortex. Naunyn-Schmiedeberg's Arch. Pharmacol. **291**, 47–54 (1975)

Terenius, L., Wahlström, A.: Morphine-like ligand for opiate receptors in human CSF. Life Sci. **16**, 1759–1764 (1975)

Teschemacher, H., Opheim, K.E., Cox, B.M., Goldstein, A.: A peptide-like substance from pituitary that acts like morphine. 1. Isolation. Life Sci. **16**, 1771–1775 (1975)

Thoa, N.B., Wooten, G.F., Axelrod, J., Kopin, I.J.: On the mechanism of release of norepinephrine from sympathetic nerves induced by depolarizing agents and sympathomimetic drugs. Molec. Pharmacol. **11**, 10–18 (1975)

Thoenen, H., Hürlimann, A., Haefely, W.: Dual site of action of phenoxybenzamine in the cat's spleen: Blockade of α-adrenergic receptors and inhibition of re-uptake of neurally released norepinephrine. Experientia (Basel) **20**, 272–273 (1964a)

Thoenen, H., Hürlimann, A., Haefely, W.: Wirkungen von Phenoxybenzamin, Phentolamin und

Azapetin auf adrenergische Synapsen der Katzenmilz. Helv. physiol. pharmacol. Acta **22**, 148–161 (1964b)

Thoenen, H., Hürlimann, A., Haefely, W.: The effect of angiotensin on the response to postganglionic sympathetic stimulation of the cat spleen; lack of facilitation of norepinephrine liberation. Med. Pharmacol. exp. **13**, 379–387 (1965)

Thoenen, H., Huerlimann, A., Haefely, W.: Interaction of phenoxybenzamine with guanethidine and bretylium at the sympathetic nerve endings of the isolated perfused spleen of the cat. J. Pharmacol. exp. Ther. **151**, 189–195 (1966a)

Thoenen, H., Tranzer, J.P., Hürlimann, A., Haefely, W.: Untersuchungen zur Frage eines cholinergischen Gliedes in der postganglionären sympathischen Transmission. Helv. physiol. pharmacol. Acta **24**, 229–246 (1966b)

Thornburg, J.E., Blake, D.E.: Effects of morphine on 5-hydroxytryptamine uptake by rat brain striatal slices. Fed. Proc. **30**, 501 Abs (1971)

Toda, N., Hojo, M., Sakae, K., Usui, H.: Comparison of the relaxing effect of dopamine with that of adenosine, isoproterenol and acetylcholine in isolated canine coronary arteries. Blood Vessels **12**, 290–301 (1975)

Tremblay, J.P., Schlapfer, W.T., Woodson, P.B.J., Barondes, S.H.: Morphine and related compounds: Evidence that they decrease available neurotransmitter in *Aplysia californica*. Brain Res. **81**, 107–118 (1974)

Trendelenburg, P.: Physiologische und pharmakologische Versuche über die Dünndarmperistaltik. Arch. exp. Path. Pharmakol. **81**, 55–129 (1917)

Trendelenburg, U.: The action of morphine on the superior cervical ganglion and on the nictitating membrane of the cat. Brit. J. Pharmacol. **12**, 79–85 (1957)

Trendelenburg, U.: Supersensitivity and subsensitivity to sympathomimetic amines. Pharmacol. Rev. **15**, 225–276 (1963)

Trendelenburg, U.: Some aspects of the pharmacology of autonomic ganglion cells. Ergebn. Physiol. **59**, 1–85 (1967)

Vanhoutte, P.M.: Inhibition by acetylcholine of adrenergic neurotransmission in vascular smooth muscle. Circulat. Res. **34**, 317–326 (1974)

Vanhoutte, P.M., Lorenz, R.R., Tyce, G.M.: Inhibition of norepinephrine-^{3}H release from sympathetic nerve endings in veins by acetylcholine. J. Pharmacol. exp. Ther. **185**, 386–394 (1973)

Vanhoutte, P.M., Verbeuren, T.J.: Inhibition of norepinephrine release in vascular smooth muscle by acetylcholine. Due to hyperpolarization of adrenergic nerve endings? Arch. int. Pharmacodyn. **213**, 332–333 (1975)

Vargas, O., Miranda, R., Orrego, F.: Effects of sodium-deficient media and of a calcium ionophore (A-23187) on the release of (^{3}H)-noradrenaline, (^{14}C)-α-aminoisobutyrate, and (^{3}H)-γ-aminobutyrate from superfused slices of rat neocortex. Neurosci. **1**, 137–145 (1976)

Vizi, S.E.: The role of Na$^+$-K$^+$-activated ATP-ase in transmitter release: acetylcholine release from basal ganglia and its inhibition by dopamine and noradrenaline. In: Frigyesi, T.L. (ed.), Subcortical Mechanisms and Sensorimotor Activities, pp. 63–87. Bern-Stuttgart-Vienna: Hans Huber 1975

Vizi, S.E., Knoll, J.: The effects of sympathetic nerve stimulation and guanethidine on parasympathetic neuroeffector transmission; the inhibition of acetylcholine release. J. Pharm. Pharmacol. **23**, 918–925 (1971)

Vizi, E.S., Rónai, A.Z., Knoll, J.: The inhibitory effect of dopamine on acetylcholine release. Naunyn-Schmiedeberg's Arch. Pharmacol. **285**, R89 (1974)

Vizi, E.S., Somogyi, G.T., Hadházy, P., Knoll, J.: Effect of duration and frequency of stimulation on the presynaptic inhibition by α-adrenoceptor stimulation of the adrenergic transmission. Naunyn-Schmiedeberg's Arch. Pharmacol. **280**, 79–91 (1973)

Vogel, S.A., Silberstein, S.D., Berv, K.R., Kopin, I.J.: Stimulation-induced release of norepinephrine from rat superior cervical ganglia in vitro. Europ. J. Pharmacol. **20**, 308–311 (1972)

Wagner, J., Endoh, M., Reinhardt, D.: Stimulation by phenylephrine of adrenergic alpha- and beta-receptors in the isolated perfused rabbit heart. Naunyn-Schmiedeberg's Arch. Pharmacol. **282**, 307–310 (1974)

Waterfield, A.A., Kosterlitz, H.W.: Stereospecific increase by narcotic antagonists of evoked acetylcholine output in guinea-pig ileum. Life Sci. **16**, 1787–1792 (1975)

Wennmalm, Å.: Quantitative evaluation of release and reuptake of adrenergic transmitter in the rabbit heart. Acta physiol. scand. **82**, 532–538 (1971)

Wennmalm, Å.: Prostaglandin release and mechanical performance in the isolated rabbit heart during induced changes in the internal environment. Acta physiol. scand. **93**, 15–24 (1975)

Wennmalm, Å., Hedqvist, P.: Prostaglandin E$_1$ as inhibitor of the sympathetic neuroeffector system in the rabbit heart. Life Sci. **9**, I, 931–937 (1970)

Wennmalm, Å., Hedqvist, P.: Inhibition by prostaglandin E$_1$ of parasympathetic neurotransmission in the rabbit heart. Life Sci. **10**, I, 465–470 (1971)

Werner, U., Starke, K., Schümann, H.J.: Wirkungen von Clonidin (ST 155) und BAY a 6781 auf das isolierte Kaninchenherz. Naunyn-Schmiedebergs Arch. Pharmakol. **266**, 474–475 (1970)

Werner, U., Starke, K., Schümann, H.J.: Actions of clonidine and 2-(2-methyl-6-ethyl-cyclohexylamino)-2-oxazoline on postganglionic autonomic nerves. Arch. int. Pharmacodyn. **195**, 282–290 (1972)

Werner, U., Wagner, J., Schümann, H.J.: Beeinflussung der Noradrenalinabgabe aus isolierten Kaninchenherzen durch β-Adrenolytica unter sympathischer Nervenreizung. Naunyn-Schmiedebergs Arch. Pharmakol. **268**, 102–113 (1971)

Westfall, T.C.: Effect of nicotine and other drugs on the release of ^{3}H-norepinephrine and ^{3}H-dopamine from rat brain slices. Neuropharmacol. **13**, 693–700 (1974a)

Westfall, T.C.: Effect of muscarinic agonists on the release of ^{3}H-norepinephrine and ^{3}H-dopamine by potassium and electrical stimulation from rat brain slices. Life Sci. **14**, 1641–1652 (1974b)

Westfall, T.C., Besson, M.J., Giorguieff, M.F., Glowinski, J.: The role of presynaptic receptors in the release and synthesis of ^{3}H-dopamine by slices of rat striatum. Naunyn-Schmiedeberg's Arch. Pharmacol. **292**, 279–287 (1976)

Westfall, T.C., Brasted, M.: The mechanism of action of nicotine on adrenergic neurons in the perfused guinea-pig heart. J. Pharmacol. exp. Ther. **182**, 409–418 (1972)

Westfall, T.C., Brasted, M.: Effect of 4,4'-biphenylenebis-[(2-oxoethylene)-bis-(2,2-diethoxyethyl)]dimethylammonium dibromide (DMAE) on accumulation and nicotine-induced release of norepinephrine in the heart. J. Pharmacol. exp. Ther. **184**, 198–204 (1973)

Westfall, T.C., Brasted, M.: Specificity of blockade of the nicotine-induced release of ^{3}H-norepinephrine from adrenergic neurons of the guinea-pig heart by various pharmacological agents. J. Pharmacol. exp. Ther. **189**, 659–664 (1974)

Westfall, T.C., Hunter, P.E.: Effect of muscarinic agonists on the release of (^{3}H)noradrenaline from the guinea-pig perfused heart. J. Pharm. Pharmacol. **26**, 458–460 (1974)

Wooten, G.F., Thoa, N.B., Kopin, I.J., Axelrod, J.: Enhanced release of dopamine β-hydroxylase and norepinephrine from sympathetic nerves by dibutyryl cyclic adenosine 3',5'-monophosphate and theophylline. Molec. Pharmacol. **9**, 178–183 (1973)

Yaksh, T.L., Yamamura, H.I.: Blockade by morphine of acetylcholine release from the caudate nucleus in the mid-pontine pretrigeminal cat. Brain Res. **83**, 520–524 (1975)

Yoshizaki, T.: Effect of histamine, bradykinin and morphine on adrenaline release from rat adrenal gland. Jap. J. Pharmacol. **23**, 695–699 (1973)

Younkin, S.G.: An analysis of the role of calcium in facilitation at the frog neuromuscular junction. J. Physiol. (Lond.) **237**, 1–14 (1974)

Zimmerman, B.G.: Effect of acute sympathectomy on responses to angiotensin and norepinephrine. Circulat. Res. **11**, 780–787 (1962)

Zimmerman, B.G.: Blockade of adrenergic potentiating effect of angiotensin by 1-Sar-8-Ala-angiotensin II. J. Pharmacol. exp. Ther. **185**, 486–492 (1973)

Zimmerman, B.G., Gisslen, J.: Pattern of renal vasoconstriction and transmitter release during sympathetic stimulation in presence of angiotensin and cocaine. J. Pharmacol. exp. Ther. **163**, 320–329 (1968)

Zimmerman, B.G., Gomer, S.K., Liao, J.C.: Action of angiotensin on vascular adrenergic nerve endings: facilitation of norepinephrine release. Fed. Proc. **31**, 1344–1350 (1972)

Zimmerman, B.G., Liao, J.C., Gisslen, J.: Effect of phenoxybenzamine and combined administration of iproniazid and tropolone on catecholamine release elicited by renal sympathetic nerve stimulation. J. Pharmacol. exp. Ther. **176**, 603–610 (1971)

Zimmerman, B.G., Ryan, M.J., Gomer, S., Kraft, E.: Effect of the prostaglandin synthesis inhibitors indomethacin and eicosa-5,8,11,14-tetraynoic acid on adrenergic responses in dog cutaneous vasculature. J. Pharmacol. exp. Ther. **187**, 315–323 (1973)

Zimmerman, B.G., Whitmore, L.: Effect of angiotensin and phenoxybenzamine on release of norepinephrine in vessels during sympathetic nerve stimulation. Int. J. Neuropharmacol. **6**, 27–38 (1967)

Zwieten, P.A. van: Antihypertensive drugs with a central action. Progr. Pharmacol. **1**, 1–63 (1975)

Rev. Physiol., Biochem. Pharmacol., Vol. 77
© by Springer-Verlag 1977

Biology of Leukotaxis

P.A. WARD and E.L. BECKER *

Contents

I. Introduction

The bulk of the fragmentary information on the mechanism of chemotaxis of inflammatory cells has come primarily from studies employing the neutrophil. Unless otherwise noted, the following discussion will deal wholly with this cell type.

The mechanisms of the chemotactic response of neutrophils have many similarities to the secretory processes of a variety of cells, to muscle contraction and the release of neurotransmitters from nerve endings (BECKER and HENSON, 1973). More specifically, chemotaxis, phagocytosis, and lysosomal enzyme secretion in the neutrophil, can all be considered examples of cell movement—either movement of the whole cell, as in chemotaxis; movement of a part of the cell membrane and contiguous structures, as in phagocytosis, or movement of an intracellular organelle from the interior of the cell to the plasma membrane, as in lysosomal enzyme release (BECKER et al., 1974). In accord with this view, chemotactic factors have been found to induce lysosomal enzyme release (BECKER et al., 1974; GOLDSTEIN et al., 1973) and when in solution to inhibit the rate of phagocytosis (MUSSON and BECKER, 1975); whereas, on a latex particle they

* Supported in part by NIH grants AI 09651, AI 11526, AI 12225, AI 09648

enhance phagocytosis (BECKER, 1976). From present work, it appears that all three functions share portions of the same biochemical sequences. Thus, study of the differences and similarities to all three functions, in some instances might be more profitable than concentrating wholly on the study of any one function.

II. Description of Neutrophil Movement

Neutrophils move only on surfaces, first, by extending onto the surface a thin, flattened protrusion called a lamellipodium, with the cell rising above and back of the lamellipodium. Then, the cell contents, cytoplasmic granules, vacuoles and nuclei flow into and fill the lamellipodium. This leads again to a protrusion of the cell above the surface, and the entire process is repeated (RAMSEY, 1974). Several lamellipodia may protrude from the cell at one time, but usually the cell contents flow only into one lamellipodium, and it is this one which determines the movement. In cells responding to a chemotactic stimulus, lamellipodia extend apparently at random, but cell contents tend to flow into lamellipodia on the side from which the attractant is diffusing (RAMSEY, 1972). However, according to ZIGMOND (1974), the frequency of pseudopod formation increases on the side closest to the source of chemoattractant. (The relation between lamellipodia and pseudopodia is not clear.) In addition to lamellipodia at the front, moving neutrophils exhibit a knob-like tail, or, uropod at the rear (RAMSEY, 1974; ZIGMOND and HIRSCH, 1972). Reportedly, the uropod is a collection of the retraction fibers formed from the lamellipodia, which had been extended and had adhered to the surface, but into which the cellular contents did not flow (RAMSEY, 1974b).

Chemotactic agents induce an increase in volume of neutrophils suspended in a liquid medium (HSU and BECKER, 1975), but, in the presence of cytochalasin B and Ca^{2+}, they cause the cell to contract (KOZA et al., 1975; HSU and BECKER, 1975). On the assumption that a surface can substitute for cytochalasin B, it was suggested that the chemotactic response involves a regulated and sequential increase in volume in the portion of the cell exposed to the bulk medium and a contraction in the portion in contact with the surface (HSU and BECKER, 1975). In moving, the neutrophil is alternately highly spread over the surface, or protruding up into the medium with relatively little area in contact with the surface (RAMSEY, 1972). It is possible that the first phase is associated with the postulated contraction and the second with the increase of cell volume.

An organism can sense a chemical gradient by at least two possible methods (ADLER, 1975). In the single receptor, or time gradient mechanism used by bacteria, the organism as it moves detects a chemical gradient by constantly assaying the medium. In the spatial, or multiple receptor model, the organism detects the gradient between at least two locations on its body at one time. The latter implies that the organism is capable of sensing a gradient across its own dimensions. Neutrophils appear to utilize a spatial mechanism (ZIGMOND, 1974; RAMSEY, 1974b).

Several studies have suggested that the rate of movement is not altered by a chemotactic stimulus, only the direction (RAMSEY, 1974a). ZIGMOND and HIRSCH (1972) have pointed out that the conditions in some of these studies, were such that a possible stimulator would have been masked.

III. The Mechanism of the Chemotactic Response

The mechanisms of how the cell moves and how the chemotactic gradient induces a direction to that movement are both essentially unknown. Chemotactic factors stimulate increased random and directed movement (ZIGMOND and HIRSCH, 1972; SHOWELL et al., 1976). However, not all agents stimulating increased neutrophil movement, are chemotactic. For example, ascorbic acid and glutathione (GOETZL et al., 1974) and K^+ (SHOWELL et al., 1976) enhance spontaneous movement and the response to chemotactic factors, but other compounds, which are not chemotactic per se, as well as the Ca^{2+} ionophore, A23187, also enhance neutrophil locomotion in the presence of Mg^{2+} (BECKER, 1975). WILKINSON (1975a), however, found that the A23187 in low concentrations enhanced the leukocyte response to casein, in the absence of Ca^{2+} and Mg^{2+}.

No biochemical sequence(s) can be written for the chemotactic process at this time, yet, something is known or suspected about certain components and elements of the process and these will be discussed below. Among these are the interaction of chemotactic factors with the neutrophil, the role of neutrophil esterases, the nature of the requirements for the external ions and metabolic energy, the role of the cyclic nucleotides (cyclic AMP and cyclic GMP), and, related to cyclic nucleotides, the effects of autonomic agents, prostaglandins, etc. A brief consideration will also be given to the contraction mechanisms of the cell, the microfilaments and microtubules.

A. Interaction of Chemotactic Factor with Cell

One attractive hypothesis as to the nature of the primary interaction of the neutrophil and chemotactic factors, is that there is a binding of a chemotactic factor to receptors on the cell surface. The term "receptor" is used to describe the molecules or group of molecules capable of specifically recognizing the factor, in a manner which, in some way, initiates the chemotactic response (HOLLENBERG and CUATRECASAS, 1975). Unfortunately, there is no direct evidence, either that the interaction of chemotactic factors occurs on the cell surface, or that specific receptors are involved. It is plausible to consider that the high molecular weight chemotactic factors (kallikrein and plasminogen activator and large peptides such as C5a) can only interact with the cell surface. One might argue that, in view of the very limited pinocytic ability of neutrophils, it is likely that even the small molecular weight chemotactic peptides (SHOWELL et al., 1976) act at the cell surface. But both of these arguments emphasize the lack of direct evidence.

Similarly, there is no direct evidence for the existence of specific chemotactic receptors. As pointed out by WILKINSON (1974a), given the very great heterogeneity of chemotactic agents, it is unlikely that there is a single species of neutrophil chemotactic receptor. In fact, both WILKINSON (1974a, b) and WISSLER et al. (1972) as well as STECHER and SORKIN (1972) argue against there being any specific chemotactic receptors. However, the recent work showing the very great specificity of chemotactically active synthetic small peptides (SHOWELL et al., 1976) and the low concentration (10^{-10} 10^{-12} M) at which some of them act, suggests that this class of compounds interacts with a specific cell receptor. Nevertheless, even accepting this hypothesis, it does not follow that this is the only species of receptor, or that all chemotactic factors interact with only one kind of receptor on the neutrophil.

In fact, there is reason to believe that not all chemotactic factors initiate a response by interacting with the cell in precisely the same way. The enzymes, kallikrein and plasminogen activator, are chemotactic and their enzymatic activity is required for their chemotactic activity (KAPLAN et al., 1973, 1974). Most chemotactic factors, studied to date, are not enzymes, so that their action must differ, in this respect, from kallikrein and the plasminogen activator.

WILKINSON (1974b) has suggested that chemotactic factors induce a response by possessing hydrophobic groups and, thus, being able to penetrate the lipid bilayer of the cell membrane. In addition to his findings with chemically modified proteins as chemoattractants, referred to above, WILKINSON (1975a) has adduced evidence in support of this hypothesis by the demonstration that the chemotactic response and the stimulated random motility of human blood monocytes are inhibited by phospholipase C and by sphingomyelinase C, but not by Cl perfringens θ toxin. Chemotaxis of human blood neutrophils is inhibited by θ toxin, but not by phospholipase C.

B. Requirement for External Cations

Extracellular Ca^{2+} and Mg^{2+} are required for the optimal spontaneous motility and chemotactic responsiveness of neutrophils (BECKER and SHOWELL, 1972; GALLIN and ROSENTHAL, 1974). Cobalt and manganese are able to substitute completely for Ca^{2+}, but not a variety of other divalent cations. A minimal chemotactic response still occurs in the presence of sufficient EDTA to chelate all extracellular divalent cations (BECKER and SHOWELL, 1972). Part of the effect of Mg^{2+} might be due to increasing granulocyte adhesiveness (BRYANT et al., 1966; ATHERTON and BORN, 1972) although arguments have been advanced to suggest that this is not the only role for Mg^{2+} (BECKER and SHOWELL, 1972). Interaction of the neutrophil with concentrations of chemotactic factors giving up to maximum stimulation of migration, leads to an inhibition of Ca^{2+}

uptake and a small dose-dependent increase in Ca^{2+} release (GALLIN and ROSEN-THAL, 1974; NACCACHE et al., 1976). Increasing the concentration of chemotactic factor to inhibitory levels, increases dramatically the release of Ca^{2+} from the cell as well as Ca^{2+} uptake (NACCACHE et al., 1976). It is possible that these Ca^{2+} shifts reflect the stimulation of increased transport of Ca^{2+} into the cell with a consequent release of Ca^{2+} into the cytoplasm from a sequestered pool, as is found in smooth and cardiac muscle. However, much further work is required before such a conclusion can be accepted.

Extracellular K^+ is not required for the chemotactic response of the neutrophil but small concentrations increase it (SHOWELL and BECKER, 1976). Spontaneous motility is also enhanced by K^+, suggesting that its effect is on neutrophil movement and not on the directional component of chemotactic stimulation. Other monovalent cations are active in the same way, the rank order of their effectiveness being $K^+ = NH_4^+ > Rb^+ > Cs^+ > Li^+ = Na^+$. The K^+ specific ionophore, valinomycin, enhances chemotactic activity by the presence of extracellular K^+ but not in its absence, whereas another K^+ specific ionophore, nigericin, inhibits chemotaxis slightly but significantly in the absence of K^+, not in its presence. Ouabain, a rather specific inhibitor of Na^+/K^+ATPase action and the monovalent cation pump activity, abolishes the enhancing effect of K^+, suggesting that this latter action depends upon the activation of the monovalent cation pump of the neutrophil. Neutrophils in buffer, devoid of Na^+, greatly increase their spontaneous movement and lose their ability to respond to chemotactic stimuli. These data suggest that chemotactic factors act, in part, at least, by stimulating a net influx of K^+ and/or a net efflux of Na^+. Some support for this suggestion has been found in work which has shown that concentrations of a synthetic peptide chemotactic factor; F-Met-Leu-Phe, which enhance migration, also accelerate K^+ uptake by neutrophils, whereas, inhibitory concentrations depress the uptake of K^+ (NACCACHE et al., 1976).

C. Esterases Involved in Chemotaxis

At least two cellular serine esterases have been implicated in the chemotactic response of the neutrophil (BECKER and HENSON, 1973; BECKER, 1974). As implied in their characterization as serine esterases, these enzymes are inhibited by organophosphorus inhibitors such as diisopropylphosphofluoridate (DFP), various series of p-nitrophenyl ethyl phosphonate esters, and inhibitors such as L-1-tosylamide-2-phenylethyl-chloromethyl ketone (TPCK), or N-α-p-tosyl-L-lysine-chloromethyl ketone (TLCK). The interaction of the complement-derived factors C5a[1], C$\overline{567}$, and C3a[1], with rabbit blood neutrophils, leads to the activation of a cell-bound serine esterase, proesterase 1 to esterase 1. Proester-

---

[1] The designations C3a and C5a formally imply anaphylatoxin activities in these fragments. Since most reports have demonstrated that preparations of C3a and C5a also contain chemotactic activity, the terms C3a and C5a are used with the implications of dual biological activities. As will become evident in a later section, there is now good evidence that certain C3 and C5 fragments, lacking in anaphylatoxin activity, are, nevertheless, chemotactically active. Thus, the terms C3a and C5a, and C3 and C5 chemotactic fragments, are used in this series without intent to imply structure.

ase 1 exists in or on the cell in an enzymatically inert form incapable of being inhibited by organophosphorus inhibitors. After activation, the resulting esterase 1 is able to hydrolyze the aromatic amino acid ester, acetyl DL phenalanine naphthyl ester, and is inhibited by 10^{-8} M to 2.5×10^{-9} M phosphonate esters. There are approximately 500 molecules of proesterase 1 per rabbit blood neutrophil, as estimated using radiolabeled DFP (unpublished experiments, BECKER and HENSON). The activation does not require divalent cations (BECKER, 1974). There is sufficient proesterase 1 in the neutrophil from rabbit blood to be demonstrable biochemically. In the peritoneal cell, its concentration is so low that its presence is inferred indirectly by its ability to become inactivated by phosphonate esters when the cell is acted upon by the chemotactic factor. Chemotactic factors from *Escherichia coli* culture filtrates do not reproducibly activate proesterase 1, however, the structure activity relationships (the inhibition profiles) of inhibition by the phosphonate esters of the chemotactic response to the factor, led to the suggestion that the bacterial factor, in causing chemotaxis activates proesterase 1 but at so low a level it cannot be detected biochemically (BECKER, 1972). The structure activity relations of the inhibition of spontaneous motility by the same phosphonate esters suggested the hypothesis that activation of proesterase 1 also is involved in the unstimulated locomotion of the neutrophil (BECKER, 1975).

Incubation of the neutrophil with $\overline{C567}$, C3a, kallikrein, or plasminogen activator in the absence of a concentration gradient, leads to "deactivation". The cell becomes unresponsive to further chemotactic stimulation when it is washed and tested in the usual way with fresh chemotactic factor (WARD, 1972). Proesterase 1 is involved in the deactivation by the complement derived factors (WARD and BECKER, 1968). The complement-derived factors not only deactivate to themselves but deactivate to all of the others as well as the chemotactic factor from *E. coli*.

Esterase 1 is produced to only a very small extent by chemotactic stimulation of the neutrophil but is present much more abundantly in the already activated state. There it exists predominately in the cytosol of rabbit peritoneal neutrophils with a molecular weight of 230,000 (TSUNG and KEGELES, unpublished observations). If the cell is lysed and extracted with buffer, or if an extract of neutrophil lysosomal granules is added to the cytosol fraction, esterase 1 activity is found associated with two molecular weight species of 170,000 and 65,000, respectively.

The chemotactic responsiveness of the rabbit neutrophil is irreversibly inhibited by organophosphorus inhibitors in the absence of contact of the cell with chemotactic factors (BECKER and WARD, 1967; WARD and BECKER, 1968). The chemotactic responsiveness of human neutrophil is also inhibited in the same way by DFP, by esterase inhibitors such as TLCK and TPCK, as well as by the globulin inhibitors, α1 antitrypsin and α2 macroglobulin, and by C1 esterase inactivator (GOETZL, 1975; SMITH et al., 1975). These findings suggest that in addition to proesterase 1 and esterase 1, there also exist one or more serine esterases in or on the neutrophil in an already activated form which are involved in chemotaxis. The nature of none of these esterases is understood nor is the place at which they act identified in the biochemical sequence or sequences leading to the chemotactic response.

D. Metabolic Requirements

The neutrophil contains few mitochondria, and 2.4 dinitrophenol (CARRUTHERS, 1967; WARD, 1968) has no effect on chemotaxis. Arsenite has little or no effect on neutrophil motility. Chemotactic factors stimulate the hexose monophosphate shunt and increase the rate of aerobic glycolysis of neutrophils (BECKER et al., 1974; GOETZL and AUSTEN, 1974b). Iodoacetate, which inhibits both the shunt and glycolytic pathways also inhibits chemotactic responsiveness (CARRUTHERS, 1966; GOETZL and AUSTEN, 1974b). Ascorbic acid exerts a parallel concentration-dependent enhancement of spontaneous motility, chemotaxis and hexose mono-phosphate shunt activity (GOETZL and AUSTEN, 1974b). However, methylene blue, which also increases shunt activity, has no effect on chemotactic responsive-ness (GOETZL and AUSTEN, 1974b). Extracellular glucose has no effect on chemo-taxis or unstimulated locomotion (CARRUTHERS, 1967) even though added insulin restores the chemotactic activity of neutrophils from diabetics (MOWAT and BAUM, 1971). Glucose does prevent the inhibition of chemotaxis by deoxyglucose, an inhibitor of the glycolytic pathway (CARRUTHERS, 1967).

These findings suggest that aerobic oxidation plays little or no role in neutro-phil locomotion or chemotaxis, but rather that the metabolic energy comes from aerobic glycolysis or, possibly, from the hexose monophosphate shunt activity, or both. Neutrophils from hypophosphatamic dogs have a chemotactic defect and a concomitant reduction in leukocyte ATP. Repletion of cellular ATP by phosphate and adenosine corrects the depression of chemotactic respon-siveness (CRADDOCK et al., 1974). This suggests that the requirement of energy from glucose metabolism is to maintain or furnish cellular ATP for the chemotac-tic process.

E. Cyclic Nucleotides

A number of investigators have reported that exogenous cyclic AMP and/or dibutyryl cyclic AMP are chemotactic; an almost equal number have reported they are not (RIVKIN and BECKER, 1975). There seems to be no reason to doubt the findings of either group of investigators and the reasons for the difference between the two groups is unknown. RIVKIN et al. (1975) tested a number of agents for their ability to affect intracellular levels of neutrophil cyclic AMP, chemotaxis and spontaneous motility. They found that prostaglan-dins E and A but not $F_{2\alpha}$ increased neutrophil cyclic AMP levels and, corre-spondingly, only the first two inhibited chemotaxis. Epinephrine, isoproterenol, and, to a much lesser extent, norepinephrine increased cyclic AMP. Only epinephrine and isoproterenol inhibited chemotaxis, but the inhibition was vari-able and not related to the ability of these catecholamines to increase intracellular cyclic AMP. The stimulation of cyclic AMP and the inhibition of chemotaxis was due to the β adrenergic action of the catecholamines. No evidence for an α adrenergic receptor was found. The effects of the catecholamines and prostaglandins were due to their interaction with different receptors on the neutrophil. Cholera toxin increased neutrophil cyclic AMP and inhibited chemotaxis and spontaneous motility after a 30-min lage period. However, the effect on chemotaxis required 50 ng/ml of toxin, whereas the effect on cyclic AMP was manifested at 2 ng/ml. Even 1250 ng/ml had no effect on either chemotaxis or cyclic AMP unless there was a 30-min, or more, preincubation. Choleragenoid (an inactivated form of cholera enterotoxin) prevented the effects of toxin on both cyclic AMP and chemotaxis. Neither the chemotactic factor from *E. coli* culture filtrates nor C5a caused any detectible change in levels of neutrophil cyclic AMP. The authors concluded that increases in intracellular cyclic AMP may inhibit neutrophil movement even though chemotactic factors seem to have no affect on adenylcyclase activity, implying cyclic AMP is not, in the main sequence of events, triggering the chemotactic response.

HILL et al. (1975) obtained results which essentially agreed with those of RIVKIN et al. (1975). TSE et al. (1972), however, tested a number of agents with supposed effect on intracellular cyclic AMP levels and obtained results which only partly agreed with the above. They did not measure the effect of these agents on intracellular levels of cyclic AMP.

The mechanism by which cyclic AMP is believed to exert its effects is through cyclic AMP-dependent protein kinases. Human neutrophils contain at least one such kinase as well as protein phosphatases (TSUNG et al., 1975a, b). The substrate or substrates in the neutrophil phosphorylated naturally by this kinase, is not known. However, demonstration of its existence does indicate how cyclic AMP may intervene in the chemotactic process.

The 8 bromo-derivative of cyclic GMP increases chemotactic responsiveness as do agents such as carbamylcholine (carbachol), phorbol myristate acetate, imidazole, etc. All of the agents are known to increase cyclic GMP levels of other cells (ESTENSON et al., 1973; HILL et al., 1975; SANDLER et al., 1975). Thus, there is some evidence that neutrophil movement and the chemotactic response may be under the reciprocal influences of cyclic AMP and cyclic GMP. It is unknown whether chemotactic stimuli increase cyclic GMP levels, and thus, there is no present basis for speculating that cyclic GMP is part of the main biochemical sequence leading to cell movement.

F. Microtubules and Microfilaments

Movement in neutrophils, as in other non-muscle cells, may involve the contractile proteins, actin and myosin. These have been isolated from neutrophils. Microfilaments, present in neutrophils, have been identified as actin polymers. Microtubules form another contractile system and they, too, have been shown to be present in neutrophils (STOSSEL, 1975).

Cytochalasin B, a fungal metabolite, which, supposedly, interferes with microfilament function, inhibits neutrophil locomotion and chemotaxis at concentrations of 1 µg/ml and above (BECKER et al., 1972; BOREL and STAEHLEIN, 1972; ZIGMOND and HIRSCH, 1972); at concentrations of 0.5 µg/ml and below, enhancement of chemotaxis occurs (BECKER et al., 1972). Due to the numerous actions of cytochalasin B on other than microfilaments, the significance of these actions of cytochalasin B on neutrophil movement is uncertain. BOXER et al.

(1974) have recently studied an infant with repeated infections whose neutrophils manifested impaired locomotion and ingestion but increased neutrophil degranulation. The patient's neutrophils contained an actin which polymerized much less completely than normal. So far, this is the most direct evidence available for a role of contractile proteins in neutrophil locomotion.

The chemotactic factor C5a induces a transient microtubule assembly in neutrophils treated with cytochalasin B (GOLDSTEIN et al., 1973), and a longer lasting one in untreated neutrophils (GALLIN and ROSENTHAL, 1974). Agents in addition to C5a that cause degranulation of human neutrophils such as Ca^{2+}, or concanavalin A, or that elevate intracellular levels of cyclic GMP such as phorbol myristate acetate or carbachol, also cause an increase in the number of microtubules (HOFFSTEIN, 1975). The argument for a functional relationship between microtubule assembly and neutrophil locomotion largely rests upon the inhibitory effect on chemotaxis of agents such as colchicine, vinblastine or vincristine, which interfere with microtubular polymerization or function. Low concentrations of colchicine, demicolchicine, podophyllic diethylhydrazide or vinblastine are reported to inhibit (CANER, 1965; PHELPS and McCARY, 1969; BANDMANN et al., 1974a; BANDMANN et al., 1974b), have no effect (WARD, 1971), or, depending on experimental circumstances, either inhibit or enhance chemotactic activity (BECKER and SHOWELL, 1974). The same agents either have essentially no effect on spontaneous movement (RAMSEY and HARRIS, 1973; BANDMANN et al., 1974b) or enhance it (EDELSON and FUDENBERG, 1973; BECKER and SHOWELL, 1974). OLIVER et al. (1975) have suggested that neutrophils from patients with Chediak-Higashi syndrome suffer from a defect in microtubule function. This suggestion is of interest in view of the defective chemotactic responsiveness and apparently normal spontaneous motility of the neutrophils from these patients (CLARK and KIMBALL, 1972).

Thus, chemotactic factors can increase microtubule number and the bulk of studies (with some disturbing discrepancies) show inhibition of chemotaxis, but not of random locomotion, by agents which interfere with microtubule function. In addition, there is evidence that microfilaments are required in random locomotion and chemotaxis. A number of workers have suggested that microtubules superimpose direction or orientation to leukocyte movement but are not required for movement itself (BHISEY and FREED, 1971; BANDMANN et al., 1974a).

It is obvious from what has just been reviewed that neither the precise reactions triggered by chemotactic factors nor their sequence is known. These reactions apparently involve activation of a proesterase and action of already activated esterase, both monovalent and divalent cations and, possibly, changes in their rate of flux across the cell membrane; an energy source, activation and regulation by cyclic nucleotides; and compartmentalized contraction and relaxation of the contractile machinery of the cell, involving both microfilaments and probably microtubules.

IV. Nature of the Chemotactic Factors

Except for the synthetic, leukotactically active small peptides (see below), there is little information regarding the structural basis for chemotactic factors. In studies with human serum albumen, casein, and globin that have been physically altered ("denatured") or chemically modified, it is the contention of WILKINSON (1974a) that chemotactic activity is attributable to tertiary structural changes in a protein/peptide rather than a reflection of the primary structure of the chemotactic factor. This theory has the advantage of providing an explanation for the large number of agents that have been reported to have chemotactic activity. One of the problems in interpreting the data of WILKINSON (1974a) relates to the high concentrations of proteins employed, and the fact that many of the chemotactic responses barely exceed the background counts by a factor of two (see below).

Recently, SCHIFFMANN et al. (1975b) reported that simple N-formyl methionyl peptides enhanced migration of neutrophils and were presumably chemotactic. SHOWELL et al. (1976) have confirmed and extended this finding to show that simple di and tripeptides, most of them N-formyl methionyl peptides, stimulated both random and directed (chemotactic) locomotion. Systematic studies of the relation of structure to activity revealed that the chemotactic activity of the peptides depends on their constituent amino acids as well as the position of the amino acids in the peptide chain. These workers have failed to find an absolute requirement for a hydrophobic character for the most active chemoattractants. The conclusion from these studies is inescapable that primary structure is a critical feature, at least in the synthetic chemotactic peptides. Nevertheless, WILKINSON (1974a) has stimulated considerable interest on the subject of structure-function relationships of chemotactic factors and no categorical statement can yet be made about the chemical basis for chemotactic activity.

One of the few nonpeptide factors with reported chemotactic activity for leukocytes is the material described by TURNER et al. (1975) who have found an oxidation product (as yet unidentified) from arachadonic acid. A few other oxidized polyenoic lipids were also found to contain chemotactic activity for neutrophils. It was stressed that prostaglandins, per se, exhibited no chemotactic activity, suggesting that some intermediate products of oxidized arachodonic acid were the responsible compounds. The authors presented the hypothesis that lipid peroxidation, which occurs in membranes, might be a mechanism by which chemotactic factors are produced following membrane damage and that this might be responsible for the inflammatory response after tissue injury. To what extent arachadonic acid derivatives represent a biologically important class of chemotactic factors has yet to be determined.

The production by bacteria of chemotactic factors (for leukocytes) has been studied by several groups (KELLER and SORKIN, 1967; WARD et al., 1968; TEMPLE et al., 1970). The general observations have been that, during the log phase of bacterial replication, chemotactic activity appears in the culture medium. The production of this chemotactic activity occurs in chemically defined media that lack proteins or peptides, indicating that the activity is not a derivative of a preexisting protein present in the culture medium. Furthermore, the in vitro elaboration of chemotactic activity is not related to the clinically determined pyogenic characteristics of the micro-organisms. However, this lack of association may simply reflect that the in vitro culture conditions do not simulate cultural conditions in vivo. There is little direct evidence that defines the chemical nature of the bacterial chemotactic factors. Most work has been done with the products from *E. coli*. Here, the chemotactic activity seems to be due to dialyzable, heat stable factors that are susceptible to inactivation by pronase, but resistant to the effects of trypsin, chymotrypsin, and sugar-cleaving enzymes (SCHIFFMAN et al., 1975a). It has been unexpectedly difficult to isolate sufficient quantities of the bacterial chemotactic factor to permit structural analysis. Preliminary studies have suggested the presence of at least one peptide with the following amino acid content—serine, glutamic, and aspartic acids and alanine. However, these studies have also provided evidence that the bacterial preparations additionally contain a nonpeptide chemotactic factor that is pronase-resistant. These studies should not be confused with evidence that some bacteria release enzymes which, upon interaction with the complement system, can generate additional chemotactic activity from the third (C3) and the fifth (C5) components of complement (WARD et al., 1973).

The most rigorously studied plasma-derived chemotactic factors relate to the complement, the kinin-forming, and fibrinolytic systems. The latter two categories include the enzymes, kallikrein, and the plasminogen activator (see above). Both of these enzymes can be generated in human plasma by the addition of activated Hageman factor (factor XII). KAPLAN et al. (1973, 1974) have demonstrated two distinctly different chemotactic agents by very careful chromatographic isolation from human plasma. They have also shown that, if the active site in either kallikrein or the plasminogen activator is blocked by diisopropylfluorophosphate, the chemotactic activity is also lost. Natural serum inhibitors of these two chemotactically active enzymes include the C1 inhibitor (C1INH) and α_2 macroglobulin. Inhibition of the enzymatic activity by these natural inhibitors is associated with concomitant loss of chemotactic activity (GOETZL and AUSTEN, 1974). It is uncertain to what extent the chemotactically active enzymes are important in biological systems. The recognition of humans with a deficiency of the kallikrein precursor (Fletcher factor deficiency) may provide some answers to this question.

The complement system (MÜLLER-EBERHARD, 1975) appears to be a major source of chemotactic factors in the human and in animals. Three complement-derived chemotactic factors have been described—fragments from the third (C3) and the fifth (C5) components of complement, and a high molecular weight compound, C567, consisting of the larger, residual portion of C5, and the essentially intact sixth (C6) and seventh (C7) components (WARD, 1974). The C567 complex can be generated in serum by the addition of immune complexes, and the amount of C5 in the mixture seems to determine whether more or less of the chemotactic activity generated will be in the form of the complex (C567) or in the form of C5 fragment (WARD, 1969). Some question has arisen regarding the significance of the C567 complex in view of the fact that it is possible to generate chemotactic activity in C6-deficient rabbit serum (KELLER and Sorkin, 1968). However, the finding of the high molecular as well as lower molecular weight chemotactic factors derived from C5 in ongoing inflammatory reactions (see below) suggests that both factors are biologically (Ward and Hill, 1972). The chemotactic activities of fragments derived from C3 and C5 have been repeatedly demonstrated. C3 chemotactic fragments have been generated by the action of plasmin (Ward, 1967), trypsin (Bokisch et al., 1969), thrombin (Bokisch et al., 1969), and neutral proteases in tissue extracts Hill and Ward (1969). None of the fragments has been chemically well characterized, although the plasmin-produced fragment is acidic in charge while, at least, one of the trypsin fragments is basic in charge. The molecular weight of all of these fragments is approximately 10,000. The C5 chemotactic fragments can be obtained as a result of cleavage produced by trypsin (Ward and Newman, 1969), neutral proteases from lysosomal granules of neutrophils (Ward and Hill, 1970) and from virusinfected tissues (Brier et al., 1970); and the acidic protease present in macrophages (Table 1). The C5 fragments seem to be more variable in their molecular weights, probably averaging 15,000 (Ward and Newman, 1969). Their charge(s) has not been determined in most cases. None of the chemotactic fragments, either C3 or C5, have been unequivocally structurally defined. It is possible to generate chemotactically active C3 or C5 fragments by appropriate activation of serum as well as by proteolysis of the isolated C3 or C5 molecules. Incubation of human serum with activators of the classical complement pathway or activation of the alternative (properdin) complement pathway by agents such as zymosan or the cobra venom factor generates the chemotactic C5 fragment (Snyderman, 1970; Till et al., 1975). Under none of these conditions has a C3 chemotactic fragment been demonstrated in serum. If, however, serum is pretreated with high concentrations of ε-aminocaproic acid, which blocks the anaphylatoxin inactivator (see below), the generation in serum of either a C3 or a C5 chemotactic fragment will occur, depending upon the method of activation of serum (Vallota and Müller-Eberhard, 1973).

Table 1. Enzymes responsible for generation of complement-dependent chemotactic factors

Chemotactic factor	Enzymes
C3 fragment	C3 activator (of alternate complement pathway)
	Plasmin
	Trypsin
	Thrombin
	Neutral protease of tissue
C5 fragment	C4̄2̄3̄
	Alternate complement pathway enzyme (not known)
	Trypsin
	Neutral protease of neutrophil lysosomal granules
	Acidic tissue protease
C5̄6̄7̄	C4̄2̄3̄
	Trypsin

Considerable controversy has arisen regarding whether the anaphylatoxin fragments (C3a and C5a) from C3 and C5 are chemotactically active. The resolution of this question is made difficult by the different ways in which C3a and C5a are produced. Limited trypsin cleavage of the purified complement components has been the obvious approach, but this has been replaced by the current practice of isolating C3a and/or C5a from treated, whole human serum in the presence of ε-aminocaproic acid (VALLOTA and MÜLLER-EBERHARD, 1973). The complete primary structure has been published for the C3a molecule isolated from whole serum (HUGLI, 1975). This peptide has a molecular weight of approximately 9,000, is very basic in charge, containing 11 residues of arginine and 7 of lysine, and has serine and arginine as the amino and the carboxyl terminal residues, respectively. C3a isolated from porcine serum is remakably similar in structure. It is not known if the trypsin-produced C3a has an identical primary structure, but the answer will probably be positive. Recent studies by FERNANDEZ et al. (1976) suggest that purified C3a is lacking in chemotactic activity. It has been shown that the generation of the C3 chemotactic factor from the purified component occurs only *after* the C3a anaphylatoxin activity has appeared and then been lost as a result of additional trypsin treatment (TILL et al., 1974). This would suggest that the C3 chemotactic fragment is a smaller portion of the larger C3a fragment, or that the C3 chemotactic fragment arises from an area of the C3 molecule different from the site giving rise to the C3a molecule.

There seems to be general agreement that anaphylatoxin preparations of C5a are chemotactically active (COCHRANE and MÜLLER-EBERHARD, 1968; JENSEN, 1972; VOGT et al., 1971; SHIN et al., 1968), although much controversy swirls around the nature of the chemotactic activity. WISSLER et al. (1972) contend that C5a preparations contain an accompanying small peptide, not detectable by the usual physical-chemical parameters, and that this cofactor causes conformational changes in the C5 fragment to permit the chemotactic activity to be expressed. It is further argued that, in the usual isolation procedures for C5a, adjustment of the mixture to an acid pH (3.5) accomplishes a tertiary confirmational change, which can also be induced by the cofactor peptide. This controversy seems unresolvable at the present time, but it does not seem unreasonable to assume that C5a, per se, must undergo a subtle change, not involving an alteration in its primary structure, before its chemotactic activity can be expressed. The ability to generate a chemotactic activity can be expressed. The ability to generate a chemotactically active C5 fragment in whole human serum (under conditions in which the anaphylatoxin inactivator has not been blocked) fails to result in the appearance of C5 anaphylatoxin activity. This means either a new C5 fragment (different from C5a) has been generated, or the bioassay for anaphylatoxin is considerable less sensitive than the assay for the chemotactic factor.

As is apparent from what has been pointed out above, tissue breakdown products, per se, do not seem to account for the presence of chemotactic activity. Most of the leukotactic activity, appearing coincident with tissue injury, is probably due to the release of C3 or C5 cleaving enzymes. The only known *chemotactically* active products, directly derived from tissue breakdown, are

Table 2. Specificity of leukotactic factors

Chemotactic factor	Leukocyte target
C5 fragment	Neutrophil, eosinophil, basophil, monocyte
C3 fragment	Neutrophil, monocyte
C567	Neutrophil
Bacterial factors	Neutrophils, monocytes
Lymphocyte products	Neutrophils[a], basophils, monocytes, lymphocytes
Kallikrein, plasminogen activator	Neutrophils, monocytes

[a] Chemotactic activity for neutrophils is quantitatively less than the amount of activity for the other leukocytes.

the fragments from collagen (CHANG and HOUCK, 1970). In addition, fibrin breakdown products have also been described as being chemotactically active (STECHER and SORKIN, 1972).

All of the chemotactic factors described above are active for neutrophils. The C3 and the C5 fragments also have chemotactic activity for monocytes and eosinophils (WARD, 1974). These observations are summarized in Table 2. Basophils, likewise, respond to the C5 fragment (WARD et al., 1975). Thus, products of the complement system are active for the entire granulocytic series, although the relative responses of each cell type to a given concentration of fragment has not been established.

A different type of chemotactic factor for neutrophils has been described by HAYASHI et al. (1974). This group has described a neutral, SH-dependent protease that acts on human IgG_2 and IgG_4, mouse IgG, and rabbit "fast" IgG, to generate chemotactic activity. All of these immunoglobulins are relatively papain-resistant, i.e., they are not readily hydrolyzed into the conventional Porter Fab fragments. The chemotactic factor of HAYASHI et al. (1974) has an estimated molecular weight of 140,000 and a sedimentation coefficient of 6.58S. Dialyzable, nonchemotactic products have been found in the IgG preparations that have been chemotactically activated, indicating a partial hydrolysis of the Ig molecule. The IgG-related chemotactic factor, termed "leukoegresin", induces neutrophil accumulation upon injection into animal skin, and has been isolated in extracts of 12-hour-old Arthus sites (HAYASHI et al., 1974). It is postulated that leukoegresin is produced by the action of a locally released SH-dependent neutral protease derived from connective tissue cells. An interesting, parallel story has been developed with a neutral SH-dependent protease, derived from lysosomal granules of the neutrophil. This enzyme degrades rabbit IgM into small fragments (mol wt approximately 14,000) which are said to be chemotactic for rat lymphocytes (HAYASHI et al., 1974).

Factors chemotactically active for monocytes, eosinophils, and basophils have been recently reported (WARD, 1974). Perhaps the most unusual factors relate to eosinophils. Two different chemotactic factors have been described with functional specificity for the eosinophil—a small peptide and a larger product of stimulated lymphoid cells. The former, termed the eosinophil chemotactic factor of anaphylaxis (ECF-A), was found in fluids after incubation of sensitized lung slices with antigen, and in perfusion fluids from antigen-stimulated sensitized lung (KAY et al., 1971). ECF-A is a tetrapeptide with one of two amino acid sequences — Val-Gly-Ser-Glu and Ala-Gly-Ser-Glu (GOETZL and AUSTEN, 1975). These peptides are active at micromolar concentrations. Although basophils were thought to be resevoirs of ECF-A, there are now data indicating that ECF-A (or its functional equivalent) also resides in lysosomal granules of neutrophils and is released during phagocytosis (KONIG et al., 1976). This would indicate that anaphylactic responses are not a unique condition leading to release of ECF-A. This fact and the two different sources of ECF-A, may be indications for dropping the anaphylactic designation (-A) in ECF-A.

It has been recently reported that, at low concentrations, histamine is a chemoattractant for eosinophil (CLARK et al., 1975). Neutrophils do not respond to this factor, and the eosinophil response is abolished by the presence of histaminase. The chemotactic response of the eosinophil cannot be blocked by H-1 or H-2 type antihistaminic drugs.

Another eosinophil-specific chemotactic factor is elaborated from antigen-stimulated lymphoid cells (COHEN and WARD, 1971; TORISU et al., 1973). As such, this factor can be considered in the category of the lymphokines. However, this eosinophil chemotactic factor (ECF is released in precursor form ECF_p) from sensitized guinea pig lymphoid cells, cultured in the presence of specific antigen. The generation of ECF_p, like delayed-type hypersensitivity reactions, is carrier specific. That is, the hapten must be presented with the protein to which it was attached when the animals were originally immunized. Linkage of the hapten to an unrelated carrier protein will fail to cause stimulation of lymphoid cells. ECF_p can be converted to the active form, ECF, by incubation with immune complexes containing homologous antibody and antigen. Immunoabsorbant columns with antibody to the antigen (which had induced the sensitized lymphoid cells to release both ECF_p and MIF) removed ECF_p but not MIF from the lymphoid cell culture fluids. Antigen coated columns affected neither activity. On the other hand, ECF, which had been activated by incubation with homologous immune complexes (see above), could not be removed by either immunoabsorbant. On the basis of these findings, it has been postulated that ECF_p coprecipitates with immune complexes because of an antigenic fragment contained within ECF_p, and that removal of this fragment is responsible for conversion of ECF_p into the active form, ECF. Precisely why the antibody must be in the form of an immune complex, and why antibody alone will not produce the same result is unclear. An additional, alternative explanation is that the ECF_p fluids contain an inhibitor that is antibodylike, similar to a rheumatoid factor, and that immune complexes remove this material. However, the data from the immunoabsorbant columns are difficult to explain if this alternative hypothesis is correct.

It should also be pointed out that an eosinophil-specific chemotactic factor can be generated in fresh guinea pig serum by incubation with immune complexes (KAY, 1970). This factor has been separated from the neutrophil-specific chemotactic factor present in the same mixture. On the basis of gel filtration, the eosinophil factor appears to be of lower molecular weight when compared with *the chemotactic factor for neutrophils*. The chemical nature of this eosinophil factor is not known and, whether it bears any structural relationship to the neutrophil chemotactic factor, cannot be answered.

These studies do suggest an amazingly complex series of events involving cell-cell interactions (lymphocytes-eosinophils), the role of at least one soluble, antigen-specific mediator (ECF_p), an interaction with immune complexes before the initial trigger (antigen), and the final event (accumulation of eosinophils). The findings might explain the requirement for an intact lymphoid system in the eosinophil accumulation (in blood and in tissues) of experimental trichinosis (BASTEN and BEESON, 1970), and the long known association between the presence of lymphoid cells and eosinophils.

V. In Vivo Demonstration of Chemotactic Factors

Chemotactic factors for neutrophils and for monocytes have been demonstrated in biological fluids from animals as well as man (Table 3). Both complement-derived, and lymphocyte-derived factors have been isolated from these fluids. In experimental animals with developing reversed passive Arthus reactions, chemotactic activity accumulates over a 2–3 h period. It is shortly after the appearance of this leukotactic activity that accumulation of neutrophils in tissues is seen. The tissue extracts contain both the C5 chemotactic fragment as well as the C$\overline{567}$ complex (WARD and HILL, 1972). Blocking of the complement system, which suppresses development of the Arthus reaction (WARD and COCHRANE, 1965), precludes the appearance of chemotactic activity in the tissue extracts, and no inflammatory reaction occurs. In mice, the C5 chemotactic fragment has been found in peritoneal fluids of animals injected intraperitoneally with bacterial endotoxin, presumably, following activation of the alternate complement pathway. In similarly treated C5 deficient mice, no chemotactic factor was found in the peritoneal fluids, and the intensity of the inflammatory cellular reaction was markedly diminished (SNYDERMAN et al., 1971). In humans, the C5 chemotactic fragment and C$\overline{567}$ have been demonstrated in a majority of synovial fluids from patients with rheumatoid arthritis (WARD and ZVAIFLER, 1971).

Table 3. Leukotactic factors in biological reactions

Reaction	Factor
Reversed passive Arthus reaction	C5 fragment, C$\overline{567}$
Human synovial fluids:	
rheumatoid arthritis	C5 fragment, C$\overline{567}$
inflammatory nonrheumatoid arthritis	C3 fragment
Delayed reaction (in skin, peritoneum)	Macrophage (monocyte) chemotactic factor
Myocardial infarct	C3 fragment

C3 chemotactic fragments have been found in vivo in two situations—in synovial fluids of human patients with inflammatory non-rheumatoid arthritis (WARD and ZVAIFLER, 1971), and in the developing myocardial infarct of the rat (HILL and WARD, 1971). In the latter case, it has been shown that the C3 fragment is generated following the release from injured tissue of a C3-cleaving enzyme. Treatment of rats with the C3 inactivator prior to ligation of the coronary artery prevents the appearance of the C3 chemotactic fragments in tissue and blocks the development of the acute (neutrophil-rich) cellular inflammatory reaction to the tissue injury. In a completely different animal model, depletion of C3 from the serum has suppressed the appearance of neutrophils at sites of surgical wounds (WEINER et al., 1973).

Eosinophil chemotactic activity (ECF-A) has been demonstrated in perfusion fluids of guinea pig lung following exposure to antigen (the lungs had been sensitized with anaphylactic-type antibody) (KAY et al., 1971), and in extracts of a malignant pulmonary tumor that was associated with the presence of eosinophils in and around the tumor (WASSERMAN et al., 1974). In this case, it was postulated that ECF-A was being produced by the tumor cells, analogous to ectopic hormone production by malignant pulmonary tumors.

The monocyte-specific chemotactic factor, which is presumably the product of antigen-stimulated lymphoid cells, has been demonstrated in developing skin reactions of delayed-type hypersensitivity in the guinea pig (COHEN et al., 1973). The tissue extracts were virtually devoid of chemotactic activity for neutrophils. The monocyte chemotactic activity in the tissue extract had physical-chemical features similar to those of monocyte chemotactic activity present in culture fluids from antigen-stimulated lymphoid cells (WARD et al., 1969). In other experiments, noncomplement-dependent chemotactic activity for monocytes has been found in peritoneal fluids of sensitized guinea pigs, following intraperitoneal challenge with antigen (POSTLETHWAITE and SNYDERMAN, 1973). Recently, in a form of graft versus host reaction, in which thoracic duct lymphoid cells are injected subcapsularly into the kidney, it has been demonstrated that the injection of parentaltype lymphocytes, but not syngeneic (F_1-hybrid) lymphocytes, results in the appearance of extractable monocyte chemotactic factor (WARD and VOLKMAN, 1976). This material is similar to that produced in mixed lymphocyte cultures, and the appearance of the chemotactic factor in the kidney roughly parallels the development of the monocyte-lymphocyte infiltrate that appears in the interstitium of the kidneys. These data provide additional support for the role of lymphocyte-derived monocyte chemotactic factors in cell-mediated immune reactions.

VI. Regulation of Chemotactic Systems

Regulation of chemotactic systems can occur indirectly by depletion of substrates necessary for generation of leukotactic factors. However, the most important *regulatory mechanisms* appear to be related to the naturally occurring leukotactic

Table 4. Comparisons between natural regulators of leukotaxis

Characteristic	Chemotactic factor inactivator (CFI)	Cell-directed inhibitor (CDI)	Neutrophil neutral-izing factor (NF)
Source	Plasma/serum	Plasma/serum	Neutrophil
Target of inhibition	Chemotactic factors	Neutrophil, monocyte	Neutrophil, eosinophil
Nature of inhibition	Amino peptidase	Unknown	Unknown
Function blocked	Leukotactic response	Leukotactic and phagocytic responses	Random and chemo-tactic motility
Electrophoretic position in serum	α- and β-globulins	Unknown	Unknown
Estimated mol wt	60,000 and 150,000	150,000 and >150,000	5000

inhibitors found in human serum. The chemotactic systems for neutrophils and monocytes are under regulatory control by chemotactic factor-directed and cell-directed inhibitors (Table 4). The former inhibitor is also known in the chemotactic factor inactivator (CFI), and it consists of a substance present in low concentration in normal human serum (BERENBERG and WARD, 1973). CFI irreversibly inactivates complement-derived (C3 and C5 fragments, $\overline{C567}$), chemotactic factors, chemotactic factors derived from bacteria, and the mono-cyte chemotactic factor present in lymphokinerich fluids (WARD and BERENBERG, 1974). CFI also inactivates migration inhibitor factors (MIF), produced by antigen-stimulated lymphoid cells (WARD and ROCKLIN, 1975). CFI has amino-peptidase activity that progressively cleaves chemotactic peptides, starting from the N-terminal amino acid (WARD and OZOLS, 1976). Direct evidence for the biochemical action has been obtained by the use of synthetic di and tripeptides that are exquisitely active (10^{-10} M) as chemotactic factors. CFI exists in human serum in two forms, as an α-globulin (sedimentation velocity of approximately 4S), and as a β-globulin (sedimentation velocity of approximately 7S). Both CFIs have the amino peptidase activity described above, although substrate specificity has been demonstrated using the C3 and C5 chemotactic fragment. The α-globulin CFI inactivates the C5 chemotactic fragment, while the β-globulin CFI inactivates the C3 fragment.

Elevations in serum levels of CFI occur in a variety of disease states, many of which are associated with defective (anergic) cellular inflammatory responses to intradermal stimuli such as *Candida* extract, streptokinase/streptodornase, mumps antigen, trichophytin extract, and nonspecific irritants, such as croton oil. Diseases associated with elevated serum levels of CFI include Hodgkin's diseas (WARD and BERENBERG, 1974), hepatic cirrhosis (MADERAZO et al., 1975), sarcoid (MADERAZO et al., 1976a), and lepromatous leprosy (WARD et al., 1976). In experimental animal studies with the Walker carcino-sarcoma tumors in rats, CFI has been found in the ascitic fluids associated with the tumor cells (BROZNA and WARD, 1975). This may explain the lack of an inflammatory response to the tumor in experimental animals.

A second regulator of chemotaxis is the cell-directed inhibitor (CDI), a substance also present in low concentrations in normal human serum (MADERAzo et al., 1976b). In contrast to CFI, CDI is heat-stable (56° C, 2 h), has a biphasic sedimentation velocity when isolated from human serum (7S and > 7S), and is cell-directed in its action. Incubation of human neutrophils as well as monocytes depresses the responses of these cells to a variety of chemotactic stimuli. Also, in the neutrophil, CDI blocks the phagocytic activity as determined by quantitative measurements of uptake' of immune precipitates. That both chemotactic and phagocytic responses are blocked by CDI implies a common, susceptible mechanism related both to chemotaxis as well as to phagocytosis.

CDI was originally discovered because of an unexplained inhibitor in the serum of a patient with *Listeria* meningitis (MADERAZO et al., 1976b). It has now been found that nearly 70% of all human patients with malignant tumors (carcinomas represent the most frequently studied class of tumor) manifest a leukotactic defect that is due to the presence of high levels of CDI (ANTON et al., 1976). The well-known propensity of cancer patients to show defects in expressing delayed-type hypersensitivity (including contact-type hypersensitivity) skin reactions may well be related to abnormally high levels of CDI in their plasmas. On the basis of what is presently known about CDI, it would be expected that both acute (neutrophil) as well as chronic (delayed-type hypersensitivity) cellular inflammatory reactions would be depressed. These findings also have implications for the use of immunotherapy, the purpose of which may be to increase the intensity of the inflammatory stimulus in an effort to overcome the leukotactic abnormality.

One additional regulator, with more restricted activity, has been described by GOETZL and AUSTEN (1972, 1974a). This inhibitor, termed the neutrophil-immobilizing factor (NIF), is contained within neutrophils, and is released with the culture medium during exposure of neutrophils to bacterial endotoxin or phagocytic particles, and during acidification. NIF is thought to be a small compound (mol wt about 5,000), is heat-stable, and can block chemotactic (as well as random) motility of neutrophils and eosinophils to a variety of factors. However, NIF has no effect on the chemotactic responses of monocytes. It is as yet uncertain what role NIF has as a natural regulator of inflammatory responses.

Alterations in leukocytes, leading to the state of "deactivation" (see above) might also be considered a form of regulation. Deactivation is the loss of chemotactic responsiveness by leukocytes, not only to the factor with which they have been in contact (WARD and BECKER, 1968), but, also, to other chemotactic factors. Deactivation requires contact of leukocytes with the chemotactic factors at relatively high concentration and for a finite period of time. Whether deactivation represents an important "shut-off" mechanism for the inflammatory response is unknown.

References

Adler, J.: Chemotaxis in bacteria. Ann. Rev. Biochem. **44**, 391 (1975)

Anton, T., Maderazo, E.G., Ward, P.A.: Leukotactic Dysfunction in Humans with Cancer. In press (1976)

Atherton, A., Born, G.V.R.: Quantitative investigations of the adhesiveness of circulating polymorphonuclear leukocytes to blood vessel walls. J. Physiol. (Lond.) **22**, 447 (1972)

Bandmann, U., Norberg, B., Rydgren, L.: Polymorphonuclear leukocyte chemotaxis in Boyden chambers. Scand. J. Haematol. **13**, 305 (1974b)

Bandmann, U., Rydgren, L., Norberg, B.: The difference between random movement and chemotaxis. Exp. Cell Res. **88**, 63 (1974a)

Basten, A., Beeson, P.B.: Mechanism of eosinophilia II. Role of the lymphocytes. J. Exp. Med. **131**, 1288 (1970)

Becker, E.L.: The relationship of the chemotactic behavior of the complement-derived factors C3a, C5a and C567 and a bacterial chemotactic factor to their ability to activate proesterase 1 of rabbit polymorphonuclear leukocytes. J. Exp. Med. **135**, 376–387 (1972)

Becker, E.L.: Enzyme activation and the mechanism of neutrophil chemotaxis. *In* Chemotaxis Its Biology and Biochemistry. Sorkin, E. (ed.). Basel: S. Korger 1974, p. 409

Becker, E.L.: Enzyme activation and the mechanism of polymorphonuclear leukocyte chemotaxis. *In* Cell in Host Resistance. Bellanti, J.A., Dayton, D.H. (eds.). New York: Raven Press, 1975, pps. 1–14

Becker, E.L.: Some inter-relationships among chemotaxis, lysosomal enzyme secretion and phagocytosis. Nobel Symposium. Molecular and Biological Aspects of the Acute Allergic Reaction. In press

Becker, E.L., Davis, A.T., Estenson, R.D., Quie, P.G.: Cytochalasin B IV. Inhibition and stimulation of chemotaxis of rabbit and human polymorphonuclear leukocytes. J. Immunol. **108**, 396 (1972)

Becker, E.L., Henson, P.M.: In vitro studies of immunologically induced secretions of mediators from cells and related phenomena. Advanc. Immunol. **17**, 93 (1973)

Becker, E.L., Showell, H.J.: The effect of Ca^{2+} and Mg^{2+} on the chemotactic responsiveness and spontaneous mobility of rabbit polymorphonuclear leukocytes. Z. Immun. Forsch. **143**, 466 (1972)

Becker, E.L., Showell, H.J.: The ability of chemotactic factors to induce lysosomal enzyme release II. The mechanism of the release. J. Immunol. **112**, 2055 (1974)

Becker, E.L., Showell, H.J., Henson, P.M., Hsu, L.S.: The ability of chemotactic factors to induce lysosomal enzyme release. 1. The characteristics of the release, the importance of surfaces and the relation of the release to chemotactic responsiveness. J. Immunol. **112**, 2047 (1974)

Becker, E.L., Ward, P.A.: Partial biochemical characterization of the activated esterase required in complement dependent chemotaxis of rabbit leukocytes. J. exp. Med. **125**, 1021 (1967)

Becker, E.L., Ward, P.A.: Enzymatic mechanisms concerned in the complement induced chemotaxis of rabbit polymorphonuclear leukocytes. In: Vth. Internat. Immunopath. Symp. Grabar, P., Miescher, P. (eds.). New York and London: Grune and Stratton, 1967, p. 189

Berenberg, J.L., Ward, P.A.: The chemotactic factor inactivator in normal human serum. J. clin. Invest. **52**, 1200 (1973)

Bhisey, A.N., Freed, J.J.: Altered movement of endosomes in colchicine-treated culture macrophages. Exp. Cell Res. **64**, 430 (1971)

Bokisch, V.A., Müller-Eberhard, H.J., Cochrane, C.G.: Isolation of a fragment (C3a) of the third component of human complement containing anaphylatoxin and chemotactic activity and description of an anaphylatoxin inactivator of human serum. J. exp. Med. **129**, 1109 (1969)

Borel, J.F., Staehlein, H.: Effects of cytochalasin B on chemotaxis and immune reactions. Experienta (Basel) **28**, 745 (1972)

Boxer, L.A., Hedley-White, E.T., Stossel, T.P.: Neutrophil actin dysfunction and abnormal neutrophil behavior. New Engl. J. Med. **291**, 1093 (1974)

Brier, A.M., Snyderman, R., Mergenhager, S.E., Notkins, A.L.: Inflammation and herpes simplex virus: release of a chemotaxis-generating factor from infected cells. Science **170**, 1104 (1970)

Brozna, J., Ward, P.A.: Anti-leukotactic properties of tumor cells. J. clin. Invest. **56**, 616 (1975)

Bryant, R.E., DePrez, R.M., van Way, M.H., Rogers, D.E.: Studies on leukocyte motility. 1. Effects of alterations in pH, electrolyte concentration and phagocytosis on leukocyte migration, adhesiveness and aggregation. J. exp. Med. **124**, 483 (1966)

Caner, J.E.Z.: Colchicine inhibition of chemotaxis. Arthr. Rheum. **8**, 757 (1965)

Carruthers, B.M.: Leukocyte motility. I. Method of study, normal variation effect of physical alterations in environment and effect of codoacetate. Canad. J. Physiol. Pharmacol. **44**, 475 (1966)

Carruthers, B.M.: Leukocyte motility II. Effect of absence of glucose in the medium: effect of presence of deoxyglucose, dinitrophenol, puromycin antimycin D and trypsin on the response to chemotactic substance; effect of segregation of cells from chemotactic substance. Canad. J. Physiol. Pharmacol. **45**, 269 (1967)

Chang, C., Houck, J.C.: Demonstration of the chemotactic properties of collagen. Proc. Soc. exp. Biol. (N.Y.) **134**, 22 (1970)

Clark, R.A.F., Gallin, J.I., Kaplan, A.P.: The selective eosinophil chemotactic activity of histamine. J. exp. Med. **142**, 1462 (1975)

Clark, R.A.F., Kimball, H.R.: Defective granulocyte chemotaxis in the Chediak-Higashi syndrome. J. clin. Invest. **51**, 649 (1972)

Cochrane, C.G., Müller-Eberhard, H.G.: The derivation of two distinct anaphylatoxin activities from the third and fifth components of human complement. J. exp. Med. **127**, 371 (1968)

Cohen, S., Ward, P.A.: In vitro and in vivo activity of a lymphocyte and immune complex-dependent chemotactic factor for eosinophils. J. exp. Med. **133**, 133–146 (1971)

Cohen, S., Ward, P.A., Yoshida, T., Burek, C.L.: Biological activity of extracts of delayed hypersensitivity skin reaction sites. Cell. Immunol. **9**, 363 (1973)

Craddock, P.R., Yawata, Y., van Santen, L., Gilberstadt, S., Silvis, S., Jacob, H.S.: Acquired phagocytic dysfunction. A complication of the hypophosphatemia of parental hyperalimentation. New Engl. J. Med. **290**, 1403 (1974)

Edelson, P.J., Fudenberg, H.F.: Effect of vinblastine on the chemotactic responsiveness of normal human neutrophils. Infect. Immun. **8**, 127 (1973)

Estenson, R.D., Hill, H.R., Quie, P.G., Hogan, N., Goldberg, N.D.: Cyclic GMP and cell movement. Nature (Lond.) **245**, 458 (1973)

Fernandez, H., Henson, P., Hugh, T.E.: A single isolation procedure for obtaining both C3a and C5a from activated human serum. J. Immunol. In press (1973)

Gallin, J.I., Rosenthal, A.S.: The regulatory role of divalent cations in human granulocyte chemotaxis: Evidence for an association between calcium exchange and microtubule assembly. J. Cell Biol. **62**, 594 (1974)

Goetzl, E.J.: Modulation of human neutrophil polymorphonuclear leucocyte migration by human plasma alpha-globulin inhibitors and synthetic esterase inhibitors. Immunol. **29**, 163 (1975)

Goetzl, E.J., Austen, K.F.: A neutrophil immobilizing factor derived from human leukocytes. I. Generation and partial characterization. J. exp. Med. **136**, 1564 (1972)

Goetzl, E.J., Austen, K.F.: Active site chemotactic factors and the regulation of the human neutrophil chemotactic response. In: Chemotaxis: Its Biology and Biochemistry. Sorkin, E. (ed.). Basel: Karger, 1974a, p. 218

Goetzl, E.J., Austen, K.F.: Stimulation of human neutrophil leukocytic aerobic glucose metabolism by purified chemotactic factors. J. clin. Invest. **53**, 591 (1974b)

Goetzl, E.J., Austen, K.F.: Purification and synthesis of eosinophilotactic tetrapeptides of human lung tissue: Identification as eosinophil chemotactic factor of anaphylaxis. Proc. nat. Acad. Sci. (Wash.) **72**, 4123 (1975)

Goetzl, E.J., Wasserman, S.I., Gigli, I., Austen, K.F.: Enhancement of random migration and chemotactic response of human leukocytes by ascorbic acid. J. clin. Invest. **53**, 813 (1974)

Goldstein, I.M., Hoffstein, S., Gallin, J., Weissmann, G.: Mechanisms of lysosomal enzyme release from human leukocytes: microtubule assembly and membrane fusion induced by a component of complement. Proc. nat. Acad. Sci. (Wash.) **70**, 2916 (1973)

Hayashi, H., Yoshinaga, M., Yamamoto, S.: The nature of a mediator of leukocyte chemotaxis in inflammation. In: Chemotaxis: Its Biology and Biochemistry, Sorkin, E. (ed.). Basel: Karger, 1974, p. 296

Hill, H.R., Estenson, R.D., Quie, P.G., Hogan, N.A., Goldberg, N.D.: Modulation of neutrophil chemotactic responses by cyclic 3′,5′-guanosine monophosphate and cyclic 3′,5′ adenosine monophosphate. Metab. **24**, 447 (1975)

Hill, J.H., Ward, P.A.: C3 Leukotactic factors produced by a tissue protease. J. exp. Med. **130**, 505 (1969)

Hill, J.H., Ward, P.A.: The phlogistic role of C3 leukotactic fragments in myocardial infarcts of rats. J. exp. Med. **133**, 885 (1971)

Hoffstein, S.: Microtubule assembly and secretion in human polymorphonuclear leukocytes. Fed. Proc. **34**, 868 (1975)

Hollenberg, M.D., Cuatrecasas, P.: Insulin: Interaction with membrane receptor and relationship to cyclic purine nucleotides and cell growth. Fed. Proc. **34**, 1556 (1975)

Hsu, L.S., Becker, E.L.: Chemotactic factor and cytochalasin B induced volume changes in rabbit polymorphonuclear leukocytes. Amer. J. Path. **81**, 1 (1975)

Hugli, T.E.: Human anaphylatoxin (C3a) from the third component of complement. J. biol. Chem. **250**, 8293 (1975)

Jensen, J.A.: Anaphylatoxins. In: Biological Activities of Complement. Ingram, D. (ed.). Basel: Karger, 1972, p. 136

Kaplan, A.P., Goetzl, E.J., Austen, K.F.: The fibrinolytic pathway of human plasma. II. The generation of chemotactic activity by activation of plasminogen proactivator. J. clin. Invest. **52**, 2591 (1973)

Kaplan, A.P., Kay, A.B., Austen, K.F.: A prealbumin activator of prekallikrein. III. Appearance of a chemotactic factor for human neutrophils by the conversion of human prekallikrein to kallikrein. J. exp. Med. **135**, 81 (1974)

Kay, A.B.: Studies on eosinophil leucocyte migration. II. Factors specifically chemotactic for eosinophils and neutrophils generated from guinea-pig serum by antigen-antibody complexes. Clin. exp. Immunol. **7**, 723 (1970)

Kay, A.B., Stechschulte, D.J., Austen, K.F.: An eosinophil leukocyte chemotactic factor of anaphylaxis. J. exp. Med. **133**, 602 (1971)

Keller, H.U., Sorkin, E.: Studies on chemotaxis. V. On the chemotactic effect of bacteria. Int. Arch. Allergy **39**, 247 (1967)

Keller, H.U., Sorkin, E.: Chemotaxis of leukocytes. Experientia (Basel) **24**, 641 (1968)

Konig, W., Czarnetzki, B.M., Lichtenstein, L.M.: Generation and release of eosinophil chemotactic factor (ECF) from human peripheral leukocytes. Fed. Proc. **35**, 515 (1976)

Koza, E.P., Wright, T.E., Becker, E.L.: Lysosomal enzyme secretion and volume contraction induced in neutrophils by cytochalasin B, chemotactic factor and A23187. Proc. Soc. exp. Biol. (N.Y.) **34**, 173 (1975)

Maderazo, E.G., Ward, P.A., Quintilliani, R.: Defective regulation of chemotaxis in cirrhosis. J. Lab. clin. Med. **85**, 621 (1975)

Maderazo, E.G., Ward, P.A., Woronick, C.L., Kubik, J., De Graff, A.C.: Leukotactic dysfunction in sarcoidosis. Ann. intern. Med. **84**, 414 (1976a)

Maderazo, E.G., Ward, P.A., Woronick, C.L., Quintilliani, R.: A cell directed inhibitor of leukotaxis in human serum. J. Lab. clin. Med. In press (1976b)

Mowat, A.G., Baum, J.: Chemotaxis of polymorphonuclear leukocytes from patients with diabetes mellitus. New Engl. J. Med. **284**, 621 (1971)

Müller-Eberhard, H.J.: Complement. Ann. Rev. Biochem. **44**, 697 (1975)

Musson, R.A., Becker, E.L.: The effect of chemotactic factors on erythrophagocytosis by human neutrophils. Fed. Proc. **34**, 1019 (1975)

Naccache, P., Freer, P.J., Showell, H.J., Becker, E.L., Sha'afi, R.I.: Cation fluxes and chemotaxis. Fed. Proc. **35**, 2190 (abst.) (1976)

Oliver, J.M., Zurier, R.B., Berlin, R.D.: Concanavalin A cap formation on polymorphonuclear leukocytes of normal and beige (Chediak-Higashi) mice. Nature (Lond.) **253**, 471 (1975)

Phelps, P., MacCarty, J., Jr.: Crystal induced arthritis. Postgrad. Med. **45**, 87 (1969)

Postlethwaite, A., Snyderman, R.: Monocyte leukocyte chemotactic factor in vivo in delayed hypersensitivity. Fed. Proc. **31**, 988 (1973)

Ramsey, W.S.: Locomotion of human polymorphonuclear leukocytes. Exp. Cell Res. **72**, 489 (1972)

Ramsey, W., Harris, A.: Leukocyte locomotion and its inhibition by antimitotic drugs. Exp. Cell Res. **82**, 262 (1973)

Ramsey, W.S.: In: Chemotaxis: Its Biology and Biochemistry. Sorkin, E. (ed.). Basel: Karger, 1974a, p. 179

Ramsey, W.: Retraction fibers and leukocyte chemotaxis. Exp. Cell Res. **86**, 184 (1974b)

Rivkin, I., Becker, E.L.: Effect of exogenous cyclic AMP and other adenine nucleotides on neutrophil chemotaxis and motility. Int. Arch. Allergy **50**, 95 (1975)

Rivkin, I., Rosenblatt, J., Becker, E.L.: The role of cyclic AMP in the chemotactic responsiveness and spontaneous motility of rabbit peritoneal neutrophils. The inhibition of neutrophil movement and elevation of cyclic AMP levels by catecholamines, prostaglandins, theophylline and cholera toxin. J. Immunol. **115**, 1114 (1975)

Sandler, J.A., Gallin, J.I., Vaughn, E.: Effects of serotonin, carbamylcholine and ascorbic acid on leukocyte cyclic GMP and chemotaxis. J. cell. Biol. **67**, 480 (1975)

Schiffmann, E., Showell, H., Corcoran, B., Ward, P.A., Smith, E., Becker, E.L.: The isolation and partial characterization of neutrophil chemotactic factor from *Escherichia coli*. J. Immunol. **114**, 1831 (1975a)

Schiffmann, E.A., Corcoran, B.A., Wall, S.M.: Formylmethionyl peptides as chemoattractants for leukocytes. Proc. nat. Acad. Sci. (Wash.) **70**, 2916 (1975b)

Shin, H.W., Snyderman, R., Friedman, E., Mellors, A., Mayer, M.M.: Chemotactic and anaphylatoxic fragment cleaved from the fifth component of guinea pig complement. Science **122**, 361 (1968)

Showell, H.J., Becker, E.L.: The effects of external K^+ and Na^+ on the chemotaxis of rabbit peritoneal neutrophils. J. Immunol. **116**, 99 (1976)

Showell, H.J., Freer, R.J., Zigmond, S.H., Schiffman, E., Aswanikumar, S., Corcoran, B., Becker, E.L.: The structure activity relations of synthetic peptides as chemotactic factors and inducers of lysosomal enzyme secretion for neutrophils. J. exp. Med. In press 1976

Smith, C.W., Hollers, J.C., Bing, D.H., Patrick, R.A.: Effects of human CI inhibitor on complement-mediated human leukocyte chemotaxis. J. Immunol. **114**, 216 (1975)

Snyderman, R., Phillips, J., Mergenhagen, S.E.: Polymorphonuclear leukocyte chemotactic activity in rabbit serum and guinea pig serum treated with immune complexes: Evidence for C5a as the major chemotactic factor. Infect. Immunol. **1**, 521 (1970)

Snyderman, R., Phillips, J., Mergenhagen, S.E.: Biological activity of complement in vivo. Role of C5 in the accumulation of polymorphonuclear leukocytes in inflammatory exudates. J. exp. Med. **134**, 1131 (1971)

Stecher, V.J., Sorkin, E.: The chemotactic activity of fibrin lysis products. Int. Arch. Allergy **43**, 879 (1972)

Stossel, T.P.: Phagocytosis: Recognition and ingestion. Semin. Hemat. **12**, 83 (1975)

Temple, T.R., Snyderman, R., Jordan, H.V., Mergenhagen, S.E.: Factors from saliva and oral bacteria, chemotactic for polymorphonuclear leukocytes: their possible role in gingivalinflammation. J. Periodont. **41**, 71 (1970)

Till, G., Ward, P.A.: Two distinct chemotactic factor inactivators in human serum. J. Immunol. **114**, 843 (1975)

Torisu, M., Yoshida, T., Ward, P.A., Cohen, S.: Lymphocyte-derived eosinophil chemotactic factor. I. Studies on the mechanism of activation of the precursor substance by immune complexes. J. Immunol. **111**, 1450–1458 (1973)

Tse, R.L., Phelps, P., Urban, D.: Polymorphonuclear leukocyte motility in vitro. VI. Effect of purine and pyrimidine analogues: Possible role of cyclic AMP. J. Lab. clin. Med. **80**, 265 (1972)

Tsung, P.K., Kegeles, S., Becker, E.L.: Isolation of an acetyl DL phenylalanine β-naphthyl esterase from rabbit peritoneal polymorphonuclear leukocytes. Biochim. biophys. Acta **403**, 98 (1975a)

Tsung, P., Sakamoto, T., Weissmann, G.: Protein kinase and phosphatase from human polymorphonuclear leukocytes. Biochem. J. **145**, 437 (1975b)

Turner, S.R., Campbell, J.A., Lynn, W.S.: Polymorphonuclear leukocyte chemotaxis toward oxidized lipid components of cell membranes. J. exp. Med. **141**, 1437 (1975)

Vallota, E., Müller-Eberhard, H.J.: Formation of C3a and C5a anaphylatoxins in whole human serum after inhibition of the anaphylatoxin inactivator. J. exp. Med. **137**, 1109 (1973)

Vogt, W., Lufft, E., Schmidt, G.: Studies of the relation between the fifth component and complement and anaphylatoxinogen. Europ. J. Immunol. **1**, 141 (1971)

Ward, P.A.: A plasmin-split fragment of C's as a new chemotactic factor. J. exp. Med. **126**, 189–206 (1967)

Ward, P.A.: Chemotaxis of polymorphonuclear leukocytes. Biochem. Pharmacol. spec. Suppl. 99 (1968)

Ward, P.A.: The heterogeneity of chemotactic factors for neutrophils generated from the complement system. In: Cellular and Humoral Mechanisms in Anaphylaxis and Allergy. Movat, H.Z. (ed.). Basel and New York: Karger, 1969, p. 279

Ward, P.A.: Leukotactic factors in health and disease. Amer. J. Path. **64**, 521 (1971)

Ward, P.A.: Complement derived chemotactic factors and their interactions with neutrophilic granulocytes. In "Biological Activities of Complement", Fifth International Symposium of the Canadian Society of Immunology, edited by D. Ingram, pg. 108–116 (1972)

Ward, P.A.: Leukotaxis and leukotactic disorders. Amer. J. Path. **77**, 519–538 (1974)

Ward, P.A., Becker, E.L.: The deactivation of rabbit neutrophils by chemotactic factor and the nature of the activatable esterase. J. exp. Med. **127**, 693 (1968)

Ward, P.A., Berenberg, J.L.: Defective regulation of inflammatory mediators in Hodgkin's disease. Supernormal levels of chemotactic factor inactivator. New Engl. J. Med. **290**, 76–80 (1974)

Ward, P.A., Chapitis, J., Conroy, M.C., Lepow, I.H.: Generation by bacterial proteinases of leukotactic factors from human serum and C3 and C5. J. Immunol. **110**, 1003–1009 (1973)

Ward, P.A., Cochrane, C.G.: Bound complement and immunologic vasculitis. J. exp. Med. **121**, 215 (1965)

Ward, P.A., Data, R., Till, G.: Regulatory control of complement derived chemotactic and anaphylatoxin mediators. In: Progress in Immunology II. Brest, L., Holborow, J. (eds.). Amsterdam: North-Holland, 1974, Vol. I, pp. 209–215

Ward, P.A., Dvorak, H.F., Cohen, S., Yoshida, T., Data, R., Selvaggio, S.: Chemotaxis of basophils by lymphocyte-dependent and lymphocyte-independent mechanisms. J.Immunol.**114**,1523–1531 (1975)

Ward, P.A., Goralnick, S., Bullock, W.E.: Defective leukotaxis in patients with lepromatous leprosy. J. Lab. clin. Med. **87**, 1025 (1976)

Ward, P.A., Hill, J.H.: C5 chemotactic fragments produced by an enzyme in lysosomal granules of neutrophils. J. Immunol. **104**, 535 (1970)

Ward, P.A., Hill, J.H.: Biological role of complement products. Complement-derived leukotactic activity extractable from lesions of immunologic vasculitis. J. Immunol. **108**, 1137 (1972)

Ward, P.A., Lepow, I.H., Newman, L.J.: Bacterial factors chemotactic for polymorphonuclear leukocytes. Amer. J. Path. **52**, 725 (1968)

Ward, P.A., Newman, L.J.: A neutrophil chemotactic factor from human C'5. J. Immunol. **102**, 93–99 (1969)

Ward, P.A., Ozols, J.: The protease activity of the chemotactic factor inactivator. J. clin. Invest. **58**, 123 (1976)

Ward, P.A., Remold, H.G., David, J.R.: A leukotactic factor produced by sensitized lymphocytes. Science **163**, 1079–1081 (1969)

Ward, P.A., Rocklin, R.: Regulation of MIF by a factor in human serum. J. Immunol. **115**, 309–311 (1975)

Ward, P.A., Volkman, A.: The elaboration of leukotactic mediators during the interaction between parental-type lymphocytes and F_1 hybrid cells. J. Immunol. **115**, 1394–1399 (1975)

Ward, P.A., Zvaifler, N.J.: Complement derived leukotactic factors in inflammatory synovial fluids of humans. J. clin. Invest. **50**, 606 (1971)

Wasserman, S.I., Goetzl, E.J., Ellman, L., Austen, K.F.: Tumor-associated eosinophilotactic factor. New Engl. J. Med. **290**, 420 (1974)

Weiner, S.L., Lendair, S., Rogers, B., Urivetzsky, M., Meilwan, E.: Non-immune chemotaxis in vivo: inhibition by complement depletion with cobra factor. Amer. J. Path. **73**, 807 (1973)

Wilkinson, P.C.: Chemotaxis and Inflammation. Edinburgh: Churchill Livingstone, 1974a, p. 94

Wilkinson, P.C.: Surface and cell membrane activities of leukocytic chemotactic factors. Nature (Lond.) **251**, 58 (1974b)

Wilkinson, P.C.: Leukocyte locomotion and chemotaxis. Exp. Cell Res. **93**, 420 (1975a)

Wilkinson, P.C.: Inhibition of leukocyte motion and chemotaxis by lipid-specific-bacterial toxins. Nature (Lond.) **255**, 485 (1975b)

Wissler, J.H., Stecher, V., Sorkin, E.: Chemistry and biology of the anaphylatoxin related peptide system. III. Evaluation of leukocyte activity as the property of a new peptide system with classical anaphylatoxin and cocytalaxin components. Europ. J. Immunol. **2**, 90 (1972)

Zigmond, S.H.: Mechanisms of sensing chemical gradients by polymorphonuclear leukocytes. Nature (Lond.) **249**, 450 (1974)

Zigmond, S.H., Hirsch, J.G.: Effects of cytochalasin B on polymorphonuclear leukocyte locomotion, phagocytosis and glycolysis. Exp. Cell Res. **73**, 383 (1972)

Rev. Physiol. Biochem. Pharmacol., Vol. 77

Capillary Structures and O_2 Supply to Tissue *

An Analysis with a Digital Diffusion Model as Applied to the Skeletal Muscle

W. A. GRUNEWALD and W. SOWA

Contents

---

* Dedicated to: Prof. Dr. D. W. Lübbers, director of the Max-Planck-Institut für Systemphysiologie in Dortmund in honor of his 60th birthday.

Supported by the Deutsche Forschungsgemeinschaft.

I. Introduction

The development of the polarographic measurement of oxygen partial pressure (PO_2) to the micromethod has made possible a direct analysis of the oxygen supply to tissue. The local PO_2 in tissue is measured with membrane-covered needle electrodes that have a tip diameter of about 1 µm (Fatt, 1964; Silver, 1965, 1966; Lübbers and Baumgärtl, 1967; Kunze, 1966, 1969; Whalen et al., 1967; Lübbers et al., 1969; Bicher and Knisely, 1970; Günther et al., 1972; Baumgärtl and Lübbers, 1973; Vaupel et al., 1973; Baumgärtl et al., 1974). Local PO_2 measurements were made in the last few years by Silver (1965) in rabbit cortex; Lübbers (1967) in guinea-pig cortex; Bicher and Knisely (1970), Bicher et al. (1973) in dog and cat brains; Erdmann et al. (1969), Heidenreich et al. (1969), Metzger et al. (1970) in the rat and cat brains. Measurements in other organs that have been reported include those of Leichtweiss et al. (1969) and Günther et al. (1974) in the rat kidney; Vaupel et al. (1972), Vaupel (1974) in tumor tissue of rat kidney; Kunze (1969) in healthy as well as pathological human skeletal muscle; Kessler (1967) in rat and dog livers; Kadatz (1967) in dog heart muscle; Schuchhardt (1971a, b, c) in heart tissue from dogs, rats and guinea-pigs; Acker et al. (1970, 1971), Acker (1975), and Whalen and Nair (1973) in the carotid body in cats; Rodenhäuser et al. (1970) in the vitreous body of cat eye; and Mendler et al. (1973) in the human heart during cardiac surgery [1].

In experiments where steady-state conditions are at hand, a series of PO_2 values along the "puncture channel" results in a so-called PO_2 *profile*. From these PO_2 profiles, the beforementioned authors obtained information concerning the frequency of the PO_2 values measured, the so-called PO_2 *frequency distribution*, PO_2 differences between two points of measurement, the so-called PO_2 *gradient*, and the position of the needle electrode with respect to near-lying capillaries.

In experiments where nonsteady-state conditions are at hand, the resulting changes of the PO_2, at the measuring point of the electrode with respect to time gives us the so-called PO_2 *transient function*.

Before the development of this micromethod, oxygen supply models for tissue were developed in order to gain an insight into the basic laws of oxygen supply. Recently, models were also used to interpret the results of the PO_2 measurements in tissue. For *nonsteady-state processes*, Reneau et al. (1969) used the PO_2 transient functions that were calculated from the Krogh cylinder model and compared the ensuing results with the PO_2 transient functions that they measured in the cat cortex after changing arterial O_2 content. The capillary model that will be described in this work allows for an interpretation of the PO_2 measurements under *steady-state conditions* using the PO_2 frequency distribution. The models known to date are too schematic for such a comparison. The new capillary model is based upon microcirculatory units (MCUs) that allow for the simulation of capillary networks of variegated form. From the

---

[1] For further literature see: Lübbers et al. (1968), Kessler et al. (1973), and Bicher and Bruley (1973).

PO_2-fields of MCUs, PO_2 distributions can be determined which correspond to the *measured* distributions.

The diffusion model, especially the calculated PO_2 distributions, allow for *general* conclusions concerning the O_2 supply to tissue as well as for an interpretation of results for individual organs. As an example, the O_2 supply of the skeletal muscle is considered using the diffusion model. Furthermore, an analysis of the O_2 supply using the measured PO_2 distribution is possible based upon a large number of measured data.

II. Historical Development of the Capillary Models [2]

The development of models describing the O_2 supply to tissue began shortly after the turn of the century. Many of these models that are known to date can only be classified in a broad sense of the word as capillary models. In most instances, the capillary is considered as an isolated part of the capillary network so that the spatial relationship with other capillaries is lost. The reason for this method of approach is based upon the fact that the O_2 supply to tissue from an isolated capillary with its respective supply volume is represented by a model that can be mathematically comprehended for the most part in a closed form. The analysis of O_2 supply to tissue, taking into account the *spatial relationship* of capillaries to one another (the capillary structure) as is found in the capillary network of the individual organs, has become only possible through the use of a high-speed digital computer.

For all supply models, the basis of the O_2 transport from the capillaries to tissue is diffusion. Diffusion explains best the, at first, paradoxical situation whereby the blood, rich in O_2, leaves the capillary network of an organ, whereas in the tissue, an O_2 demand and even anoxic areas can exist. In capillary blood, the PO_2 level must be high enough to maintain a PO_2 gradient from capillary to tissue which is necessary for the O_2 transport by diffusion. The basis of all models that describe the O_2 supply to tissue is therefore Fick's first law and the derived diffusion equation. This equation is a partial differential equation. The method which is used to solve this diffusion equation allows for a subdivision of the models into three groups. In the *first* group, the boundary conditions are formulated in such a fashion that the solution of the differential equation in a closed form can be given as an analytical expression. For this group, the *cylinder model* from KROGH (1918/19 b) initiates the systematic analysis of the O_2 supply. KROGH used measurements from the diffusion properties of biological media and histological investigations of the capillary networks of different organs as the foundation of his model theory (KROGH, 1918/19a). He describes his model—represented in literature as the *Krogh cylinder* or *standard cylinder*—using the following assumptions: each capillary supplies its respective surrounding circular tissue cylinder; the oxygen diffusion from the capillary into

---

[2] For a more detailed synopsis of the mathematical description of many of the diffusion models noted in this chapter, see LEONARD and JØRGENSEN (1974).

the tissue is only dependent upon the radius; the PO_2 at the boundary between capillary and tissue is known and equals the mean PO_2 of the capillary cross-section; the diffusion coefficient and the O_2 consumption of the tissue are distributed homogeneously throughout the tissue cylinder and are constant; steady-state conditions are at hand. With these conditions, Krogh (1918/19b) and his mathematician Erlang were able to derive the following equation:

$$P(r_c) - P(r) = \frac{AR^2}{2D_K} \left(\ln \frac{r}{r_c} - \frac{r^2 - r_c^2}{2R^2} \right) \tag{1}$$

where $P(r_c)$ and $P(r)$ are the PO_2 at the capillary wall and at point r in the tissue cylinder respectively; A is the O_2 consumption of the tissue; D_k is the Krogh diffusion constant (later used as the "diffusion conductivity" and annotated as "K"); r_c is the capillary radius, R is the tissue cylinder radius.

Up until today, the isolated capillary with its tissue cylinder is the basis for most models. This Krogh tissue cylinder is modified and special conditions are assumed, whereby the mathematical description of the model becomes more complicated. From the multiple variations of the Krogh cylinder, only the most important will be considered.

Hill (1929) expanded the Krogh cylinder model in that he included to the constant O_2 consumption in muscle tissue an O_2 deficit of the tissue, that linearly increases with respect to time. He handled this nonsteady-state problem by considering the *advancing diffusion front* without solving the diffusion equation itself. Furthermore, he combined an O_2 diffusion into tissue with a lactic acid diffusion out of the tissue during constant lactic acid production. He modified the Krogh model by assuming that the tissue cylinder is supplied from the periphery (see also: Jacobs, 1935). In case the O_2 consumption of the tissue can be neglected, he calculates the nonsteady-state solution of the diffusion equation for the Krogh model.

Opitz (1948) and Opitz and Schneider (1950) transfer the Krogh model that was developed for muscle tissue to the O_2 supply of brain tissue. They indicated the influence of the longitudinal diffusion in the tissue cylinder by referring to the results published later by Thews (1953).

Roughton (1952) described the O_2 supply of the Krogh cylinder for nonsteady-state conditions without Hill's (1929) limitation of a neglectable O_2 consumption of the tissue. In addition to the constant O_2 consumption, he considered a first order reversible reaction between a diffusable and a nondiffusable substance in tissue (O_2 and myoglobin respectively).

Opitz and Thews (1952) calculated the O_2 supply of myocard fiber in nonsteady-state conditions. They considered that the myocard fiber is represented by a cylinder with a constant O_2 consumption and that such a cylinder is supplied from its periphery.

The longitudinal diffusion in the Krogh cylinder which Roughton (1952) neglects is dealt with by Thews (1953) in his mathematical essay concerning the diffusion processes in cylindrical objects. However, he limited this expansion of the Krogh model by assuming that the concentration of oxygen in blood decreases linearly from the arterial to the venous end of the capillary. Blum (1960) considers the Krogh cylinder assuming a finite permeability of the diffusable substance through the vessel wall. The concentration of the diffusable substance on the inner wall of the capillary is assumed to be equal to the mean concentration of the cross section of the capillary. Because of a difference in the arrangement of the cells in radial and longitudinal direction in the tissue cylinder and because of a difference in the diffusion properties of the intracellular and extracellular areas, Blum (1960) assumes an anisotropic diffusion coefficient in radial and longitudinal direction. He describes the consumption of the diffusable substance using the Michaelis-Menten-equation. The Blum modification of the Krogh cylinder is considered to be more applicable for large molecules. However, a transcription of his model to O_2 diffusion is possible if the PO_2 along the length of the capillary is assumed to be linear.

Thews (1960) in his consideration of O_2 diffusion in the brain modifies the radius of the Krogh tissue cylinder and calculates the PO_2 decrease across the capillary cross-section. In doing so, he assumes that the PO_2 gradients at the boundary blood/tissue are reciprocal to their respective diffusion coefficients (THEWS and NIESEL, 1959). GROTE and THEWS (1962) and THEWS (1962) transcribe this modification of the Krogh cylinder to the O_2 supply in myocardium.

HUDSON and CATER (1964) expanded the Krogh model by introducing in their considerations a PO_2 dependent O_2 consumption for the case where the tissue PO_2 drops below a critical level. They thereby considered a linear and exponential dependence of the O_2 consumption upon PO_2 and such a dependence described by TANG (1958), BÄNDER and KIESE (1955), and LONGMUIR (1957). They also assumed that, along with diffusion, an O_2 transport through the extracellular fluid flow from the arterial to the venous areas of the tissue cylinder occurs. They considered the nonstationary diffusion with the assumption that the PO_2 courses exponentially along the length of the capillary. For economical reasons, they give a calculation method for the O_2 supply of a hexagonal tissue cylinder instead of for the Krogh circular cylinder. A hexagonal tissue cylinder is also discussed by THEWS (1960).

Further modifications of the Krogh cylinder tend towards models that are no longer capable of being described in a closed mathematical form. Before we consider such models, a few other supply models should be mentioned which can only be considered in the broadest sense as capillary models. They belong to the group that still can be calculated in the closed form. WARBURG (1923), HILL (1929), and JACOBS (1935) investigated the O_2 supply in tissue layers. CALIGARA and ROOTH (1961) have in their *golf ball model* the idea that the capillaries in the subcutaneous tissue form a network which encompasses a spherical tissue volume. They assume a mean intracapillary PO_2 value which is evenly "spread" over the surface area of the sphere.

DIEMER (1963, 1965) leads to a new concept of the model form of O_2 supply due to his anatomical investigations of the capillary course in the brain. He found that along with the arterial part of the capillary, for the most part a venous part was at hand in the neighboring capillary. In his model, he simulates this fact in that he assumes two parallel-running capillaries which do not flow in the same direction (concurrent flow direction), but rather in *opposite* directions (countercurrent flow direction). From the assumption that the arterial capillary end supplies the tissue up to the neighboring venous capillary end, he concludes that under normal supply conditions the lowest PO_2 in the tissue is identical to the PO_2 at the venous capillary end. Assuming a radial symmetry, DIEMER arrives at a *supply cone*. Under special conditions, this supply cone can change to a *supply-truncated cone*. In this case, the venous end also takes part in the O_2 supply of the tissue. Disregarding further development of the Krogh model, DIEMER goes back with his assumptions to the original form of the tissue cylinder and uses Equation 1 for his calculations. The important new idea from DIEMER's conical model is that he considers the arrangement of the capillaries, although he limits himself to two capillaries and thereby remains with a two-dimensional problem. BAILEY (1967) had the same idea of a countercurrent flow direction in neighboring capillaries with his *double-layer model*. In this model, two capillary layers lie parallel to one another and have countercurrent flow directions. He assumes that an O_2 diffusion in the tissue between the blood-transporting layers is only possible perpendicular to the blood layers. BAILEY approximates the O_2 dissociation curve of hemoglobin by using two straight lines and considers the O_2 consumption of the tissue from the 0-th and the 1-st order. With this model, BAILEY analyzed the O_2 supply of tissue between the blood transporting capillary layers with concurrent and countercurrent flow directions.

A *second* group of O_2 supply models is found in the analog models. In these models, the tissue is replaced by a network of electrically switching elements (NIESEL and THEWS, 1959, 1962). THEWS (1962) simulates in this fashion the O_2 supply in myocardium during systole and diastole. The myocard fibers are thereby considered to be a Krogh cylinder with a time-dependent O_2 consumption.

EVANS and NAYLOR (1964) describe a two-dimensional analog model whereby they consider the change of the O_2 supply in tissue between two capillaries with concurrent and countercurrent flow directions after a change in the arterial PO_2 and at different values of the blood flow velocity.

A *third* group of supply models are the digital models. All of these models have in common that the differential equation for diffusion is replaced by a difference equation whose solution is approximatively calculated in an iteration process. With such an approximation, the boundary conditions for a capillary model can be generalized. The diffusion model that will be described in this paper uses this possibility by taking particularly into account the spatial position of the capillaries to one another, as well as the direction of capillary blood flow under general conditions. The development of this model began in 1965 (GRUNEWALD and LÜBBERS, 1966; GRUNEWALD, 1968, 1969, 1971).

Before this model is described in detail, a few other digital models should be considered. They are, for the most part, further modifications of the Krogh tissue cylinder. BAILEY (1967) calculated the O_2 supply in the Krogh cylinder taking into account the longitudinal diffusion but without the assumption of a linear PO_2 course along the length of the capillaries, to which THEWS and BLUM were forced in their models. BAILEY approximates the O_2 dissociation curve of hemoglobin using polynomes of the first and second order. The approximative calculation of the O_2 supply is done by using a two-dimensional grid point system. The error of the iteration process is controlled by special cases which can also be calculated in the closed form. RENEAU et al. (1967), KNISELY et al. (1969), and BRULEY and KNISELY (1970) expand the Krogh cylinder model by solving in an approximative fashion the diffusion equation for the combined blood/tissue system. They take into account that, in capillaries as well as in tissue, there is an axial as well as a radial diffusion gradient and they calculate the intracapillary PO_2 profile using the O_2 dissociation curve described by BARCROFT et al. (1922). With this model, the authors simulate the O_2 supply in the brain. RENEAU and KNISELY (1971) use this modification of the Krogh cylinder to consider the O_2 supply between two capillaries with countercurrent blood flow direction (similar to DIEMER). GONZALES-FERNANDES and ATTA (1968) expand the THEWS, BLUM and BAILEY-calculated Krogh tissue cylinder with longitudinal diffusion by describing the O_2 consumption of the tissue with the Michaelis-Menten equation. RENEAU et al. (1970) expand their digital model for the Krogh cylinder to include nonstationary conditions. In this model, they calculate PO_2 transient functions and investigate the behavior of the O_2 supply in brain cortex after changes in the arterial PO_2.

A digital model that is not derived from the Krogh tissue cylinder is described by METZGER (1967, 1972). He considers the O_2 supply in a quadratic capillary net in which, depending upon the arrangement of the arterial inflows and venous outflows, various flow behaviors can be realized. The blood flow in the capillaries is described by the Kirchhoff laws. The dependence of the oxyhemoglobin concentration upon the PO_2 is described by the Hill equation and the O_2 consumption in tissue by the Michaelis-Menten equation.

IWANOW and KISLIAKOW (1974) and KISLIAKOW and IWANOW (1974) modify the idea of spatially arranged parallel-running capillaries with differing blood flow directions (GRUNEWALD and LÜBBERS, 1968). They impose a spherical cell (nerve cell) with a high O_2 consumption between the spatially arranged capillaries. Using the capillary structure with a countercurrent blood flow direction, IWANOW and KISLIAKOW (1974) investigate the O_2 supply of nerve cells of the brain under normoxy and hypoxy conditions at differing values of the O_2 consumption, the capillary blood flow velocity, and the cell diameter.

III. Diffusion Model for Comprehension of the O_2 Supply from 3-Dimensional Capillary Structures

A. Model Concept and Assumptions

The model describes the O_2 supply to tissue from three-dimensional (in space) *capillary structures*. On one hand, the morphology of the tissue and its capillary network has to be taken into account; on the other hand, a certain abstraction is necessary for the mathematical description. This results in a compromise which defines the assumptions of the model. These assumptions can be divided into anatomical, physiological, and mathematical ones.

In order to comprehend the *anatomical* aspects of this problem, one has to consider the structure of the capillary network, i.e., the positioning of the capillaries to one another, the arrangement of their inflows and outflows, and thus the resulting flow directions. It is assumed that a capillary network and the tissue supplied by it, is divisible into *basic elements* (GRUNEWALD, 1971), also called *microcirculatory units* (LEONARD and JØRGENSEN, 1974). Such a microcirculatory unit (MCU) consists of a *tissue fragment* and of four capillaries of the length $2 \cdot l$ (Fig. 1). The length l is identical to the capillary length, i.e., the distance between arterial inflow and venous outflow of the capillary. The capillaries run rectilinear and parallel to one another. The shortest distance between two capillaries is the capillary distance d. The tissue fragment of a MCU is that volume V which is supplied by all four capillaries *together*. It has the form of a squared column with the edge lengths $2 \cdot l$ and d, and yields:

$$V = 2 \cdot d^2 \cdot l,$$

more exactly: (2)

$$V = 2 \cdot (d^2 - r_c^2 \pi) \cdot l.$$

r_c is the capillary radius; l, d, and r_c are the so-called structure parameters.

Assuming that the capillary network with its correspondingly supplied tissue is built from such MCUs, it follows that each capillary participates in the O_2 supply of four such tissue fragments. If $V_c^{(i)}$ ($i = 1 \ldots 4$) are those volume parts of a tissue fragment V which is commonly supplied by the four capillaries, then:

$$V = \sum_{i=1}^{4} V_c^{(i)}.$$ (3)

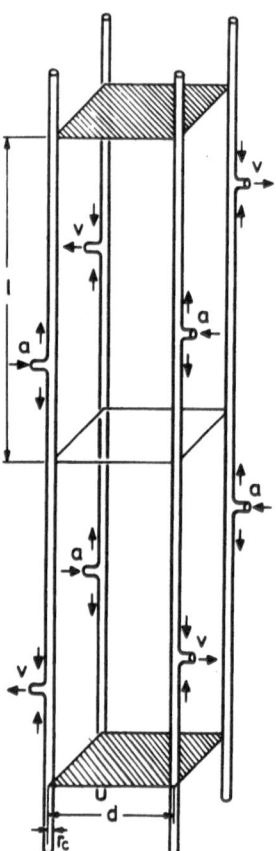

Fig. 1. Basic element (GRUNEWALD, 1971) also called microcirculatory unit (LEONARD and JØRGENSEN, 1974). A MCU consists of tissue fragment and of 4 rectilinear and parallel-running capillaries of length $2 \cdot l$, whereby l is identical to capillary length, i.e., distance between arterial inflow a and venous outflow v. Shortest distance between two capillaries is capillary distance d. r_c is capillary radius. SHORT *arrows* denote direction of blood flow in capillaries. Tissue fragment of MCU is that volume supplied by all 4 capillaries together. ▨ Upper and lower surface areas of tissue fragment

The capillary branchings, i.e., the arterial inflows a and the venous outflows v can be distributed arbitrarily along the length $2 \cdot l$ of the four capillaries that run on the edges of a tissue fragment.

It is assumed that on each capillary an arterial inflow always lies the length l away from a venous outflow. Hence, the natural asymmetry of the capillary network is almost completely reproduced.

If the capillaries in the MCU are so arranged that they run with the same direction of flow (MCU 1 in Fig. 2), we have the MCU with a concurrent flow (also called *Krogh capillary structure*). The capillary ends are situated in such a way that always arterial-arterial and venous-venous ends face one another at the distance d. If in two neighboring capillaries the flow direction is concurrent and in the two other capillaries is countercurrent (MCU 7 in Fig. 2), we have a MCU with a partial concurrent and countercurrent flow direction respectively.

However, if the capillary flow in two diagonally neighboring capillaries is opposite in direction (MCU 19 in Fig. 2), we have a MCU with a total counter-current flow. Thus, an arterial inflow always lies opposing a venous outflow with the capillary distance d. If this model is enlargened into a three-dimensional one, the MCU 19 likens DIEMER's model.

If the arterial inflows in a MCU are positioned in such a fashion that a spiral form with a $1/2$ shift results (MCU 16 in Fig. 2), then we have the so-called *helix structure* which is a particularly efficient MCU.

The above-named MCUs show a symmetry in their capillary arrangement. However, such symmetrical conditions are not a general assumption for the

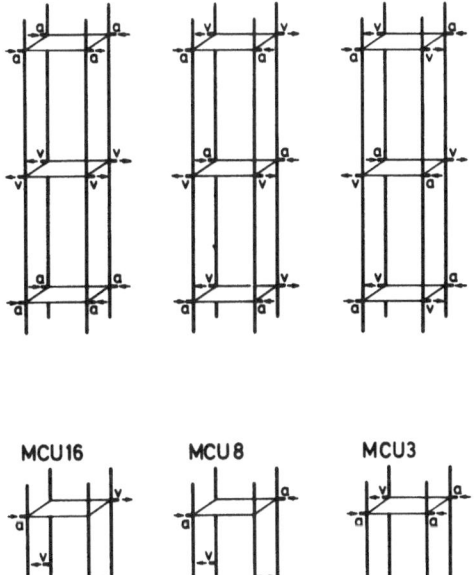

Fig. 2. Six microcirculatory units:

MCU 1   with concurrent capillary blood flow direction (Krogh capillary structure)
MCU 7   with partial concurrent or countercurrent capillary blood flow directions respectively
MCU 19   with total countercurrent capillary blood flow direction (capillary structure according to DIEMER (1963), enlargened into 3-dimensional structure)
MCU 16   with spirally arranged arterial inflows (and venous outflows respectively) shifted against one another by $1/2$. (Helix structure, according to GRUNEWALD, 1969 b, 1971)
MCU 3 } with a capillary blood flow direction without specific geometry in arrangement of capillary
MCU 8 } ends (asymmetric capillary structures)

a = arterial inflow b = venous outflow

model. Other MCUs can be constructed by the above-named model with an asymmetry in the flow direction. Two such MCUs are depicted in Figure 2 (MCUs 3 and 8).

The *physiological* assumptions of the model refer to the O_2 transport mechanism and the O_2 supply parameters. It is assumed that the O_2 transport from the capillaries to tissue is exclusively carried out by diffusion. Thereby the transport can be described by Fick's first law:

$$v = -K \cdot \text{grad} \, P; \quad K = \alpha \cdot D \tag{4}$$

where:

v {ml/cm²/s} = the flow density of the O_2 molecules
P {mm Hg} = the oxygen partial pressure (PO_2)
K {ml/cm/min/atm} = the diffusion conductivity
α {ml/g/atm} = the Bunsen solubility coefficient
D {cm²/s} = the tissue diffusion coeffient.

It is assumed that K, α, and D are independent of place and time. For the O_2 dissociation curve of blood in general, standard conditions are assumed. If necessary, the standard dissociation curve can be adjusted to the actual pH and temperature, according to OPITZ and BARTELS (1955) and GROTE (1968). A pH alteration along the length of the capillary and its influence upon the HbO_2 saturation is not considered in the model.

For the O_2 consumption $A(p)$ of the tissue, it is assumed that it can be described by the Michaelis-Menten equation

$$A(p) = A \cdot \frac{P}{P + P_{50}} \tag{5}$$

where:

A {ml O_2/100 g/min} = maximal O_2 consumption of tissue
P_{50} {mm Hg} = the PO_2 for which the O_2 consumption is $A/2$.

The volume with $P \simeq P_{50}$ should be small compared to the volume of the tissue fragment of a MCU.

The arterial and the mean venous PO_2 are associated with the blood flow and the O_2 consumption of the tissue by the following equation

$$A = W \cdot AVDO_2 \tag{6}$$

where:

W {ml$_{bl}$/100 g/min} = the blood flow
$AVDO_2$ {ml O_2/ml$_{bl}$} = the arteriovenous O_2 difference.

From the relation between the $AVDO_2$, dissociation curve, and PO_2 (THEWS, 1957), the following equation results:

$$A = W \cdot \left\{ 1.34 \frac{c_{Hb} \{s(P_a) - s(\bar{P}_v)\}}{100 \cdot 100} + \alpha \frac{(P_a - \bar{P}_v)}{760} \right\} \tag{7}$$

where:

1.34 {ml O$_2$/g$_{Hb}$} = the Hüfner constant
c_{Hb} {g/100 ml$_{bl}$} = the hemoglobin concentration in blood
$s(P)$ = the relative HbO$_2$ saturation at the oxygen partial pressure P
    (O$_2$ dissociation curve)
α {ml/ml$_{bl}$/atm} = the Bunsen solubility coefficient
P_a {mm Hg} = arterial PO_2
P_v {mm Hg} = mean venous PO_2.

Because of the anatomical assumptions that a capillary network and its corresponding tissue consists of MCUs, Equations 6 and 7 can be further specified.

For the volume which is supplied by the i-th capillary of a MCU, Equation 6 can be written accordingly:

$$A = \frac{\phi_c^{(i)}}{V_c^{(i)}} \cdot AVDO_2^{(i)}, \quad i = 1 \dots 4 \tag{8}$$

with:

$\phi_c^{(i)}$ {ml/min} = the blood flow through the i-th capillary of a MCU (capillary blood flow)
$V_c^{(i)}$ {cm^3} = the tissue volume supplied by the i-th capillary (capillary supply volume)
$AVDO_2^{(i)}$ {ml O$_2$/ml$_{bl}$} = arteriovenous O$_2$ difference of i-th capillary.

From Equation 7, the following equation is applicable for the i-th ($i = 1 \dots 4$) capillary:

$$A = \frac{\phi_c^{(i)}}{V_c^{(i)}} \cdot \left\{ 1.34 \frac{c_{Hb} s(P_a) - s(P_v^{(i)})}{100 \cdot 100} + \alpha \frac{P_a - P_v^{(i)}}{760} \right\} \tag{9}$$

where:

$s(P_v^{(i)})$ = the venous HbO$_2$ saturation of the i-th capillary
$P_v^{(i)}$ = the venous PO_2 of the i-th capillary.

The tissue volume $V_c^{(i)}$ which is supplied by the i-th capillary does not have to be distributed equally from the arterial end along the length of the capillary to the venous end. Furthermore, $V_c^{(i)}$ of all four capillaries of a MCU can differ. Different values $s(P_v^{(i)})$ and $P_v^{(i)}$ for all four capillaries result from these differing $V_c^{(i)}$ values. From the capillary venous HbO$_2$ saturation values $s(P_v^{(i)})$, the mean venous HbO$_2$ saturation $\bar{s}_v$ of the MCU is calculated by the following:

$$\bar{s}_v = \sum_{i=1}^{4} \frac{\phi_c^{(i)}}{\phi} s(P_v^{(i)}) \tag{10}$$

where $\phi_c^{(i)}/\phi$ are the weighting factors. Thereby:

$$\phi = \sum_{i=1}^{4} \phi_c^{(i)} \tag{11}$$

is the capillary blood flow of the MCU. From the mean venous HbO_2 saturation $\bar{s}_v$ and the HbO_2 dissociation curve, the mean venous PO_2 value $\bar{P}_v$ of the MCU results.

From the capillary blood flows $\phi_c^{(i)}$, the capillary supply volumes $V_c^{(i)}$, and the Equations 3 and 11, the following results:

$$W_c^{(i)} = \frac{\phi_c^{(i)}}{V_c^{(i)}}, \qquad i = 1 \ldots 4. \tag{12}$$

Thus the mean *local blood flow* of the MCU:

$$W = \frac{\sum\limits_{i=1}^{4} \phi_c^{(i)}}{\sum\limits_{i=1}^{4} V_c^{(i)}} = \sum\limits_{i=1}^{4} \frac{V_c^{(i)}}{V} W_c^{(i)}. \tag{13}$$

$W_c^{(i)}$ is the local blood flow of the supply volume $V_c^{(i)}$ of the i-th capillary.

A special case results from the assumption of a *homogeneous* capillary blood flow. Then $\phi_c^{(i)}$ for $i = 1 \ldots 4$ are equal. Since the capillary supply volumes $V_c^{(i)}$ generally differ from one another (e.g., in the MCUs 8 and 3 in Fig. 2), from Equation 9 and the assumption of a constant O_2 consumption and of a constant arterial PO_2, differences in the capillary-venous PO_2 values $P_v^{(i)}$ result. On the other hand, if, in all MCUs with the same structure and supply parameters, the same capillary venous PO_2 values are assumed ($\bar{P}_v = P_v^{(i)}$ for $i = 1 \ldots 4$), then for asymmetrical MCUs an inhomogeneous capillary blood flow, i.e., different values of $\phi_c^{(i)}$ result. In the symmetrically arranged MCUs (e.g., MCU 1, 7, 16, 19) no differences in the capillary-venous PO_2 values $P_v^{(i)}$ and in the capillary blood flow values $\phi_c^{(i)}$ result because of the equal capillary supply volumes $V_c^{(i)}$ of these MCUs.

The *mathematical* assumptions refer to the partial differential equation for diffusion. This equation can be derived from Fick's first law (Eq. 4) and the continuity equation (e.g., Thews, 1957). Since the model is limited to steady-state conditions, the following equation holds true:

$$\nabla^2 P = \frac{A(P)}{K} \tag{14}$$

where:

∇^2 = the Laplace operator
$P = P(r)$ = the O_2 partial pressure
r = the location vector
$A(P)$ = the O_2 consumption of tissue

whereby:

$$A(P) = A \qquad\qquad \text{for } P \gg P_{50};$$

$$A(P) = A \frac{P}{P + P_{50}} \qquad \text{for } P \simeq P_{50}.$$

The solution $P(r)$ from the partial differential equation is the *oxygen partial pressure field* (PO_2 field). The PO_2 field of a MCU is only calculated outside of the capillaries. It is assumed that the intracapillary PO_2 (i.c. PO_2) in a capillary cross-section is constant (mean i.c. PO_2), and coincides with the PO_2 of the outer wall of the capillary. For the calculation of a PO_2 field, i.e., the solution of the partial differential Equation 14, boundary conditions are necessary. On the lateral surface areas of a MCU outside of the capillaries, it is assumed that oxygen transport from the outside—into, or from the inside—out of the MCU does not occur. The i.c. PO_2 along the length of the capillary is calculated by a special iteration process before the solution of the partial differential Equation 14 is calculated. Boundary conditions for the upper and lower surface area of the MCU (Fig. 1) are not necessary because the upper surface area of the MCU is identical with its lower surface area.

Particular assumptions concerning the mathematical form of the HbO_2 dissociation curve $s(P)$ are not necessary. However, careful tabulation must be carried out (BARTELS and HARMS, 1958; THOMAS, 1972; BORK et al., 1975).

B. Mathematical Method

1. Boundary Conditions

The PO_2 along the length of the capillaries of a MCU, the i.c. PO_2, is used to solve the partial differential Equation 14. The i.c. PO_2 of the i-th capillary $(i = 1 \dots 4)$ is denoted as $P_c^{(i)}$ and must be first calculated. According to our assumption, $P_c^{(i)}$ changes only along the length of the capillary, but remains constant

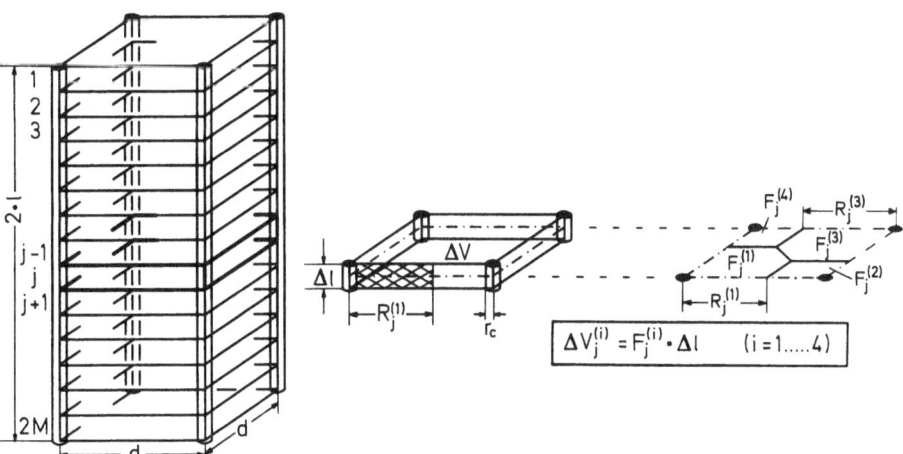

Fig. 3. Left: MCU subdivided into sections by $2 \cdot M$ sectioning perpendicular to the four capillaries of the length $2 \cdot l$ and the distance d. Center: j-th section, Δl from each capillary $(i = 1 \dots 4)$ supplies partial volume $\Delta V_j^{(i)}$ of total section volume ΔV. $R_j^{(i)}$ are supply boundaries of 2 capillary parts on lateral surfaces of j-th section. Right: Determination of partial areas $F_j^{(i)}$ from $R_j^{(i)}$: supply volume $\Delta V_j^{(i)}$ is calculated from areas $F_j^{(i)}$ and thickness Δl of the 4 capillaries $(i = 1 \dots 4)$ of j-th section (encircled Eq.)

across the cross-section of the capillary. The arterial PO_2 and the mean venous PO_2 of a MCU are associated with the HbO_2 saturation values s_a and $\bar{s}_v$ through the O_2 dissociation curve. Except for a constant factor, the saturation difference $\Delta s = s_a - \bar{s}_v$ gives the O_2 quantity that the four capillaries have at their disposal to supply the tissue fragment of the MCU. The portion that each capillary has at its disposal to supply the tissue fragment (Eq. 3) is the capillary supply volume. Generally, the supply volume is indiscriminantly extended along the length of the four capillaries of a MCU, depending upon the arrangement of the in- and outflows of the capillaries. In an iteration process, the capillary supply volume and the i.c. PO_2 (and i.c. HbO_2 saturation) are adjusted to one another (Fig. 3).

By $2M$-sectioning (perpendicular to the capillary), the MCU is subdivided into $2M$ sections of thickness $\Delta l \, (=h_y)$ and volume ΔV. As a representative for all sections we can consider the j-th section. Each Δl of the four capillaries in this section supplies a partial volume $\Delta V_j^{(i)}$ of the total ΔV with oxygen. As a good approximation we can say:

$$\Delta V_j^{(i)} = F_j^{(i)} \cdot \Delta l, \quad i = 1 \ldots 4. \tag{15}$$

$F_j^{(i)}$ is determined from the supply boundary $R_j^{(i)}$ between two adjoining capillaries on the lateral surface of the section (Fig. 3). $R_j^{(i)}$ is determined from the mean i.c. oxygen partial pressures $P_c^{(i)}(j)$ of the section under consideration and the diffusion equation.

Let us examine one of the four lateral surface areas of the j-th section of the MCU (Fig. 3, center). The supply volume of the two capillary parts ($i = 1$ and 2) on the lateral surface area extends to $R_j^{(1)}$. It is assumed that the PO_2 distribution is cylindrically symmetrical at the lateral surface area of the j-th section. Since $P_c^{(i)}$ at the length Δl of the capillary is nearly constant if $\Delta l \ll l$, the diffusion equation for the striated as well as for the nonstriated part of the examined lateral surface area (Fig. 3) takes the form:

$$\frac{d^2 P}{dr^2} + \frac{1}{r} \frac{dP}{dr} = \frac{A}{K}; \tag{16}$$

$P = P_j^{(1)}(r)$ is the solution for the striated part and $P = P_j^{(2)}(r)$ for the nonstriated part. They have to fulfill the following boundary conditions:

$$P_j^{(1)}(r_c) = P_c^{(1)}(j); \qquad P_j^{(2)}(r_c) = P_c^{(2)}(j)$$

$$\left. \frac{dP_j^{(1)}(r)}{dr} \right|_{r = R_j^{(1)}} = 0; \qquad \left. \frac{dP_j^{(2)}(r)}{dr} \right|_{r = d - R_j^{(1)}} = 0 \tag{17}$$

and,

$$P_j^{(1)}(R_j^{(1)}) = P_j^{(2)}(d - R_j^{(1)}). \tag{18}$$

Likewise, the supply boundaries $(R_j^{(1)} - R_j^{(4)})$ for all j are calculated, whereby using Equation 15, all $\Delta V_j^{(i)}$ are determined.

At the beginning of the iteration process, we assume that the i.c. HbO_2 saturation decreases *linearly* from the arterial to the venous capillary end. Then the HbO_2 saturation at j is:

$$s_j^{(i)} = s_a - \frac{\Delta s}{2M} \cdot j \quad (i = 1 \ldots 4) \tag{19}$$

if we start at the arterial end of each capillary.

Furthermore the four i.c. HbO_2 saturation changes $\Delta s_j^{(i)} = s_j^{(i)} - s_{j-1}^{(i)}$ in the j-th section are equal and are caused by the O_2 consumption of the four partial volumes $\Delta V_j^{(i)}$. These partial volumes $\Delta V_j^{(i)}$ are equal to one another and equal to one-fourth of the section volume ΔV. The assumed linear HbO_2 saturation course $s_j^{(i)}$ is correlated with an i.c. $P_c^{(i)}(j)$ through the O_2 dissociation curve. For each j, the $R_j^{(i)}$ and the $\Delta V_j^{(i)}$ (Eq. 15) of the first iteration step are calculated. Depending upon the shape of the MCU under consideration, the newly calculated partial volumes $\Delta V_j^{(i)}$ no longer equal one another. This is only the case in the MCU 1 where—as assumed at the beginning of the iteration process—the $R_j^{(i)}$ are equal to one half of the capillary distance. In general, the i.c. HbO₂ saturation change $\Delta s_j^{(i)}$ approaches:

$$\Delta s_j^{(i)} = \frac{\Delta V_j^{(i)}}{\Delta V/4} \cdot \frac{\Delta s}{2M}. \tag{20}$$

The i.c. HbO₂ saturation course of the first iteration step takes the form:

$$s_j^{(i)} = s_a - \sum_{\rho=1}^{j} \Delta s_\rho^{(i)} \quad \text{for all } j \text{ and } i = 1 \ldots 4. \tag{21}$$

Using Equation 21 and the O₂ dissociation curve yields the $P_c^{(i)}(j)$ of the *first* iteration step. This procedure is continued up to the iteration step n, where the maximal deviation of the $P_c^{(i)}(j)$ of the n-th and the $n-1$-th iteration step is smaller than a given barrier ε_c. This iteration process converges with an increasing number of iteration steps (GRUNEWALD, 1971); e.g., with $2M = 120$ sections through the MCU and with a truncation barrier of $\varepsilon_c = 0.02$, the error of the $P_c^{(i)}(j)$ is smaller than 0.4 mm Hg.

The capillary venous PO₂ values $P_v^{(i)}$ calculated in the iteration process generally differ from one another. The mean venous PO₂ of a MCU, the $\bar{P}_v(\bar{s}_v)$, results according to Equation 10 and the O₂ dissociation curve. Furthermore, the capillary blood flows $\phi_c^{(i)}$, the capillary supply volumes $V_c^{(i)}$ for all $i = 1 \ldots 4$, and the local blood flow W of the MCU are calculated in the iteration process using Equations 2, 3, 9, 11, 12, and 13.

2. Solution of Diffusion Equation

After the assumptions and the boundary conditions are known, the 3-dimensional PO₂ field in a MCU can be calculated. It is the solution of the partial differential Equation 14. Due to the geometrical form of the MCU, the partial differential equation can be written in the form:

$$\frac{\partial^2 P}{\partial x^2} + \frac{\partial^2 P}{\partial y^2} + \frac{\partial^2 P}{\partial z^2} = \frac{A(P)}{K} \tag{22}$$

where: $P - P(x, y, z)$ is the PO₂ field.

The partial differential Equation 22 cannot be solved in a closed form due to the boundary conditions. The PO₂ field can only be approximatively calculated. In the MCU, a grid is set up parallel to the capillaries. The intersecting lines of the grid are called *grid points* and the distance between two grid points is the *interstice length* (Fig. 4). Then, in the direction of the coordinate axes x and z, the grid has L discrete steps of the interstice length h_x, and in the y direction, $2M$ discrete steps of the interstice length h_y. Furthermore, each capillary end must coincide with a grid point. For each capillary, a value $MA^{(i)}$ is defined—if j is a grid point coinciding with an arterial capillary end, then $MA^{(i)} = j$ for $i = 1 \ldots 4$. Accordingly, the MCUs of Figure 2 can be characterized by the following:

MCU 1: 0/0/0/0/   MCU 8: $0\left/\dfrac{M}{2}\right/0\left/\dfrac{M}{2}\right.$

MCU 3: 0/0/0/M   MCU 16: $0\left/\dfrac{M}{2}\right/M\left/\dfrac{3M}{2}\right/$

MCU 7: 0/0/M/M/   MCU 19: 0/M/0/M/.

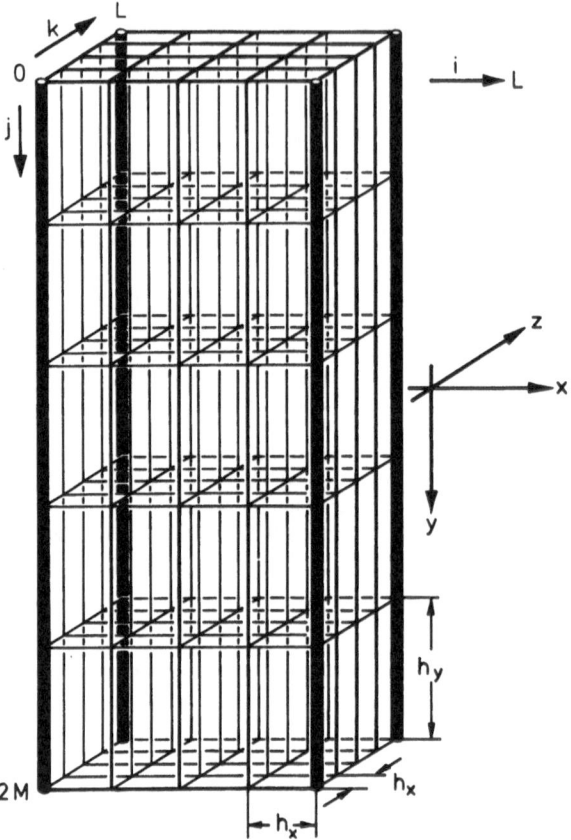

Fig. 4. Grid for approximate calculation of PO_2 field in a MCU. Intersecting *lines* of grid are called *grid-points*. 4 darkly drawn *lines* symbolize 4 capillaries of length $2 \cdot l$ in MCU. Grid has in x and z direction L $(0 \leq i \leq L, 0 \leq k \leq L)$ steps of interstice length h_x and in y direction $2M$ $(0 \leq j \leq 2M)$ steps of interstice length h_y. Each capillary end must coincide with grid point

The iteration process that is used calculates an approximate PO_2 value $P(i, j, k)$ for each grid point (i, j, k). This process can be easily understood (see Appendix 1).

IV. Results: Documentation, Interpretation, and Discussion

A. General Results—Information Content and Documentation

The results that were calculated from the described model are divided into two parts. Section A contains generally valid results that are not organ specific. Section B contains results that relate specifically to the O_2 supply of *skeletal muscle*.

In order to document the results, a form is necessary that contains as much information as possible, and that at the same time reconstructs the spatial relationship of the model results. For a documentation of the results, the PO_2 *frequency distribution* is primarily used. This suits the stated prerequisites best. With the calculated PO_2 frequency distribution, not only is the influence of the capillary structure on the O_2 supply of tissue distinguished but also general aspects of the O_2 transport to tissue. At the same time, it also conforms to the presentation form of the experimental results.

1. Capillary Supply Volume and i.c. PO_2

The *capillary supply volume* is defined as that part of the tissue fragment of a MCU which is supplied from a capillary. This supply volume can be distributed along the length of the capillary in various fashions. According to Fick's first law (Eq. 14), the association of the diffusion pathway from which the capillary supply volume is calculated to the i.c. PO_2 is of particular importance. If a high i.c. PO_2 is associated with a short diffusion path, it is just as ineffective as associat-

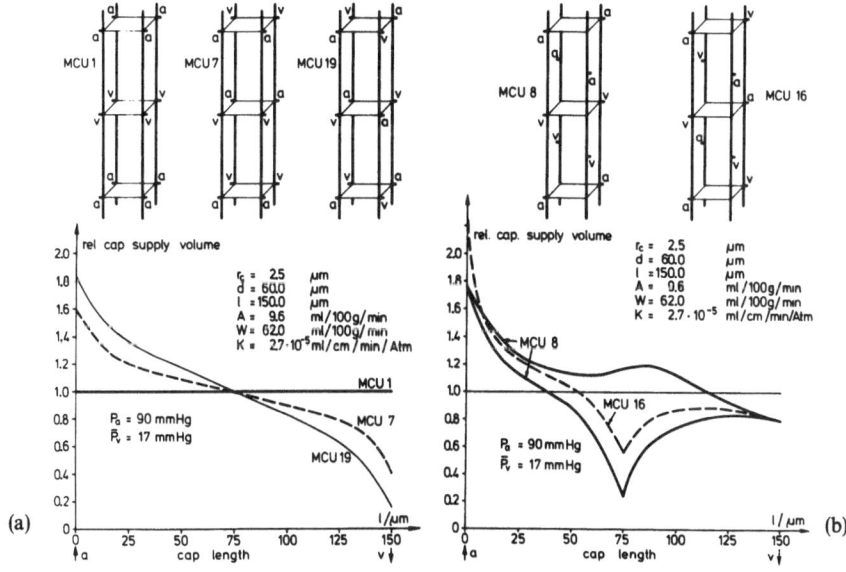

Fig. 5a and b. MCUs and capillary supply volume: a) In MCU 1, capillary supply volume does not change from arterial to venous end. If structure of MCU changes to MCU 7 with partial countercurrent flow direction and then in turn to MCU 19 with total countercurrent flow direction, arterial portion of capillary supply volume increases. Venous portion decreases accordingly. In all four capillaries of each of MCUs 1, 7, and 19, capillary supply volume has same form. b) In the MCU 16, form of capillary supply volume of all 4 capillaries is same (correspondingly, by MCUs 1, 7, 19). Minimum of supply volume no longer lies at venous capillary end rather it is to be found in the middle, between arterial inflow and venous outflow. Arterial portion of supply volume is particularly large, larger than that in MCUs 7 and 19. In MCU 8, supply volumes of 2 capillaries in major diagonal differ from supply volumes of 2 capillaries in minor diagonal by 30 %. In both figures, multifold of capillary supply volume of MCU 1 is shown on ordinate (relative capillary supply volume; parameter from gray matter of human brain under venous hypoxic conditions)

ing a low i.c. PO_2 with a long diffusion path. In the first case, more O_2 diffuses into tissue than is necessary, and in the second case too little O_2 reaches the tissue. The O_2 supply of the tissue from capillaries is most propitious when a long diffusion path is associated with a high i.c. PO_2 and likewise when a short diffusion path is associated with a low i.c. PO_2.

The PO_2 gradient and thus the flow density of the diffusing O_2 molecules (O_2 diffusion flow) into tissue can be equal in both cases. In the MCUs, the association of the diffusion path with the i.c. PO_2 is different (Fig. 5). In the MCU 1 with concurrent flow direction the diffusion pathway and thus the supply volume of each capillary part from its arterial to its venous end is equally large. The arterial end with its high i.c. PO_2 has the same supply function as the venous capillary end with a substantially lower i.c. PO_2. In contrast, in both of the MCUs 7 and 19, a larger supply volume is associated with the arterial part of the capillary as compared to the supply volume at the venous part of the same capillary (Fig. 5 a).

The arterial part of the capillary supply volume increases if the MCU changes over from the concurrent flow direction (MCU 1) via the partial countercurrent flow direction (MCU 7) to the total countercurrent flow direction (MCU 19). Compared to the other MCUs, the capillary supply volume of the MCU 16 is at its largest at the arterial part of the capillary. Halfway between the arterial and the venous ends of the capillary, the capillary supply volume reaches a minimum and then increases towards the venous end of the capillary (Fig. 5 b).

In the MCUs 1, 7, 19, and 16, the capillary supply volumes along the length of the capillaries show a difference in form, but the capillary supply volume of all four capillaries of a MCU has the same size. On the other hand, MCUs exist with (e.g., MCU 8) differences in the size of the capillary supply volumes. In the MCU 8, the supply volumes of the capillaries in the major diagonal differ from those supply volumes of the capillaries in the minor diagonal by 30% (Fig. 5 b).

The size of the supply volume along the length of the capillary under consideration has an influence upon the i.c. PO_2 decrease and upon the i.c. HbO_2 saturation decrease from arterial to venous end respectively (Fig. 6 a).

With the same structure and supply parameters, the i.c. PO_2 at the arterial capillary end decreases all the more, the larger the capillary supply volume is (MCUs 1, 7, 19 in Fig. 5 a). In the MCU 16 this decrease is most distinctive. As the capillary supply volume in the MCU 16 between arterial and venous end reaches a minimum, the i.c. PO_2 decrease along the venous part of the capillary is smaller than, for instance, the PO_2 decrease in the MCU 19. Thus, the i.c. PO_2 in the venous capillary part of the MCU 16 is higher than that in the MCU 19 (Fig. 6 a). This is advantageous to the O_2 supply of the tissue. However, at the venous capillary end all i.c. PO_2 values in the MCUs 1, 7, 19, and 16 are equal to one another. With the same structure and supply parameters the above does not hold true for the MCU 8. Since there are differences in the size of the capillary supply volumes, for the most part differing capillary venous PO_2 values result (Fig. 6 b). The mean venous PO_2 of the MCU results from these values according to Equation 10 and from the O_2 dissociation curve.

In MCUs with a short capillary distance and a short capillary length, the arterial capillary end can have – even under concurrent flow conditions – a larger

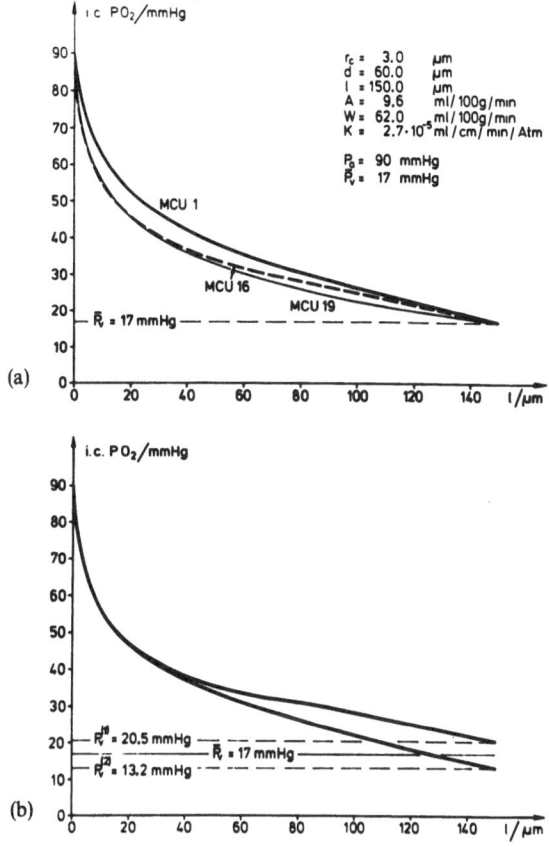

Fig. 6a and b. i.c. PO_2 decrease along capillary length l: a) MCU 1, MCU 19, MCU 16. Larger the capillary supply volume (Figs. 5a, b), larger the decrease of i.c. PO_2 at arterial capillary portion. Minimum of capillary supply volume between capillary ends in MCU 16 causes a smaller decrease of i.c. PO_2 at venous half of capillary of this MCU. Capillary venous PO_2 values are equal in MCUs mentioned above as well as in MCU 7. b) Because of differences in capillary supply volumes (Fig. 5b), i.c. PO_2 of 2 capillaries in minor diagonal and of 2 capillaries in major diagonal in MCU 8 has differing decrease. This results in different venous PO_2 values, and holds true for all asymmetric MCUs. If one calculates corresponding HbO_2 saturations from capillary venous PO_2 values $P_v^{(i)}$ and takes the mean value (Eq. 10), then from mean venous HbO_2 saturations and O_2 dissociation curve, mean venous PO_2 of MCU ($\bar{P}_v$) results

supply volume than that at the venous capillary end (Fig. 7). This holds true because then the influence of the longitudinal diffusion upon the i.c. PO_2 can no longer be neglected [3]. As shown in Figure 7a, the longitudinal diffusion along the first 10% of the capillary length in the MCU 1 can cause a 5-times larger supply volume as compared to that of the same MCU without taking the longitudinal diffusion into consideration. Thus, the i.c. PO_2 in this part of the capillary has a correspondingly steeper decrease. With an increase in capillary length (Fig. 7b),

---

[3] By taking into account the longitudinal diffusion, the calculation of the i.c. PO_2 course requires an iteration process other than that which has been described (SOWA and GRUNEWALD, in preparation).

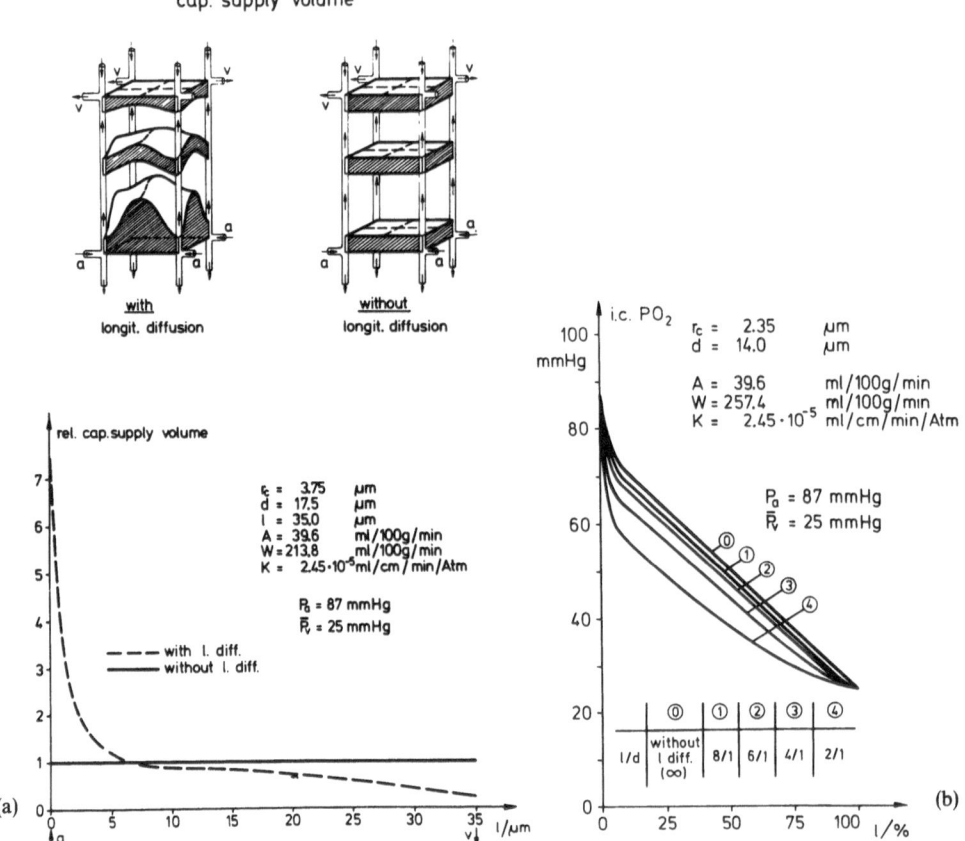

Fig. 7a and b. a) Capillary supply volumes in MCU 1 with and without longitudinal diffusion. By taking longitudinal diffusion into account, in MCUs with small capillary distance d and short capillary length l (e.g., $d=17.5$ μm and $l=35.0$ μm, $d/l=1/2$), capillary supply volume along first 10 % of the capillary length l can increase by a factor 5 times larger than corresponding supply volume without taking longitudinal diffusion into account. Upper part of figure shows supply volumes of three capillary portions equal in size ($\Delta l=l/10$). Left, with longitudinal diffusion; right, without longitudinal diffusion. Lower part of figure shows supply volume for total capillary length l with and without taking longitudinal diffusion into account. (Except for l, parameters of rat myocardium under normoxic conditions were chosen.) b) i.c. PO_2 decrease along relative length of capillary in MCU 1 with and without longitudinal diffusion. Longitudinal diffusion causes a sharper decrease of i.c. PO_2 along arterial capillary portion. Influence of longitudinal diffusion becomes smaller the longer the MCU is. With same parameters (Fig. 7a) and a $d/l \ll 1/10$, longitudinal diffusion can be neglected (e.g., in myocardium and skeletal muscle)

the influence of the longitudinal diffusion decreases. Since in the following only MCUs with comparable large capillary lengths are to be considered, the influence of the longitudinal diffusion upon the i.c. PO_2 course will be neglected. THEWS (1953) already refers to the neglectable influence of the longitudinal diffusion, for instance, on the O_2 supply of brain tissue.

2. O_2 Diffusion Shunt

The supply volume of an arterial capillary end can, under certain supply conditions, include a part of a neighboring capillary. Then an O_2 diffusion shunt exists (O_2 shunt) (EVANS and NAYLOR, 1964; GRUNEWALD and LÜBBERS, 1968; GRUNEWALD and SOWA, 1975).

In the neighboring capillary, the O_2 shunt is shown by the fact that the supply volume along the length of the capillary part under consideration is equal to 0. In this part of the capillary, the O_2 shunt leads to an increase in the i.c. PO_2. In the neighboring capillary from which the shunted O_2 comes, a correspondingly larger decrease of the i.c. PO_2 results. In the MCUs, the capillary parts are particularly predestined for an O_2 shunt in the surrounding area of the point where the capillary supply volume reaches a minimum (Figs. 5a, b). In the MCU 1 with a concurrent flow direction, no minima in the capillary supply volume exist, thus no O_2 shunt can exist in this MCU. Furthermore, this MCU 1 is the only MCU where an O_2 shunt cannot exist. In the MCUs 7 and 19 with a partial or a total countercurrent flow direction respectively, the venous part of the capillary end is always advantageous for an O_2 shunt. In the MCU 16, the capillary part in the middle between the arterial and venous capillary end is responsible for the O_2 shunt. Not only the point where the capillary supply volume reaches a minimum but also the value of the i.c. PO_2 at this same point is responsible for an O_2 shunt.

The higher the i.c. PO_2 value at the point of the minimal capillary supply volume, the later an O_2 shunt arises and vice versa. From the above it is clear, that in the MCU 19 with a total countercurrent flow direction under the same structure and supply parameters, an O_2 shunt can most easily arise. At the minimum of the capillary supply volume, the i.c. PO_2 reaches in this MCU its smallest possible value, the capillary venous PO_2. The reason for the coincidence of the minimum of the capillary supply volume and the minimum of the i.c. PO_2 lies in the fact that in this MCU all arterial and venous capillary ends lie opposing one another at the capillary distance. The largest possible PO_2 difference that can exist in a MCU (the arterio-venous PO_2 difference) exists at the shortest possible distance between two capillaries (the capillary distance d).

The i.c. PO_2 course in MCUs with an O_2 shunt can take varying forms (Fig. 8). In the MCU 16, a plateau arises in the middle between arterial and venous end with an (increasing) O_2 shunt. In the MCU 19 a plateau does not exist. Under the same conditions, the i.c. PO_2 can be less than the capillary venous PO_2. Figure 9 shows the O_2 shunt in the MCU 16 in a 3-dimensional depiction. In the MCU 19, the danger of a *non-nutritive* O_2 shunt exists, i.e., O_2 is shunted from one capillary to the venous end of a neighboring capillary, without remaining at the disposal of the O_2 supply. It is even possible that an O_2 shunt can arise even though hypoxic or anoxic areas exist in the tissue (GRUNEWALD and LÜBBERS, 1968). A part of the diffusing O_2 passes by the undersupplied area. On the other hand, the O_2 shunt in the MCU 16 is always nutritive.

Even with an O_2 shunt, the capillary venous PO_2 values in the MCUs 7, 16, and 19 remain unchanged. The O_2 shunt in these MCUs can be interpreted as a *redistribution of the O_2* from the arterial capillary part to its middle part (MCU 16)

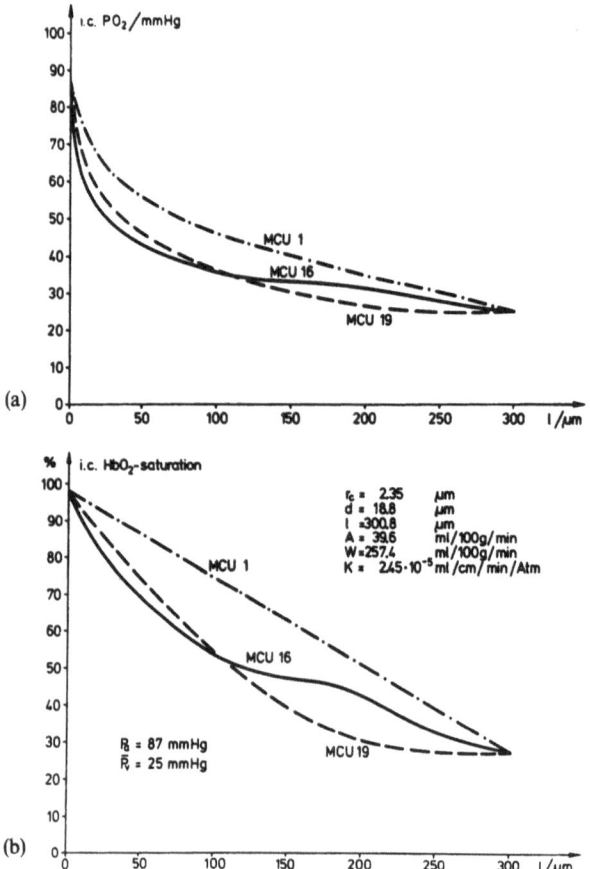

Fig. 8 a and b. i.c. HbO_2 saturation (below) and i.c. PO_2 (above) along length of capillary from arterial to venous end at an O_2 shunt in MCUs 16 and 19. For comparison, MCU 1 is shown in which O_2 shunt cannot exist. While i.c. PO_2 (i.c. HbO_2 saturation) in MCU 16 between arterial and venous capillary end takes on a plateau, i.c. PO_2 (i.c. HbO_2 saturation) in MCU 19 can even be lower than the capillary venous PO_2 (capillary venous HbO_2 saturation) (parameter: rat myocardium, normoxy)

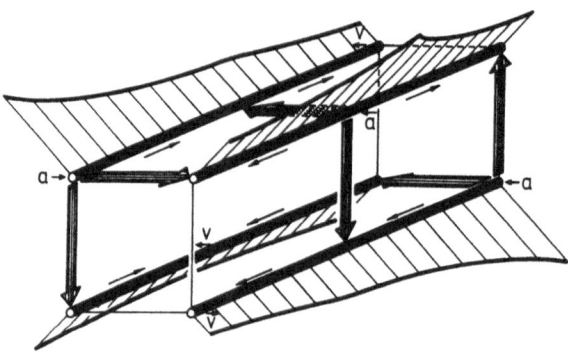

Fig. 9. 3-dimensional view of O_2 shunt and i.c. PO_2 of MCU 16. For a better view, only half of MCU 16 is shown. Thickly striped *arrows* indicate direction of O_2 shunt from each arterial capillary end to neighboring capillary. Above the capillaries, i.c. PO_2 is shown. In the middle between arterial and venous end of each capillary, a plateau in i.c. PO_2 can be seen

or to its venous part (MCUs 7 and 19) respectively. The larger the O_2 shunt, the more the O_2 is accordingly shifted from the arterial to the middle or venous capillary part.

The conditions whereby an O_2 shunt appears can be described by structure and supply parameters. Let us first consider the influence of oxygen consumption and blood flow using Equations 6, 7, and 22. If the blood flow changes proportionally to the O_2 consumption so that the $AVDO_2$, especially the venous PO_2, remains unchanged, then the probability of an O_2 shunt increases with a decreasing O_2 consumption (Fig. 10a).

If the blood flow decreases during constant O_2 consumption so that the $AVDO_2$ increases and the venous PO_2 decreases, then the probability of an O_2 shunt increases (Fig. 10b). These two cases are combined when a decreasing venous PO_2 along with an increasing O_2 consumption exists with a constant

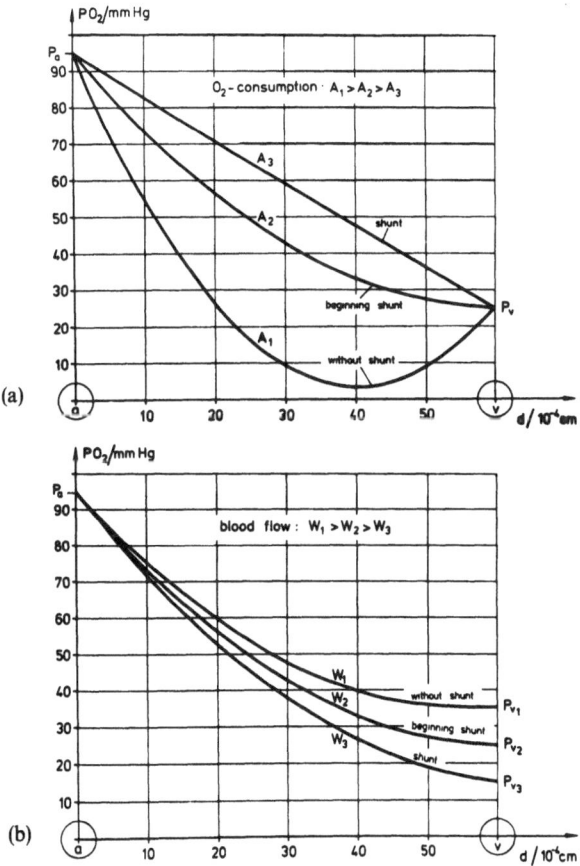

Fig. 10a and b. O_2 shunt between arterial and venous capillary ends (e.g., in MCU 19) dependent upon O_2 consumption and blood flow: a) If O_2 consumption decreases ($A_1 > A_2 > A_3$) and if blood flow diminishes proportionally so venous PO_2 remains constant, then the minimum of the PO_2 curve is shifted to the venous end. The probability of O_2 shunt increases. At $A = A_2$, O_2 shunt begins. b) If at constant O_2 consumption blood flow decreases so that the venous PO_2 sinks, then minimum of PO_2 curve also tends towards venous end. Probability of O_2 shunt increases. At $W = W_2$, O_2 shunt begins blood flow. An increasing O_2 consumption minimizes the probability of an O_2 shunt as shown in Figure 10a (the minimum in the PO_2 curve shifts away from the venous end). Competitively, the concomitant decrease of the venous PO_2 causes an increase in the O_2 shunt probability as shown in Figure 10b (the minimum in the PO_2 curve shifts towards the venous end). The organism has the ability to avoid the O_2 shunt by appropriately regulating these three supply parameters, if the shunt is non nutritive.

From all the structure parameters, the capillary distance d particularly influences the O_2 shunt. The probability of an O_2 shunt is greater with a smaller capillary distance than with a larger one. The myocardium with its capillary distance of approximately 20 µm is the organ which is particularly predestined for an O_2 shunt (Fig. 11). In the myocardium, the O_2 shunt is after a fashion obligatory.

3. PO_2 Field

a) Minimum PO_2 and Plot of Isobars

The result of a model calculation is (with sufficient convergence of the approximation process) the PO_2 field of a MCU in the form of a matrix of about 50,000–100,000 PO_2 values. This three dimensional matrix, called P matrix, contains the information concerning the O_2 supply that is specific for each MCU and its structure and supply parameters. There are necessary criteria with which this information can be gained and documented from such a large number of PO_2 values. The documentation must allow for a comparison with the results of the PO_2 measurements. The P matrix of a PO_2 field has an upper and lower limit. The upper limit of the PO_2 value is the arterial PO_2; the lower limit is P_{min}, the lowest PO_2 value that exists in the P matrix. The value P_{min} has a particular importance for the description of the O_2 supply in a MCU. This value is the PO_2 at the poorest supplied area of the tissue. The MCUs described above differ from one another under the same supply conditions in the magnitude of the value P_{min}, as is shown in Table 1, using a characteristic example.

Table 1. P_{min} of the MCUs 1, 3, 7, 8, 16, 19 under *same* supply conditions (human skeletal muscle, exercise conditions; parameters; see Fig. 13)

MCU	1	3	7	8	16	19
P_{min}/mm Hg	0.8	6.9	9.8	1.4	10.7	7.9

The MCU 1 with a concurrent flow direction has the lowest P_{min} value. It is the MCU with the poorest oxygen supply. By comparison, the P_{min} value is much larger than in, for instance, the MCU 19 with a total countercurrent flow direction. The highest P_{min} value is to be found in the MCU 16. If P_{min} is lower than the critical mitochondrial PO_2 (Starlinger and Lübbers, 1973), then anoxy exists in the tissue. Since the P_{min} values in the individual MCUs are different, it is possible that, e.g., the MCU 1 has an anoxic area, while the P_{min} value from other MCUs still lies above the limit of the critical mitochondrial PO_2; in other words, here there are no anoxic areas.

Even with an O_2 shunt, the P_{min} of a MCU shows differing values. For example, MARTINI and HONIG (1969), and HONIG and BOURDEAU-MARTINI (1973) were able to measure capillary distances in the rat myocardium under normoxy conditions of 12 to approx. 30 μm. Except for high values of d, an O_2 shunt in the MCU 16 exists for the above capillary distances. In the MCU 1, no O_2 shunt can exist as we have shown above. Let us assume that with the same structure and supply parameters the venous PO_2 remains the same despite differing capillary distance. Thus, Figure 11 a shows that the P_{min} in the MCU 1 decreases with an increasing capillary

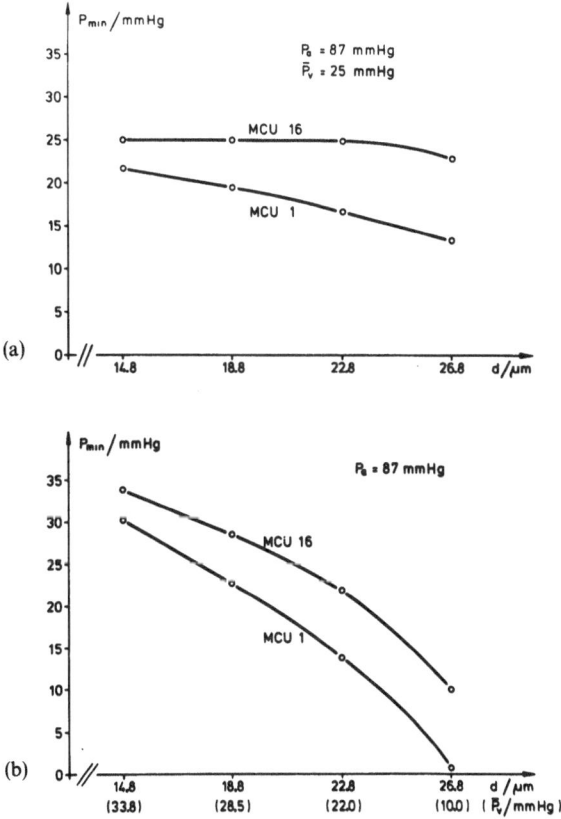

Fig. 11a and b. a) Dependence of minimal (tissue) PO_2 from capillary distance for MCUs 1 and 16. In MCU 16, P_{min} remains nearly constant between $d=14.8$ μm and 26.8 μm because of O_2 shunt. It was assumed that mean venous PO_2 remained constant ($P_v=25$ mm Hg). In MCU 1 with concurrent flow direction, P_{min} decreases by about 10 mm Hg under same conditions when $d=26.8$ μm. b) Dependence of minimal PO_2 from capillary distance as in Figure 11 a but without assumption that the $\bar{P}_v$ remains constant. It is taken into account that the $\bar{P}_v$ decreases with increasing capillary distance (increasing capillary supply volume) and vice versa. For capillary distance $d=26.8$ μm a $\bar{P}_v=10$ mm Hg is assumed. For each d on *abscissa*, corresponding mean venous PO_2 given in parenthesis. Mean venous PO_2 for all capillary distances under consideration is 25 mm Hg. In MCU 16 because of O_2 shunt, P_{min} is identical to $\bar{P}_v$ of MCU and decreases with $\bar{P}_v$ to 10 mm Hg. In MCU1 under the same conditions, P_{min} is smaller than $\bar{P}_v$ and decreases to $PO_2 \cong 1$ mm Hg (parameters: rat myocardium, normoxy; Fig. 8)

distance by approximately 10 mm Hg, whereas in the MCU 16, the P_{min} remains nearly constant and does not sink significantly below the venous PO_2. If one assumes that with an increasing capillary distance the venous PO_2 decreases correspondingly with the increasing capillary supply volume to a value of 10 mm Hg, then the P_{min} is identical to the venous PO_2 in the MCU 16 because of the O_2 shunt. Under the same conditions, the P_{min} in the MCU 1 is less than 1 mm Hg (Fig. 11 b). In the other MCUs, P_{min} can show a similar behavior even if it is less pronounced than in the MCU 16.

From the above considerations and from the results of the supply volumes and i.c. PO_2, it follows that the organism has the opportunity to optimalize the oxygen supply of tissue by developing certain capillary structures similar to those found in MCU 16 or other MCUs with a similarly favorable asymmetrical arrangement of their capillary ends.

Furthermore, one can use the P_{min} value to determine the existence of an O_2 shunt. When the P_{min} value is larger than the smallest capillary venous PO_2 of a MCU, then an O_2 shunt exists. However, an O_2 shunt can also exist when P_{min} is equal to or smaller than the lowest capillary venous PO_2. This is a result of the three dimensional shape of the PO_2 field. In his consideration, DIEMER (1965) postulated that the capillary arrangement of the brain is countercurrently perfused and that, under normoxic conditions, the PO_2 minimum in tissue is identical to the venous PO_2. Accordingly under such conditions, the brain structure would find itself constantly at the limit of an O_2 shunt, especially a nonnutritive O_2 shunt. The three-dimensional PO_2 field of the MCU 19 with total countercurrent flow direction shows however, that under the above conditions the minimal PO_2 in tissue lies far below the venous PO_2 values and that no O_2 shunt exists.

The minimal PO_2 value in the MCU is suited in considering the O_2 supply from the impropitious viewpoint. However, this value is not specific for the MCUs. More information is given by the location of the P_{min} value in the MCU. In the MCU 1, the lowest PO_2 value lies between the venous capillary ends, i.e., equidistant from each venous end. In the MCU 19, the lowest PO_2 value is located in the center of the MCU. Even though the P_{min} also has a characteristic location in the other MCUs, this value is not suited for a comparison with the experimentally obtained PO_2 values. When PO_2 values are measured for instance with a needle electrode in tissue, there is no certainty that also at *that* particular location the poorest O_2 supply is being measured.

An isobar plot gives us a view of the specific shape of the spatial PO_2 field of a MCU. Such a plot occurs when one sections the MCU perpendicular to the flow direction of the capillary (Fig. 3), and gives the PO_2 values in the sectioning plane characteristic symbols; then one connects the same symbols to one another to form isobars. From the sequence of the sections, one can complete a spatial picture of the PO_2 field for each MCU (Fig. 12).

An isobar plot is optimal for the localization of the individual PO_2 values in the PO_2 field of a MCU. However, no quantitative parameters can be named as proxy for the P matrix or the symbol arrangement that specifically characterizes the structure of the PO_2 field. Besides, the isobar plot cannot be used for comparison with the measured PO_2 fields. Since a PO_2 measurement occurs without allowing for a control over the location in the tissue, the measured PO_2 values

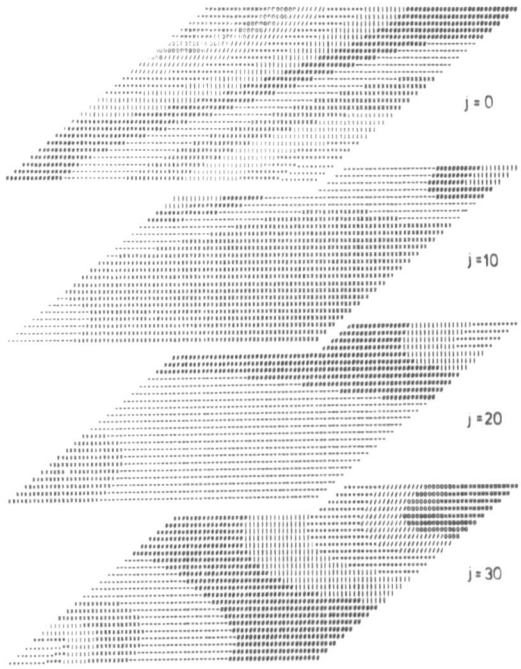

Fig. 12. Isobar plot of sections of MCU 16 with helix structure. The MCU is divided into sections that have been sectioned perpendicular to flow direction of capillaries (Fig. 3). j indicates number of sections ($0 \leq j \leq 2M$, $2M = 120$). In a section, PO_2 values from PO_2 intervals are marked with same symbol

cannot be related to one another in the same sense as in an isobar plot. The comparison between the measured and the calculated PO_2 fields succeeds only if one also disregards the location of the PO_2 values in the theoretical description of the PO_2 fields. The key to the desired comparison and to the analysis of the oxygen supply in tissue is the PO_2 *frequency distribution* (the PO_2 distribution).

b) PO_2 Frequency Distribution and its Specificity

The interval $P_{min} \leq P \leq P_a$ that includes all the values of the P matrix of a PO_2 field, is divided into PO_2 classes ΔP. The frequency with which the P values are distributed in these P classes results in the PO_2 frequency distribution $\varphi(P)$ (Appendix 2).

If one investigates the influence of structure and supply parameters upon the PO_2 distribution in the MCUs, one sees that the arrangement of capillary ends characteristically influences the form of the PO_2 distribution (Fig. 13). The PO_2 distribution in the MCU 1 with a concurrent flow direction distinguishes itself by a low frequency maximum and by a slow decrease in the frequency in the direction of not only lower but also higher PO_2 values. The low PO_2 values correspond to the *lethal corner* in the Krogh cylinder. In the MCUs 16 and 19, *which were drawn as representatives of other MCUs*, this lethal corner is cut off.

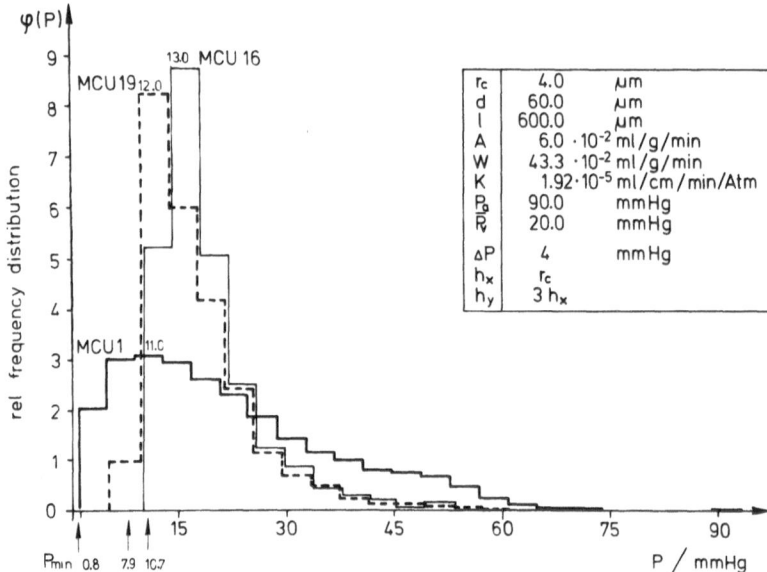

Fig. 13. PO_2 distributions $\varphi(P)$ of MCUs 1, 16, and 19: Distributions of chosen MCUs show that arrangement of capillary ends in MCU characteristically influences form of PO_2 distribution. Besides the position of the frequency maxima and P_{min} values, parameters that are necessary for calculating the respective PO_2 field are indicated (human skeletal muscle, exercise conditions). These parameters are indicated in same fashion in Figures 14–30

The low PO_2 values in these structures are much higher than in the MCU 1. These MCUs distinguish themselves through higher frequency maxima and— compared to the MCU 1—fewer high PO_2 values. The position of the frequency maximum is different in the MCU 19 with a total countercurrent flow direction as well as in the MCU 16 with Helix structure. The frequency in the lowest P class is in the MCU 16 smaller than in the MCU 19. If the O_2 supply deteriorates in these MCUs, then anoxic areas appear in the MCU 19 as well as in the MCU 16 much later than in the MCU 1. In the MCU 16 this effect appears even later than in the MCU 19.

The PO_2 frequency distributions of the other MCUs shown in Figure 2 also have a characteristic form corresponding to the arrangement of their capillary ends. The MCUs that are not taken into consideration here, also have structure specific PO_2 distributions.

From histological investigations we know for instance that the capillary distance differs from organ to organ as well as within an organ itself (Martini and Honig, 1969; Henquell and Honig, 1976; Günther et al., 1974). In general, an interval can be given for an organ in which the structure parameters vary. Therefore, it is necessary to investigate the influence of the supply parameters as well as of the structure parameters upon the PO_2 distribution in the MCUs. This can be accomplished by using the MCU 16 as an example for all the other MCUs.

To begin with, let us consider the influence of the capillary distance and length upon the PO_2 field using Equations 22 and 9. First, it is assumed that the

O_2 consumption is constant and that the capillary blood flow maintains a constant $AVDO_2$ and especially a constant venous PO_2 despite the variances in capillary distance and length. Under these assumptions, an enlargening (reduction) of the capillary distance by 10 % of its mean value ($d = 60\,\mu m$) causes a displacement of the PO_2 distribution towards lower (higher) PO_2 values in the example chosen. A massive change in capillary length (from 480 to 720 µm) shows no shift in the frequency maxima.

The explanation for the above results can be found in the form of the MCU. A change in the capillary distance means a change in the diffusion path and therefore a change in the PO_2 gradient which causes the diffusion flow into the tissue. A change in the capillary length means, at a constant $AVDO_2$ (constant venous PO_2), a change in the steepness of the PO_2 course along the length of the capillary. The diffusion flow into the tissue is influenced here by a change in the i.c. PO_2. Under the above assumptions this influence is smaller in contrast to that influence by changing the diffusion path.

Secondly, we disregard the assumption of a constant $AVDO_2$ and consider the influence of capillary distance and length only with a constant O_2 consumption and a constant capillary blood flow. Then the venous PO_2 varies with the change in capillary distance and length. Both structure parameters influence the capillary supply volume.

According to Equation 2, a change in the capillary distance affects, in the second power, the capillary supply volume $V_c^{(i)}$. However, a change in the capillary length affects the capillary supply volume $V_c^{(i)}$ only linearly [4].

Since the capillary blood flow $\phi_c^{(i)}$ should be constant, then according to Equations 8 and 9, the $AVDO_2$ must change in accordance with the capillary supply volume $V_c^{(i)}$, so that the O_2 consumption of the tissue, as assumed, remains constant. Since the O_2 capacity of blood and the arterial PO_2 should also remain constant, the venous PO_2 must change accordingly. Thus, an increase in the capillary distance along with the above-mentioned displacement of the PO_2 distribution towards lower PO_2 values leads to an additional shift to the left because of the decrease in the venous PO_2 (Fig. 14). By assuming a constant venous PO_2, an increase (reduction) of the capillary distance d of 10% leads to a shift in the frequency maximum of the PO_2 distribution of 3 mm Hg to the left (3 mm Hg to the right). By assuming a constant O_2 consumption and a constant capillary blood flow, the displacement however is 7 mm Hg in the chosen example. An increase in the capillary length only leads to a small shift of the PO_2 distribution to the left. This small shift results from the fact that, in the first case with a constant venous PO_2, no shift in the PO_2 distribution occurs and in the second case of a constant capillary blood flow, the capillary supply volume is only linearly altered. An increase (decrease) of 20% in the capillary length (Fig. 15) causes a shift of the PO_2 distribution of 3 mm Hg to the left (5 mm Hg to the right).

The O_2 consumption of the tissue influences the PO_2 distribution according to the diffusion Equation 22 as well as to Equation 9. With an assumed homo-

---

[4] This relationship also holds true for MCUs in which the capillary supply volumes do not have the same size (e.g., in MCU 8, Fig. 5b).

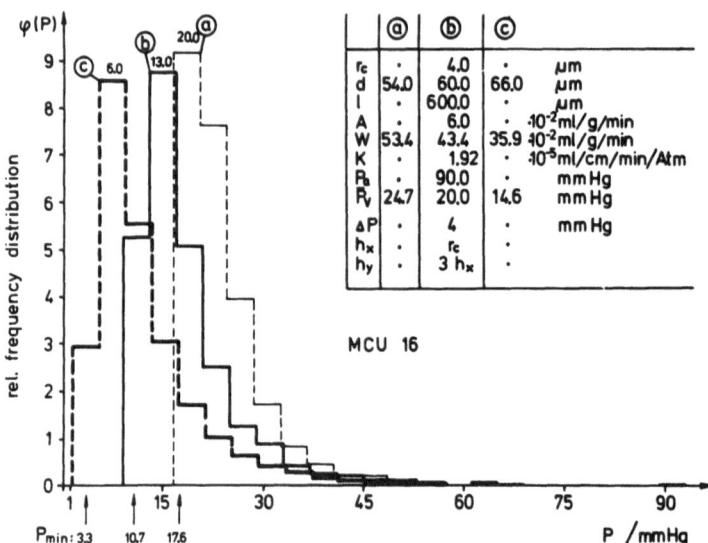

Fig. 14. PO_2 frequency distribution $\varphi(P)$ of MCU 16 at different values of capillary distances d and constant capillary blood flow ϕ_c. Under the chosen conditions an increase (reduction) of the capillary distance d of 10 % of its mean value ($d = 60 \, \mu m$) leads to a shift in the frequency maximum of PO_2 distribution of 7 mm Hg to left (7 mm Hg to right)

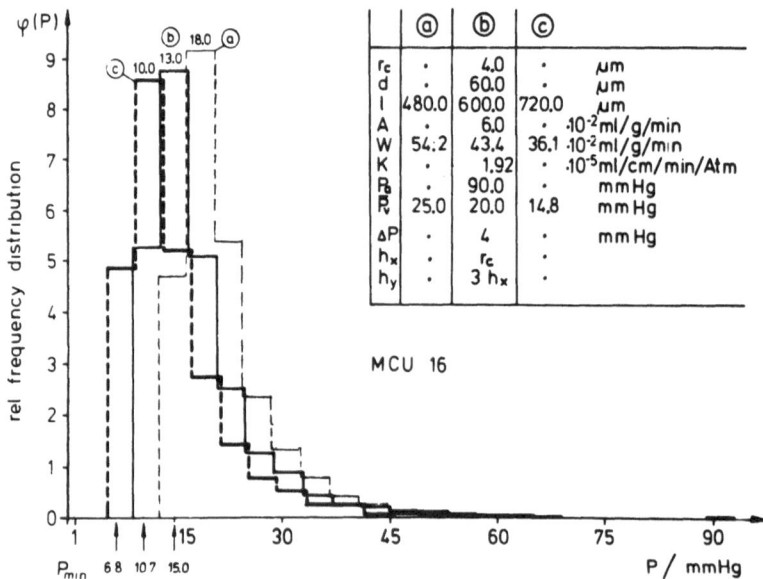

Fig. 15. PO_2 distribution $\varphi(P)$ at different values of capillary length l and constant capillary blood flow ϕ_c. Under the chosen conditions, an increase (reduction) of the capillary length l of 20 % of mean value ($l = 600.0 \, \mu m$) leads to shift in frequency maximum of PO_2 distribution of 3 mm Hg to left (5 mm Hg to right)

geneous O_2 consumption, there are two cases that must be differentiated. In the first case, the capillary blood flow varies according to the O_2 consumption so that the $AVDO_2$, particularly the venous PO_2, remains constant. Again it is assumed that the O_2 capacity of blood and the arterial PO_2 remain constant. Under such conditions, if the O_2 consumption increases (decreases) from 6 ml $O_2/100$ g/min to 8 ml $O_2/100$ g/min (to 4 ml $O_2/100$ g/min) then a resultant shift in the frequency maximum of 5 mm Hg to the left (5 mm Hg to the right) can be noted (Fig. 16).

If, in the second case, the capillary blood flow remains constant, then an increase in O_2 consumption according to Equations 8 and 9 causes an increase in the $AVDO_2$ or, in other words, it causes a decrease in the venous PO_2 and vice versa. In this case the O_2 consumption and the changes in the venous PO_2 together influence the PO_2 distribution. Thus under otherwise equal supply conditions, an increased shift of the PO_2 distribution to the left results with an increasing O_2 consumption, or an increased shift to the right with a decreasing O_2 consumption. A change of 2 ml $O_2/100$ g/min in the O_2 consumption with no change in the capillary blood flow causes a shift to the left (right) of the frequency maximum of 11 mm Hg (13 mm Hg). Furthermore, one can recognize that with a decrease in O_2 consumption, the height of the frequency maxima increases and vice versa. With a decrease in O_2 consumption the PO_2 field flattens out so that the interval in which the PO_2 values appear becomes all the more limited. If the tissue does not use any oxygen, then all of the PO_2 values fall into the class of arterial PO_2.

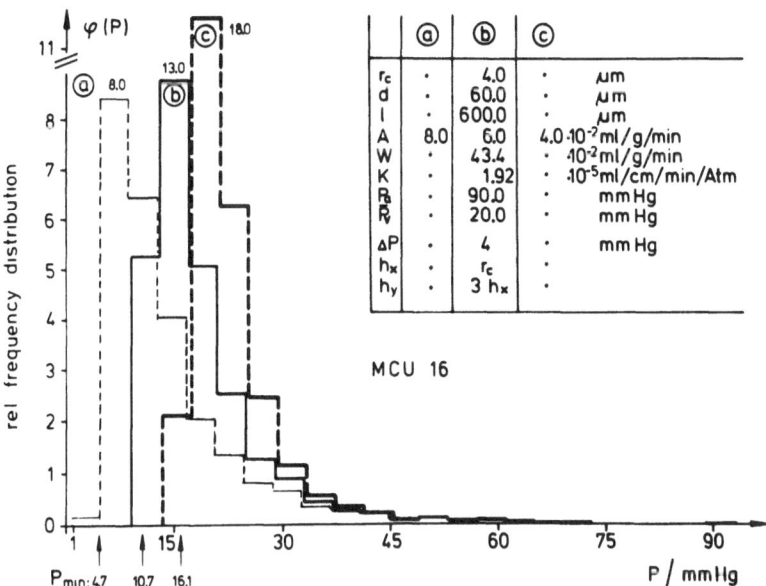

Fig. 16. PO_2 distribution $\varphi(P)$ after changing the capillary blood flow ϕ_c according to O_2 consumption A so that $AVDO_2$ remains constant (constant $\bar{P}_c$). An O_2 consumption increase (decrease) from 6 ml $O_2/100$ g/min to 8 ml $O_2/100$ g/min (to 4 ml $O_2/100$ g/min) leads to shift in frequency maximum of 5 mm Hg to the left (5 mm Hg to the right)

The capillary blood flow $\phi_c^{(i)}$ has only an indirect influence upon the diffusion equation. In order to simplify the situation, only the case of a homogeneous blood flow in the MCU 16 ($\phi_c = \phi_c^{(i)}$, $i = 1 \ldots 4$) will be considered here. A translation onto the case of an inhomogeneous capillary blood flow is possible. If the capillary blood flow $\phi_c^{(i)}$ changes with the capillary supply volume $V_c^{(i)}$ (in the case considered all $V_c^{(i)}$ are equal to one another for $i = 1 \ldots 4$) so that the quotient from both, i.e., the *local* blood flow W of the MCU remains constant, it appears that at a constant O_2 consumption an influence upon the PO_2 distribution does not come about either through Equation 22 or Equation 8. A change in the capillary supply volume means a change in either the capillary length or the capillary distance. If the capillary distance changes, then at a constant mean venous PO_2, the PO_2 distribution shifts. The latter can play a role in the mobilization or demobilization of the capillary reserve in skeletal muscle (see below) and likewise in heart muscle. If the capillary supply volume does not change then the local blood flow W of the MCU changes with a change in the capillary blood flow $\phi_c^{(i)}$. Then W is either linked to the O_2 consumption or to the $AVDO_2$ and thus to the venous PO_2. The first case was dealt with together with the change in the O_2 consumption; in the second case, the increase in local blood flow W (constant O_2 consumption) causes a decrease in the $AVDO_2$. With no changes in the oxygen capacity of blood and in the arterial PO_2, an increase in the venous PO_2 results, which causes a shift in the PO_2 frequency to the right and vice versa. Under conditions such as those chosen in Figure 17, a changing of the local blood flow W of ± 10 ml/100 g/min causes a shift in the maximum of the PO_2 frequency of 3 mm Hg to the right (6 mm Hg to the left, respectively).

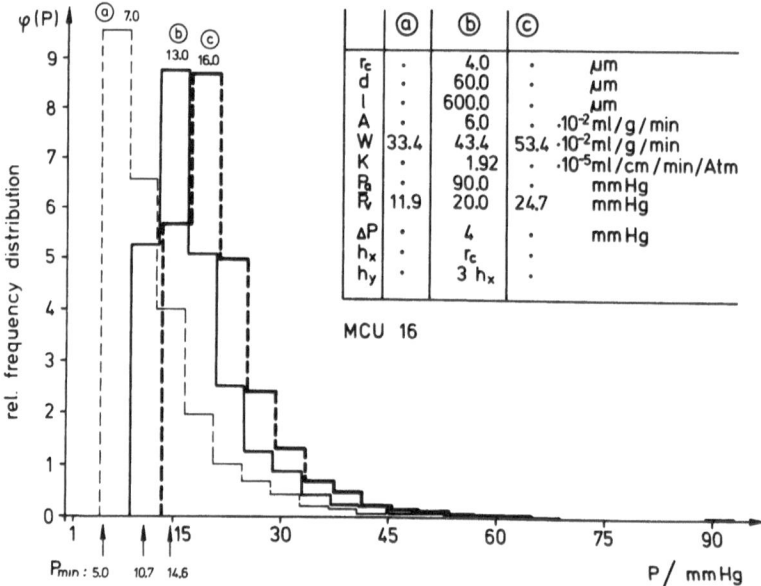

Fig. 17. PO_2 distribution $\varphi(P)$ at different values of local blood flow W and constant O_2 consumption A. Under the chosen conditions, an increase in local blood flow causes a decrease in $AVDO_2$ (increase in $\bar{P}_v$) and vice versa. Changing of local blood flow by ± 10 ml/100 g/min causes shift in frequency maximum of 3 mm Hg to right (6 mm Hg to left)

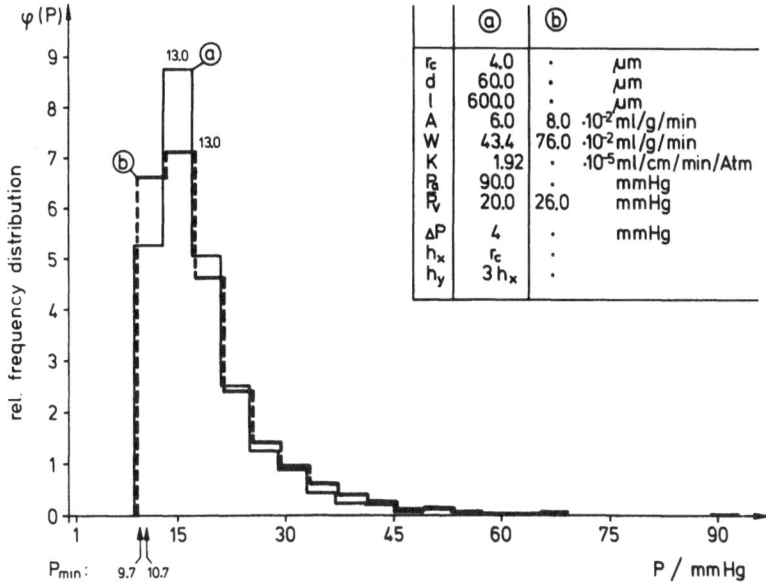

Fig. 18. *PO₂* distribution $\varphi(P)$ at increasing values of local blood flow *W* as well as of O₂ consumption *A*. Value pairs can be found (i.e. *A*=6.0 ml O₂/100 g/min, *W*=43.4 ml bl/100 g/min and *A*= 8.0 ml O₂/100 g/min, *W*=76.0 ml bl/100 g/min) having similar O₂ supply (similar *PO₂* distribution if locations of frequency maxima are compared)

If the local blood flow *W* as well as the O₂ consumption changes in such a way that *both* supply parameters increase, then *value pairs* can be noted for which the *PO₂* distribution is similar if the frequency maxima of the *PO₂* distribution are compared (Fig. 18). Under the above conditions, the value pairs *A*=6.0 ml O₂/100 g/min, *W*=43.4 ml blood/100 g/min, and *A*=8.0 ml O₂/100 g/min, *W*= 76,0 ml blood/100 g/min have a similar O₂ supply. Thus the organism has the ability to change supply parameters without having to change to any large degree the O₂ supply of the tissue.

Even though a *PO₂* distribution has a form that is specific for a MCU and moreover, changes in structure and supply parameters can be quantified (e.g., in changes in the frequency maxima), no comparison with the measured *PO₂* frequency distributions is possible at this point. In the considerations hitherto, it was tacitly assumed that the complete capillary network of an organ is composed of *equal* MCUs. Accordingly, the complete capillary network would consist, for instance, of either only MCUs with a concurrent flow direction or only MCUs with a countercurrent flow direction or only MCUs with the same asymmetric capillary arrangement.

Even though some organs appear to favor particular structural principles, it is generally the case that the capillary is composed of different MCUs and the *PO₂* electrode is located in differently structured MCUs during measurements. Only a combination of the diversely formed MCUs can allow for a comparison between measured and calculated *PO₂* distributions. Then an analysis of the O₂ supply to tissue is possible using the measured *PO₂* distribution.

c) Superimposed PO_2 Distribution

We shall assume that the capillary network of an organ is composed of such MCUs that have their genesis through a systematic displacement of the capillary ends of δl along the length of the four capillaries in the MCU. Figure 19 shows 5 of the 256 MCUs, which are formed for $\delta l = l/2$. The initial structure for the displacement of the capillary ends is the MCU 1. The PO_2 field resulting from such a capillary network is represented by the collective P matrices of the individual MCUs. The total PO_2 distribution is a resultant from the summation of the frequencies of the elements of all P matrices. The number of the MCUs with a systematic displacement of the capillary ends of δl results in:

$$\kappa = \left(\frac{2 \cdot l}{\delta l}\right)^4 \tag{23}$$

and the total PO_2 distribution of the capillary network results in:

$$\psi(P) = \frac{1}{\kappa} \sum_{i=1}^{\kappa} \varphi_i(P). \tag{24}$$

$\psi(P)$ is called a *superimposed distribution* and is defined by the interval $\hat{P}_{\min} \le P \le P_a$. The superimposed distribution is normed like the distribution $\varphi_i(P)$ of the MCUs. $\hat{P}_{\min}$ is the minimum PO_2 for *all* P matrices. Figure 20 shows the superimposed distribution for the above-named example with $\delta l = l/2$ and $\kappa = 256$. From these $\kappa = 256$ MCUs, only $\kappa' = 21$ are *non identical* MCUs. From now on, only these will be taken into consideration.

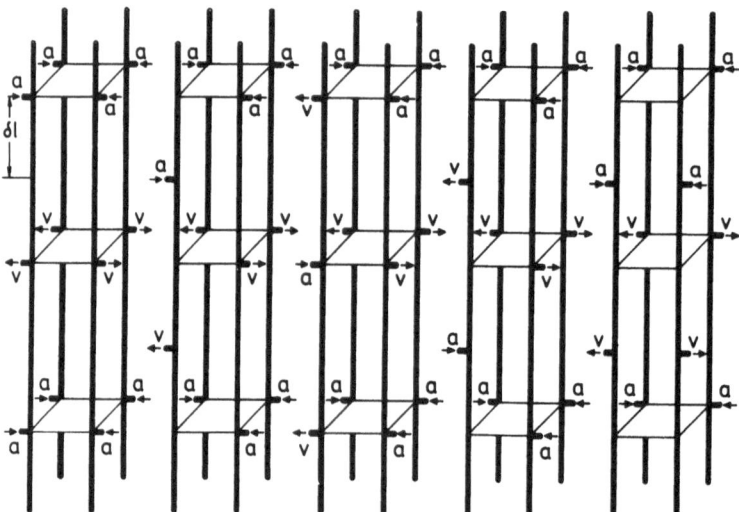

Fig. 19. Genesis of MCUs through systematic displacement of capillary ends of δl along length of the four capillaries in the MCU. Initial MCU for displacement of capillary ends is MCU 1. For $\delta l = l/2$, a number of $\kappa = 256$ MCUs results. The first 5 MCUs of these 256 MCUs are shown. From these 256 MCUs only $\kappa' = 21$ are nonidentical MCUs

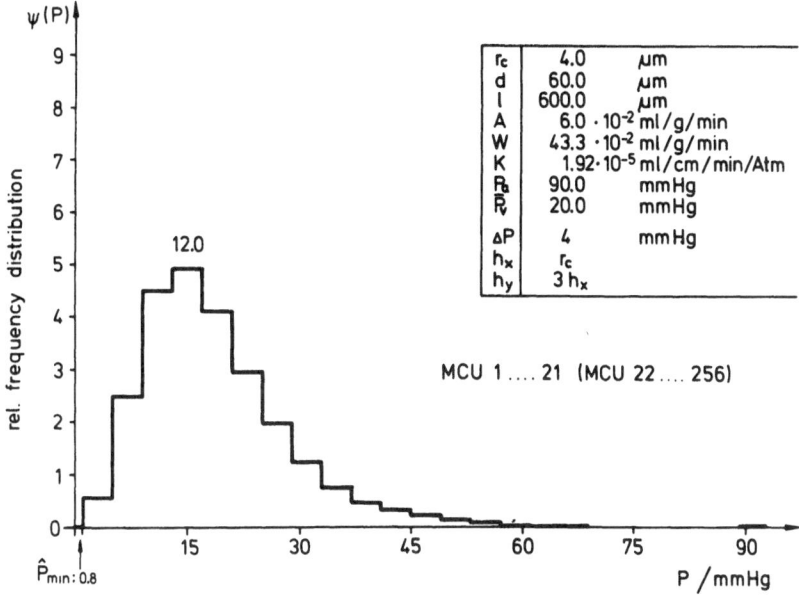

Fig. 20. Superimposed PO_2 distribution $\psi(P)$ of $\kappa'=21$ nonidentical PO_2 distributions $\varphi_i(P)$ of the MCUs, which are generated by systematic displacement of capillary ends of $\delta l = l/2$ (Fig. 19)

The values of the parameters such as the capillary distance, are statistically distributed (GÜNTHER et al., 1974; HENQUELL and HONIG, 1976; HENQUELL et al., in press) and we assume that κ'' different values exist. The PO_2 distribution $\varphi_i(P)$ for *each* of the $j=1\ldots\kappa''$ values of the parameters in Equation 24 must be provided with a weighting factor c_j. This weighting factor c_j corresponds to the frequency with which a particular capillary distance can be found. Thus the following results from Equation 24:

$$\psi(P)=\sum_{j=1}^{\kappa''}\frac{c_j}{\kappa'}\sum_{i=1}^{\kappa'}\varphi_i(P). \tag{25}$$

Furthermore, if one considers that also the κ' MCUs and therefore the $\varphi_i(P)$ can have differing weighting factors, the following results from Equation 25:

$$\psi(P)=\sum_{i=1}^{\nu}a_i\,\varphi_i(P) \tag{26}$$

with $\nu=\kappa'\cdot\kappa''$, $\sum_{i=1}^{\nu}a_i=1$ and $a_i\geq0$ the relative portion of the MCU (with the distribution $\varphi_i(P)$ and with one of the κ'' parameter values) on the whole of the capillary network.

The superimposed PO_2 distribution enables a direct comparison between a calculated and a measured PO_2 distribution. Thus the assumption that a capillary network is built up of uniform MCUs can be ignored while, at the same time, biological variations in the structure and supply parameters can be brought into the analysis. In the analysis of a measured PO_2 distribution, one starts with the known structure and supply parameters and forms parameter classes according to their statistical distribution. From this κ'', different values of parameters result. By displacing the capillary ends by the length δl for each value of the parameters, one determines the κ' non identical MCU and calculates the PO_2 distribution $\varphi_i(P)$ for $i=1$ to κ' and for each of the κ'' parameter values. The measured PO_2 distribution is then considered as the superimposed distribution $\psi_m(P)$. According to Equation 26 it follows that (with $v = \kappa' \cdot \kappa''$):

$$\psi_m(P) = \sum_{i=1}^{v} a_i \, \varphi_i(P); \quad \sum_{i=1}^{v} a_i = 1; \quad a_i \geq 0 \tag{27}$$

and using the least square conditions:

$$\int_0^{P_m} |\psi_m(P) - \sum_{i=1}^{v} a_i \, \varphi_i(P)|^2 \, dP = \text{Min} \tag{28}$$

resulting in the weighted a_i with which the basic distributions $\varphi_i(P)$ for a defined value of the κ'' parameters participate in the measured distribution $\psi_m(P)$. P_m denotes the largest measured PO_2 in $\psi_m(P)$.

As a measure of the degree of accuracy for this analysis, we use the difference between the areas under the measured distribution $\psi_m(P)$ and under the superimposed distribution $\psi(P)$ (calculated from a_i and $\varphi_i(P)$ above). In the next part of this chapter, such an analysis will be carried out for the skeletal muscle so that the details of the analysis become elucidated.

B. Special Results—O_2 Supply in Skeletal Muscle

1. Capillary Network and Special Features of the O_2 Supply

The generally relevant results concerning the O_2 supply that were gained by the PO_2 frequency distributions will be supplemented in this chapter by special results obtained from skeletal muscle. Skeletal muscle is particularly useful for an analysis of O_2 supply with the capillary model. The capillary network of the skeletal muscle is characterized by parallel-running capillaries. One muscle fiber is accompanied by several parallel capillaries. The flow direction in the capillaries and the arrangement of the in- and outflows show no particular regularity. This coincides with the anatomical assumptions of our capillary model. We therefore assume that the capillary network is built up from varying MCUs.

A special feature in the O_2 supply of skeletal muscle is the fact that skeletal muscle can increase its O_2 consumption from 0.1 ml O_2/100 g/min (rest conditions) to values above 10.0 ml O_2/100 g/min (exercise conditions). Under rest conditions and with diffusion paths that correspond to the capillary distance of 60–100 μm an enormous oxygen oversupply of the tissue exists. The skeletal muscle seems to counter this oversupply with the mechanism of demobilization of the capillary reserve. A capillary in the anatomical sense does not have to be a functional capillary, i.e., a blood vessel that transports erythrocytes and takes part in the O_2 supply. With respect to the O_2 supply, a change-over from a nonfunctional into a functional state and vice versa is the mechanism of the mobilization and the demobilization of the capillary reserve, respectively.

2. Rest Conditions

Under rest conditions (O_2 consumption between 0.1 and 0.3 ml O_2/100 g/min) and with a (anatomical) capillary distance of 60–100 µm, an O_2 diffusion shunt exists as soon as the capillary network deviates from MCUs with concurrent flow direction (MCU 1). Under these conditions, at a capillary distance of 60 µm (or 100 µm), a PO_2 difference of about 3 mm Hg (or about 13 mm Hg) between two neighboring capillaries leads to an O_2 shunt. This excessive supply of O_2 in the resting skeletal muscle fiber can be avoided in several fashions. A high capillary blood flow for example would (Eqs. 8 and 9) lead to a small $AVDO_2$ while the venous PO_2 would be high enough in order to avoid an O_2 shunt. However, an organ needs an increase in capillary blood flow in order to prevent possible anoxy in the tissue when—by an increasing O_2 consumption—the $AVDO_2$ decreases severely. Another way to prevent an excessive O_2 supply is to enlargen the diffusion path by demobilizing the capillary reserve. Under rest conditions, an increase in the functional capillary distance to 4 times the anatomical capillary distance is possible without the minimal PO_2 reaching values below 1 mm Hg.

3. Capillary Reserve and Muscle Exercise

We assume that under rest conditions the functional capillary distance is so large that the minimal PO_2 in tissue lies at about 1–5 mm Hg. This assumption is in good agreement with measurements, for example, of KUNZE (1966, 1969) and COBOURN and MAYERS (1971). If the O_2 consumption increases with an unchanged local blood flow W, then the $AVDO_2$ becomes larger and the venous PO_2 sinks. The PO_2 distribution shifts to the left (cf. Fig. 16). This shift of the PO_2 distribution towards lower PO_2 values is checked by mobilization of the capillary reserve in that the functional capillary distance is reduced to the anatomical capillary distance. However, a shorter diffusion path means a larger PO_2 gradient and thus a larger diffusion flow of O_2 to the muscle tissue. If the total capillary reserve is mobilized and the functional capillary distance is reduced to the anatomical capillary distance, then the skeletal muscle still has the possibility of countering a further sinking of the $AVDO_2$ by increasing the local blood flow of the muscle tissue. The two steps in vivo, the mobilization of the capillary reserve and the increase of the local blood flow, do not occur in succession.

To begin with, it is incomprehensible that the mobilization of the capillary reserve can exist without a change in the local blood flow W. Since even more capillaries are filled with blood with a decrease in the capillary distance, the inflowing blood is distributed into more capillaries and the capillary blood flow of the individual capillaries decreases. According to Equations 8 and 9, the local blood flow W is the quotient from the capillary blood flow $\phi_c^{(i)}$ and the capillary supply volume $V_c^{(i)}$. If $\phi_c^{(i)}$ and $V_c^{(i)}$ vary during mobilization of the capillary reserve, then the local blood flow W can remain the same. That this is possible can be demonstrated on the capillary model at hand.

For a better understanding we shall also consider here the MCU 16 with Helix structure. In this MCU, the capillary blood flows and the capillary supply volumes of the four capillaries are equal to one another respectively. Such a consideration can be made for all other MCUs.

In the model, four capillaries are ordered to the MCU in such a fashion that they *collectively* supply the tissue fragment with oxygen. If during mobilization of the capillary reserve a further capillary arises at the intersection of the diagonal lines between four capillaries in a MCU (Fig. 21), then the number of functional capillaries is doubled. Since the capillary blood flow is equally distributed amongst the capillaries, then, with the additional capillary, the capillary blood flow is reduced to one half of its previous values. The capillary distance is reduced by a factor of $1/\sqrt{2}$. The capillary distance quadratically changes the capillary supply volume (Eq. 2). Thus, the capillary supply volume and the capillary blood flow were decreased to one half of their value before the diffusion path was decreased. Hence the local blood flow W remains un-

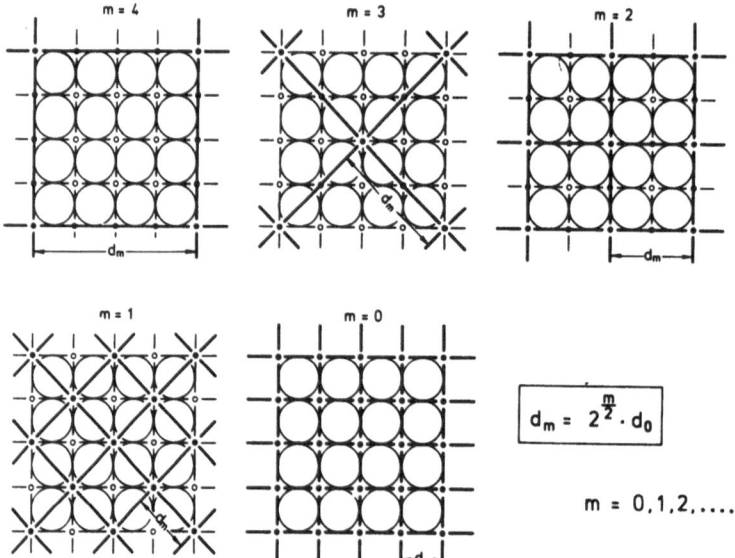

$$d_m = 2^{\frac{m}{2}} \cdot d_0$$

$$m = 0, 1, 2, \ldots$$

Fig. 21. Mobilization and demobilization of capillary reserve. Filled dots and hollow dots indicate functional and nonfunctional capillaries (capillary reserve) respectively. Darkened lines indicate tissue fragments of MCUs, the circles indicate muscle fibers (in cross-section). At $m=0$, all anatomical capillaries are equal to functional capillaries (total capillary reserve mobilized). Four capillaries collectively supply one muscle fiber. At $m=1$ at intersection of diagonal lines between four capillaries, functional capillary changes over to nonfunctional capillary. Then number of functional capillaries is halved and four capillaries collectively supply two muscle fibers (step 1 of demobilization of capillary reserve). Capillary blood flow and capillary supply volume are doubled. In figure, step 0 to step 4 of demobilization (mobilization) of capillary reserve is shown. For each step m, functional capillary distance d_m can be calculated from anatomical capillary distance d_0 using the encircled equation

changed. Likewise, demobilization of capillary reserve can function with a decreasing O_2 consumption and a decrease in the number of functional capillaries.

The principle of mobilization and demobilization of capillary reserve can be described as follows: if d_0 is the anatomical capillary distance and ϕ_{co} the capillary blood flow, given the case where the anatomical capillary distance is concomitant with the functional capillary distance, then the functional capillary distance and capillary flow of the m-th step in the demobilization of the capillary reserve are given by the following:

$$d_m = 2^{m/2} \cdot d_0 \quad \text{and} \quad \phi_{cm} = 2^m \cdot \phi_{co}. \tag{29}$$

Because of Equation 2:

$$V_{cm} = \tfrac{1}{4}(d_m^2 \cdot l) = \tfrac{1}{4}(2^m \cdot d_0^2 \cdot l). \tag{30}$$

This is the capillary supply volume of this step. According to Equation 12 it follows

$$W_c = \frac{\phi_{cm}}{V_{cm}} = \frac{\phi_{co}}{V_{co}} = \text{constant} \tag{31}$$

for all $m = 0, 1, 2, 3, \ldots$, i.e., in the course of mobilization and demobilization of the capillary reserve (Fig. 21).

With the mobilization and demobilization of the capillary reserve, the skeletal muscle can not only change the functional capillary distance but also the structure of the capillary network. As shown in Figure 22, it is possible that through mobilization of the capillary reserve, the functional structure of the MCU 1 with concurrent flow direction can change into a MCU 19 with counter-current flow direction and, finally, can change over into a MCU 16 with Helix structure. If, though, the functional structure of the capillary network coincides with the anatomical structure, i.e., if the total capillary reserve is mobilized, then the capillary network with its MCUs 16 has taken on its optimal structure for the O_2 supply (Fig. 13)[5].

In combination with the mobilization of the capillary reserve, the hypoxy experiments from STAINSBY and OTIS (1964) shall be considered. The authors caused hypoxy in the gastrocnemius muscle of a dog under rest conditions by decreasing the arterial PO_2 along with the venous PO_2, so that the $AVDO_2$ remained constant. The O_2 consumption was 0.4 ml O_2/100 g/min, the blood flow 3.5 ml bl/100 g/min. Both values remained unchanged up to the *critical* arterial $PO_2 = 60$ mm Hg and the correlated venous $PO_2 = 25$ mm Hg. From this point on, the O_2 consumption was diminished. For these critical supply conditions, the functional capillary distance for the M. gastrocnemius is calcu-lated in the model, whereby it is assumed that the lowest PO_2 in tissue (under

---

[5] At this point, however, we neglect that the constancy of the local blood flow W is, for hemodynamic reasons, only possible under certain limits. The capillary blood flow cannot increase indefinitely with an increase in functional capillary distance. Then the blood flow of skeletal muscle decreases with a reduction in heart activity. The same holds true in the opposite case by a decreasing functional capillary distance.

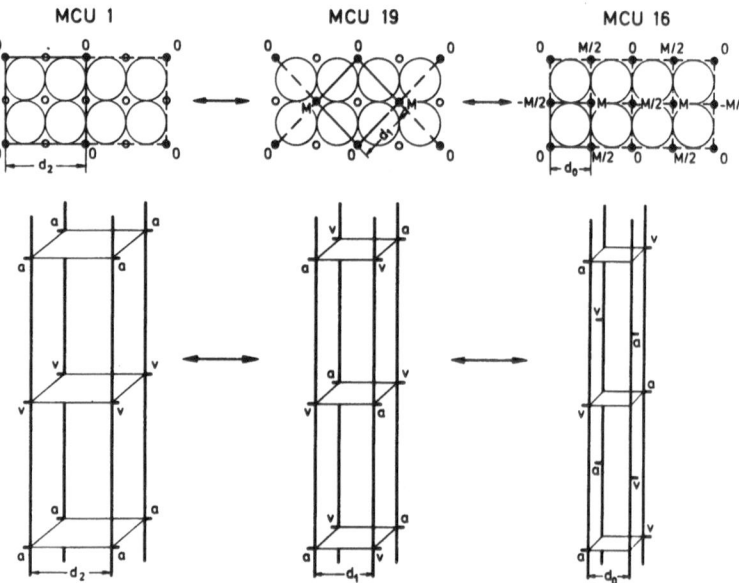

Fig. 22. Changing of functional structure of capillary network by demobilization and mobilization of capillary reserve. E.g., at step $m=2$ of capillary reserve mobilization (cf. Fig. 21), capillary network is composed of MCUs with structure as in MCU 1. At step $m=1$, arising capillaries can change structure of capillary network so it is composed of MCUs with structure as in MCU 19. Finally if all capillaries are mobilized ($m=0$), capillary network can be composed of MCUs 16. O_2 supply of muscle fiber is then better than if at $m=0$ capillary network were composed of structure as in MCU 1 (Fig. 13)

the above conditions) is about 1 mm Hg. Since in these experiments no PO_2 measurements in tissue were made, this assumption is necessary. Here, too, the assumptions correspond with the experimental results on skeletal muscle described by COBOURN and MAYERS (1971). If one again uses the MCU 16 as the basis for the calculation, then the functional capillary distance is calculated up to about 210 µm (Fig. 23, *curve b*).

In further experiments from STAINSBY and OTIS (1964) on exercised muscle, the O_2 consumption was higher by a factor 8 than under rest conditions. A critical arterial PO_2 of 40 mm Hg and the correlating venous PO_2 of 10 mm Hg were reached. These lower critical values, compared to the values under rest conditions, indicated a further mobilization of the capillary reserve with increasing exercise of the skeletal muscle. With a shorter functional capillary distance, the venous PO_2 can be smaller without changing the PO_2 gradient that is necessary for the O_2 transport. Under the above conditions, the functional capillary distance is calculated to be about 66 µm (Fig. 23, *curve c*). This value lies within the normal limits of the anatomical capillary distances. The quotient derived from the functional capillary distance under rest conditions and under exercise conditions allows us to draw some conclusions concerning the step m in the mobilization of the capillary reserve (cf. Fig. 21). The relationship of the two calculated capillary distances results in a value of 3.2. According to Equation 29, the relationship of the functional capillary distance d_m of the m-th step to the anatomical capillary distance d_0 $(m=0)$ is:

$$\frac{d_m}{d_0} = 2^{m/2}.$$ (32)

This value is, for $m=3$, approximately as large as the comparable relationship of the capillary distances that have been determined with experimental data. In the experiments under consideration, three steps of the mobilization of the capillary reserve are fully exhausted in the change-over from rest conditions to exercise conditions in skeletal muscle.

If an analogous calculation for skeletal muscle under rest conditions with normoxy ($P_a=95$ mm Hg, $P_v=38$ mm Hg) is carried out (Fig. 23), a functional capillary distance of approximately 330 µm results. This results in a step $m=5$ of the demobilization of the capillary reserve. If in step $m=4$, the functional capillary distance d_m is (Eq. 29) 4 times that of the anatomical distance ($d_0=60$ µm), neither anoxic areas nor an O_2 shunt result in tissue ($P_{min}=20$ mm Hg) under the chosen conditions. With smaller values d_m ($m=3$) an O_2 shunt arises. These results indicate that the functional capillary distance in skeletal muscle can always be maintained as large as possible. Then even under rest conditions, the minimal PO_2 values in tissue lie between 1–5 mm Hg. With an increasing

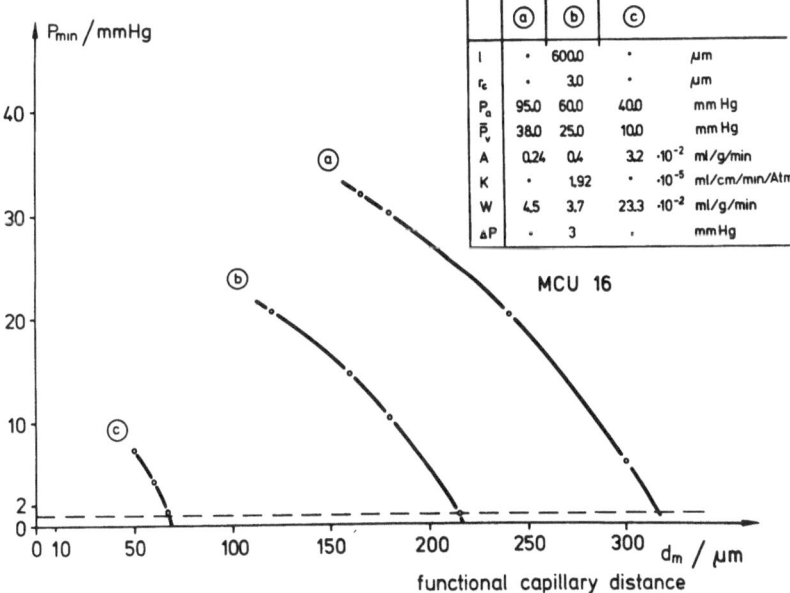

Fig. 23 a–c. Dependence of minimal tissue PO_2 (P_{min}) from functional capillary distance d_m in MCU 16. a) at normoxic conditions in resting skeletal muscle. b) at diminished arterial PO_2 (and mean venous PO_2) under rest conditions corresponding to experiments from STAINSBY and OTIS (1964). c) at further diminished arterial PO_2 (and mean venous PO_2) under exercise conditions corresponding to same experiments. If P_{min} curve in case a–c reaches value of about 1 mm Hg, functional capillary distance (d_m) results. In case c, functional capillary distance of 66 µm results. This value lies within normal limits of anatomical capillary distance

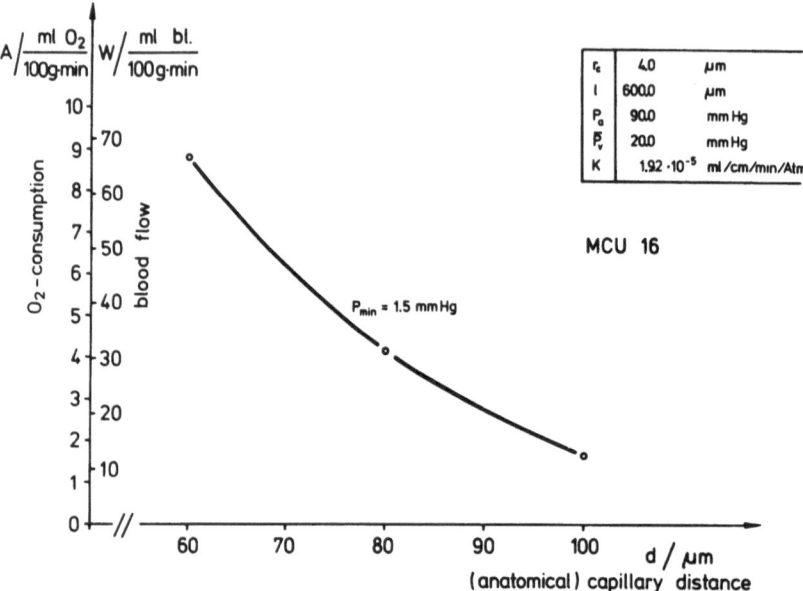

Fig. 24. Dependence of O_2 consumption and local blood flow of MCU 16 from (anatomical) capillary distance (d) under muscle exercise conditions and at constant $AVDO_2$ ($P_a=90$ mm Hg, $\bar{P}_v=20$ mm Hg), where minimal tissue PO_2 (P_{min}) is about 1,5 mm Hg. Under the chosen conditions at O_2 consumption of $A=9$ ml $O_2/100$ g/min (2 ml $O_2/100$ g/min), at local blood flow $W=70$ ml bl/100 g/min (15 ml bl/100 g/min) and at capillary distance $d=60$ μm (100 μm), a minimal tissue PO_2 of 1.5 mm Hg results

exercise the capillary reserve is mobilized in a stepwise fashion. Thereby the functional capillary distance is held as large as possible for each step until it finally reaches the anatomical capillary distance. Up to this point it is still possible despite an increase in O_2 consumption to maintain the lowest PO_2 values between 1–5 mm Hg. Then only by further increasing the O_2 consumption does the danger of anoxic areas in muscle tissue exist. The PO_2 distribution shifts to the left (Fig. 16).

Thus, if the blood flow in the skeletal muscle increases such that the $AVDO_2$ remains unchanged, then the shift to the left of the PO_2 distribution is not hindered but is only lessened. A shift to the left can only be hindered when the increase in blood flow is so high that the $AVDO_2$ sinks. Because of the elevated i.c. PO_2, the PO_2 gradient between capillary and tissue is then steep enough and the O_2 diffusion flow to tissue is then sufficient.

Under extreme exercise of the skeletal muscle (Keul, 1970; Doll, 1971), the venous PO_2 can sink to 19 or 20 mm Hg. If one assumes that at this point the functional capillary distance (equal to the anatomical capillary distance) is 60 μm, then, at an O_2 consumption rate of approximately 9 ml $O_2/100$ g/min, the PO_2 distribution in the capillary network with MCUs 16 reaches a minimal tissue PO_2 of 1.5 mm Hg (Fig. 24). From Equations 9 and 12, a local blood flow of the skeletal muscle of approximately 70 ml bl/100 g/min can be calculated.

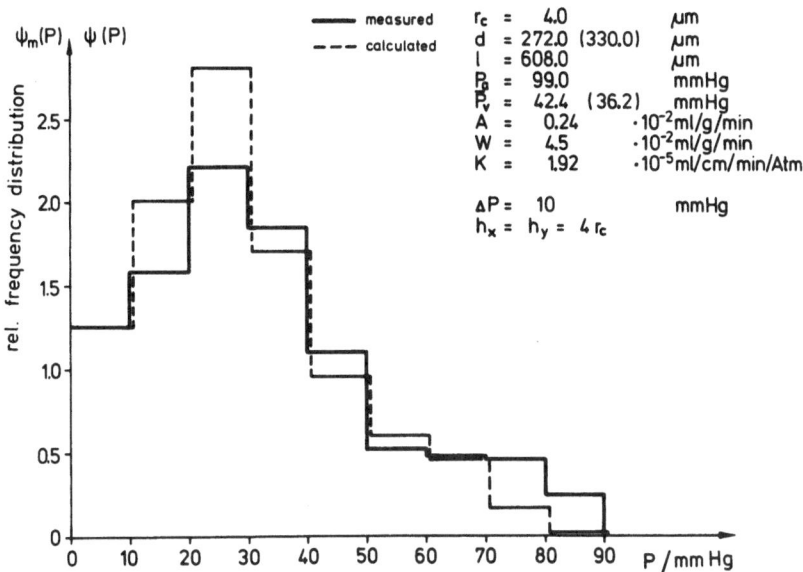

Fig. 25. PO_2 distribution $\psi_m(P)$ measured in resting human M. tibialis anterior (KUNZE, 1969) and recalculated PO_2 distribution $\psi(P)$ (normed such that the areas under the respective PO_2 distributions equal 100%). $\psi(P)$ is calculated from basic distributions $\varphi_i(P)$ and their weighting factors a_i which result from analysis of the measured distribution $\psi_m(P)$. Sum of absolute deviations from $\psi_m(P)$ and $\psi(P)$ (absolute difference in areas in individual P classes) is 20% of total area

4. Analysis of the PO_2-Distribution Measured in Skeletal Muscle

a) Problem Disposition and Analysis

If one assumes that a PO_2 distribution is measured in tissue with a Pt electrode, and any parameters that influence the O_2 supply are known, then the problem disposition for the analysis reads as follows: which information concerning the unknown structure and supply parameters and concerning the structure of the capillary network can be ascertained from the measured PO_2 distribution? The measured PO_2 distribution is (Eq. 27) considered as a superimposed distribution $\psi_m(P)$. KUNZE (1969) described PO_2 measurements with Pt electrodes in the healthy M. tibiales anterior of patients under rest conditions. The PO_2 distribution formed from 815 measurements (Fig. 25) has between 0-100 mm Hg a P class of $\Delta P = 10$ mm Hg. The following supply parameters are known:

O$_2$ consumption　　　　$A = 0.24$ ml O_2/100 g/min (KUNZE, 1969)
Blood flow　　　　　　$W = 4.5$ ml bl/100 g/min (RENKIN, 1970)
Diffusion conductivity $K = 1.92 \times 10^{-5}$ ml O_2/cm/min/atm (KUNZE, 1969)
Arterial PO_2　　　　　$= 99$ mm Hg [6]
Mean venous PO_2　　　$= 42.4$ mm Hg (calculated from A, W, and Eq. 7).

---

[6] The arterial PO_2 is chosen relatively high since the measured PO_2 distribution has values at about 100 mm Hg.

The capillary length is assumed to be 600 μm. Along with the mean value of the capillary distance d, the capillary distances of $d \pm 20\%$ were also considered in the analysis. The mean capillary distance d is determined with respect to the capillary reserve such that the minimal PO_2 lies at about 1 mm Hg for the MCUs under consideration. Thus under rest conditions, a mean functional capillary distance of $d = 275$ μm results. Furthermore, it is assumed that the capillary network in the skeletal muscle under consideration is composed of four MCUs [7]: the MCU 1 with concurrent flow direction, the MCU 16 with helix structure, the MCU 19 with total countercurrent flow direction, and the MCU 10 as a representative of one of the many asymmetric MCUs. From the above assumption of four MCUs and the differing values of the capillary distance, 12 basic distributions $\varphi_i(P)$, $i = 1 \ldots 12$ are calculated. Since the measured PO_2 distribution $\psi_m(P)$ is subdivided into 10 P classes and has only 10 frequency values, the 12 basic distributions cannot be concomitantly considered in an analysis. Therefore a number $\nu = 10$ of basic distributions $\varphi_i(P)$ is used per analysis and the coefficients a_i are chosen such that (Eq. 28) the coefficient leads to the best congruence between the measured distribution $\psi_m(P)$ and the recalculated distribution $\psi(P)$. The result of the analysis is that the following MCUs result in significant coefficients a_i and produce a good congruence:

$\varphi_1(P)$ of the MCU 1 with concurrent flow and a mean capillary distance d
     to 61% of the total weight
$\varphi_2(P)$ of the MCU 16 with helix structure and a mean capillary distance d
     to 15% of the total weight
$\varphi_3(P)$ of the same MCU 16 with a capillary distance $d + 20\%$ to 24% of the
     total weight
$\varphi_4(P)$ of the MCU 19 with countercurrent flow with a neglected part of the
     total weight.

Like before, the distributions $\psi_m(P)$ and $\psi(P)$ are normed and the areas under the distribution curves differ from one another by approximately 20% (Fig. 25). From these results one can conclude that in the capillary network of resting skeletal muscle, the MCU 19 with countercurrent flow directions are avoided and the MCU 1 with a concurrent flow direction along with the MCU 16 are preferred in a ratio of 3:2.

Because of the low O_2 consumption in skeletal muscle under rest conditions, the probability of an O_2 shunt is particularly large. Since the MCU 1 (which is not prone to an O_2 shunt) is preferred and the MCU 19 (which is most prone to an O_2 shunt) is avoided, the O_2 shunt probability is minimized for the capillary network. Even though the MCU 1—in comparison to the other MCUs—is at a disadvantage in its supply situation, there is no problem in the propitious O_2 supply of the skeletal muscle under rest conditions. This result cannot be transfered to skeletal muscle under exercise conditions. The position of the capillary ends in the neighboring capillaries can change at the mobilization of the capillary reserve, and other MCUs can participate in the arrangement of

---

[7] Because of the necessary computer time, the analysis was restricted to the four MCUs. The results of the analysis are thus not detailed enough and, therefore, the analysis has only an exemplary character.

the capillary network (Fig. 22). Thus it is better to speak of the *functional capillary network* for skeletal muscle under rest conditions.

b) Unequivocality and Error Sources of the Analysis

Because of the large P classes ($\Delta P = 10$ mm Hg), the measured PO_2 distribution is given only by 10 frequency values. This small number of frequency values limits the analysis to maximally 10 basic distributions. The analysis shows that for maximally 4 basic distributions significant coefficients a_i ($a_i \geq 0$) result. This does not preclude the fact that other combinations of basic distributions exist that have a similar accuracy of measured and recalculated frequency distributions $\psi_m(P)$ and $\psi(P)$. The analysis of the PO_2 distribution is *not unequivocal*. If the P classes are chosen such that they are smaller, then the number of frequency values of a PO_2 distribution increases, and more basic distributions can be taken into consideration in the analysis. The number of possible combinations of basic distributions that result in the same superimposed distribution decreases, however an unequivocality in the analysis does not result. Thus the results of the analysis must be interpreted with caution and can only be considered as an hypothesis.

Along with the problem of the unequivocality is the problem of error which influences such an analysis. The sources of error can be found in the abstraction of the morphological data of the model, in the size of the random sampling of the PO_2 measurements, and in the systematic error of the Pt electrode. The errors that result from the abstraction of the model cannot be quantified. Compared to other diffusion models, the restricting limitations of the diffusion model under discussion with its possibilities of simulating a 3-dimensional asymmetric capillary network are relatively small. This is particularly true with skeletal muscle with its mostly parallel-running capillaries. The number of the random samplings of the measured PO_2 distribution is, with 815 individual measurements and a ΔP class of 10 mm Hg (Fig. 29, App. 2), large enough so that the resulting error from the random sampling can be neglected. The influence of the systematical errors of the Pt electrode (e.g., O$_2$ consumption, diffusion error) particularly depends upon the tip diameter of the electrode (GRUNEWALD, 1969 a, 1970, 1971; BAUMGÄRTL et al., 1974). The results from such an error analysis show that all the measured PO_2 values maximally lie 1–2 mm Hg below the nominal value. With a P class of $\Delta P = 10$ mm Hg, the resulting shift to the left of the measured PO_2 distribution however has practically no influence upon the results of the analysis.

Appendix 1

Starting out with the above-defined grid system (Fig. 4), the differential terms of the partial differential Equation 22 are replaced by difference terms derived through a Taylor series (BATSCHELET, 1952; COLLATZ, 1955; KANTOROWITSCH and KRYLOW, 1956). The resulting difference equation takes, for each grid-point that lies outside the capillaries and does not lie on the surface area of the MCU, and, if $h_x = h_y = h$, the following form:

$$
\begin{aligned}
P^{(n+1)}(i, j, k) = \frac{\omega}{6} \Big(& P^{(n)}(i-1, j, k) + P^{(n)}(i+1, j, k) \\
& + P^{(n)}(i, j-1, k) + P^{(n)}(i, j+1, k) \\
& + P^{(n)}(i, j, k-1) + P^{(n)}(i, j, k+1) - \frac{h^2 A}{K} \Big) \\
& + (1-\omega) P^{(n)}(i, j, k).
\end{aligned} \tag{33}[8]
$$

Because of the mathematical assumptions and, if $h \leq r_c$, the difference equation on the surface area of the MCU outside of the capillary takes the following form: At the upper and lower surface (both surfaces are identical, see Fig. 1):

$$
\begin{aligned}
P^{(n+1)}(i, 0, k) = \frac{\omega}{6} \Big(& P^{(n)}(i-1, 0, k) + P^{(n)}(i+1, 0, k) \\
& + P^{(n)}(i, 2M-1, k) + P^{(n)}(i, 1, k) \\
& + P^{(n)}(i, 0, k-1) + P^{(n)}(i, 0, k+1) - \frac{h^2 A}{K} \Big) \\
& + (1-\omega) P^{(n)}(i, 0, k).
\end{aligned} \tag{34}
$$

At the lateral surfaces:

$$
\begin{aligned}
P^{(n+1)}(0, j, k) = \frac{\omega}{6} \Big(& 2 P^{(n)}(1, j, k) + P^{(n)}(0, j-1, k) + P^{(n)}(0, j+1, k) + P^{(n)}(0, j, k-1) \\
& + P^{(n)}(0, j, k+1) - \frac{h^2 A}{K} \Big) + (1-\omega) P^{(n)}(0, j, k)
\end{aligned}
$$

$$
\begin{aligned}
P^{(n+1)}(L, j, k) = \frac{\omega}{6} \Big(& 2 P^{(n)}(L-1, j, k) + P^{(n)}(L, j-1, k) + P^{(n)}(L, j+1, k) \\
& + P^{(n)}(L, j, k-1) + P^{(n)}(L, j, k+1) - \frac{h^2 A}{K} \Big) + (1-\omega) P^{(n)}(L, j, k)
\end{aligned} \tag{35}
$$

$$
\begin{aligned}
P^{(n+1)}(i, j, 0) = \frac{\omega}{6} \Big(& P^{(n)}(i-1, j, 0) + P^{(n)}(i+1, j, 0) + P^{(n)}(i, j-1, 0) + P^{(n)}(i, j+1, 0) \\
& + 2 P^{(n)}(i, j, 1) - \frac{h^2 A}{K} \Big) + (1-\omega) P^{(n)}(i, j, 0)
\end{aligned}
$$

$$
\begin{aligned}
P^{(n+1)}(i, j, L) = \frac{\omega}{6} \Big(& P^{(n)}(i-1, j, L) + P^{(n)}(i+1, j, L) + P^{(n)}(i, j-1, L) + P^{(n)}(i, j+1, L) \\
& + 2 P^{(n)}(i, j, L-1) - \frac{h^2 A}{K} \Big) + (1-\omega) P^{(n)}(i, j, L).
\end{aligned}
$$

---

[8] For an area in a MCU with $P \cong P_{50}$ (assumed to be small), for each gridpoint an additional algebraic equation must be solved because of $A = A(P)$ (Eq. 5) (SOWA and GRUNEWALD, in preparation).

Equation 35 corresponds to Neumann's boundary conditions for the differential Equation 22.

If $h > r_c$, then instead of Equation 35 at the lateral surface areas the PO$_2$ values are calculated for each j according to Equations 16, 17, and 18 (Dirichlet's boundary conditions). This method of approach is necessary for capillary distances larger than 100 µm (Figs. 23 and 25) in order to keep the computer time within reasonable limits. Both approximation processes yield similar PO$_2$ distributions. The validity of the general and the special results is thereby not restricted.

Equations 33 and 34 and Neumann's or Dirichlet's boundary conditions are a linear algebraic equation system, which can be solved by relaxation (FORSYTHE and WASOW, 1967), by using the Liebmann process (LIEBMANN, 1918). This process replaces each value $P(i, j, k)$ outside the capillaries by a new value which is calculated from its neighboring points according to the equation system; ω is the overrelaxation factor. By choosing a suitable overrelaxation factor, the convergence of the iteration process can be accelerated (Fig. 26). Arbitrary values can be assigned to the grid points outside the capillaries as starting values.

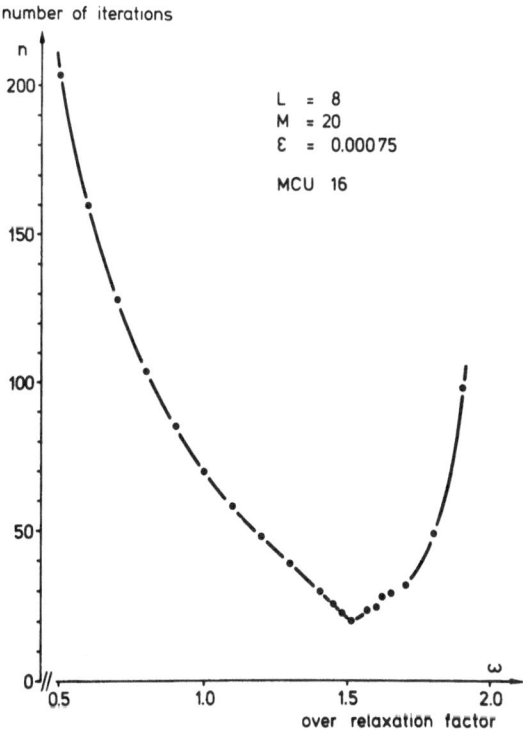

Fig. 26. Number of iteration steps n, dependent upon overrelaxation factor ω (MCU 16, $L=8$ and $M=20$, truncation barrier $\varepsilon=0.00075$), the number n of the iteration steps for $\omega=1.52$ in example chosen takes on a minimal value. Without overrelaxation factor, i.e., $\omega=1$, 5 times more iteration steps are necessary to reach same truncation barrier

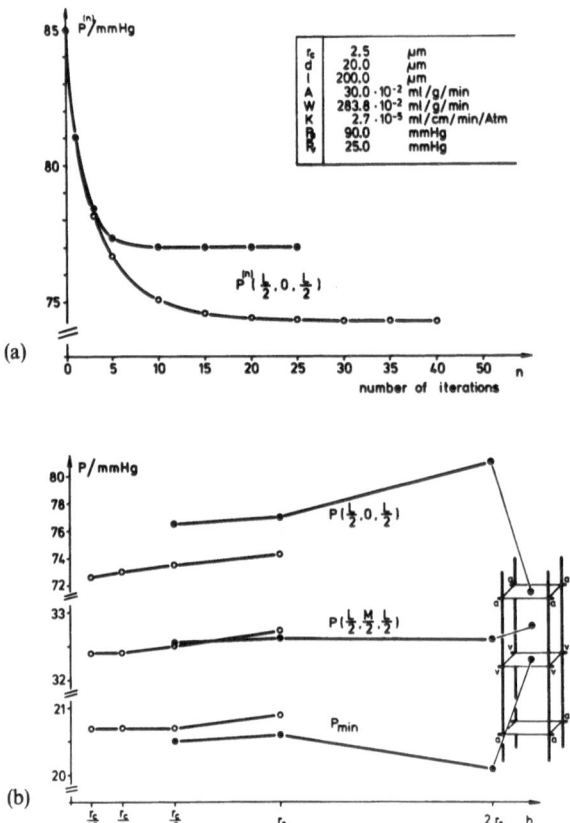

Fig. 27a and b. a) Convergence of approximative solution $P^{(n)}(i, j, k)$ of difference equation system at interstice length $h_x = h_y = h = r_c$, depending upon number of iterations (truncation error). Behavior of approximative value $P^{(n)}\left(\dfrac{L}{2}, 0, \dfrac{L}{2}\right)$ of PO_2 field is shown (MCU 1) for Neumann's boundary conditions (o–o–o) and Dirichlet's boundary conditions (●–●–●). b) Convergence of approximative solution $P(i, j, k)$ of differential equation, depending upon interstice length $h(=h_x=h_y)$ (discretization error). For this, truncation barrier ε was chosen large enough so that $P(i, j, k)$ is independent of number of iterations n. $P(i, j, k)$ was calculated according to Eqs. 33–35 (Neumann's boundary conditions o–o–o) and Eqs. 33, 34 and Eqs. 6–18 for determination of PO_2 on lateral surface areas of tissue fragment (Dirichlet's boundary conditions ●–●–●). Convergence of $P(i, j, k)$ is shown for three points of PO_2 field (see MCU 1 on right side of Fig. 27b). For $h_x = h_y = h = r_c$ and Neumann's boundary conditions a sufficient approximation of PO_2 field results (discretization error ≤ 1.5 mm Hg at high PO_2 values and ≤ 0.5 mm Hg at PO_2 values in range of P_{min}). For $h > r_c$ and Dirichlet's boundary conditions, deviations of more than 5 mm Hg result in range of high PO_2 values

The iteration process converges, independent of the choice of starting values (COURANT et al., 1928). In any case, the process calculates from the boundary conditions the approximative values of the PO_2 field except for a given error (Fig. 27a). If $P^{(n)}(i, j, k)$ is the approximative solution of the n-th iteration step, and if $P(i, j, k)$ is the solution of the difference equation system (Eq. 33) and

Dirichlet's boundary conditions, the following is valid for $h=r_c$ (COLLATZ, 1955):

$$|P^{(n)}(i,j,k)-P(i,j,k)|\leq\tfrac{1}{6}(\delta/2)^2\cdot\frac{|\text{Max }RS^{(n-1)}|}{h^2}=\varepsilon \tag{36}$$

where:

ε = truncation error
δ = the diameter of the smallest sphere which includes all grid points
$RS^{(n-1)}$ = the difference between the PO_2 values of the $(n-1)$-th and the n-th iteration step.

Except for a controllable error, $P^{(n)}(i,j,k)$ is the solution of the difference equation system, but not the solution of the differential Equation 22, i.e., the PO_2 field in question. This solution is further dependent upon the interstice length h ($h_x=h_y=h$) of the introduced grid. COURANT et al. (1928) showed that a solution of the difference equation system for each h unequivocally exists and that for $h\to0$ this solution also converges towards the solution of the differential equation, i.e., towards the PO_2 field in question. Figure 27b shows that for the equation system (Eq. 33) and Neumann's boundary conditions and for $h=r_c$, a sufficient approximation to the sought PO_2 field results. Diminishing the h practically leads to no further improvement in the approximation. Only in the range of high PO_2 values, the improvement is about 1.5 mm Hg if h is diminished from $r_c-r_c/10$. However, high PO_2 values seldom arise so that the error of about 1.5 mm Hg at high PO_2 values practically does not influence the PO_2 distribution. At great capillary distances ($d>100\,\mu$m) where in addition to the equation system (Eqs. 33, 34), PO_2 values at the lateral surface areas are calculated (Dirichlet's boundary conditions), the figure shows that in the range of high PO_2 values (at $h=2r_c$), errors of more than 5 mm Hg can arise and that, in addition, a deviation to the approximation solution of the equation system with Neumann's boundary conditions exists. By reducing the interstice length h, the number of grid points increases by a power of 3 and the computing time for the iteration process increases accordingly. Moreover, according to estimation (Eq. 36), Max $RS^{(n-1)}$ has to be reduced when h is reduced in order to remain under the truncation error. In turn, this means an increase in the number of iterations.

The computer time can be shortened considerably if $h_y>h_x$ is chosen. Figure 28 shows that for $h_y:h_x=3:1$, practically no difference in the value P_{min} exists compared to $h_y=h_x=h$. The same holds true for the PO_2 frequency distribution $\varphi(P)$. At $h_y:h_x=3:1$, the computer time is shortened by a factor of 3. For $h_y\neq h_x$, the Equations 33, 34, and 35 change over to the form:

$$P^{(n+1)}(i,j,k)=\Big(P^{(n)}(i-1,j,k)+P^{(n)}(i+1,j,k)$$
$$+\big(P^{(n)}(i,j-1,k)+P^{(n)}(i,j+1,k)\big)\cdot(h_x/h_y)^2$$
$$+P^{(n)}(i,j,k-1)+P^{(n)}(i,j,k+1)-\frac{h_x^2A}{K}\Big) \tag{37}$$
$$\cdot\frac{\omega}{4+2(h_x/h_y)^2}+(1-\omega)\,P^{(n)}(i,j,k)$$

for each grid point that lies outside the capillaries and does not lie on the surface area of the MCU;

$$
\begin{aligned}
P^{(n+1)}(i,0,k) = & \Big(P^{(n)}(i-1,0,k) + P^{(n)}(i+1,0,k) \\
& + \big(P^{(n)}(i,2M-1,k) + P^{(n)}(i,1,k)\big)\cdot(h_x/h_y)^2 \\
& + P^{(n)}(i,2M,k-1) + P^{(n)}(i,2M,k+1) - \frac{h_x^2 A}{K} \Big) \qquad (38) \\
& \cdot \frac{\omega}{4+2(h_x/h_y)^2} + (1-\omega)\,P^{(n)}(i,0,k)
\end{aligned}
$$

at the upper and lower surface outside the capillaries;

$$
\begin{aligned}
P^{(n+1)}(0,j,k) = & \Big(2P^{(n)}(1,j,k) + \\
& + \big(P^{(n)}(0,j-1,k) + P^{(n)}(0,j+1,k)\big)\cdot(h_x/h_y)^2 \\
& + P^{(n)}(0,j,k-1) + P^{(n)}(0,j,k+1) - \frac{h_x^2 A}{K} \Big) \\
& \cdot \frac{\omega}{4+2(h_x/h_y)^2} + (1-\omega)\,P^{(n)}(0,j,k)
\end{aligned}
$$

$$
\begin{aligned}
P^{(n+1)}(L,j,k) = & \Big(2P^{(n)}(L-1,j,k) \\
& + \big(P^{(n)}(L,j-1,k) + P^{(n)}(L,j+1,k)\big)\cdot(h_x/h_y)^2 \\
& + P^{(n)}(L,j,k-1) + P^{(n)}(L,j,k+1) - \frac{h_x^2 A}{K} \Big) \\
& \cdot \frac{\omega}{4+2(h_x/h_y)^2} + (1-\omega)\,P^{(n)}(L,j,k)
\end{aligned}
$$

$$
\qquad (39)
$$

$$
\begin{aligned}
P^{(n+1)}(i,j,0) = & \Big(P^{(n)}(i-1,j,0) + P^{(n)}(i+1,j,0) \\
& + \big(P^{(n)}(i,j-1,0) + P^{(n)}(i,j+1,0)\big)\cdot(h_x/h_y)^2 \\
& + 2P^{(n)}(i,j,1) - \frac{h_x^2 A}{K} \Big) \\
& \cdot \frac{\omega}{4+2(h_x/h_y)^2} + (1-\omega)\,P^{(n)}(i,j,0)
\end{aligned}
$$

$$
\begin{aligned}
P^{(n+1)}(i,j,L) = & \Big(P^{(n)}(i-1,j,L) + P^{(n)}(i+1,j,L) \\
& + \big(P^{(n)}(i,j-1,L) + P^{(n)}(i,j+1,L)\big)\cdot(h_x/h_y)^2 \\
& + 2P^{(n)}(i,j,L-1) - \frac{h_x^2 A}{K} \Big) \\
& \cdot \frac{\omega}{4+2(h_x/h_y)^2} + (1-\omega)\,P^{(n)}(i,j,L)
\end{aligned}
$$

at the lateral surfaces outside the capillaries.

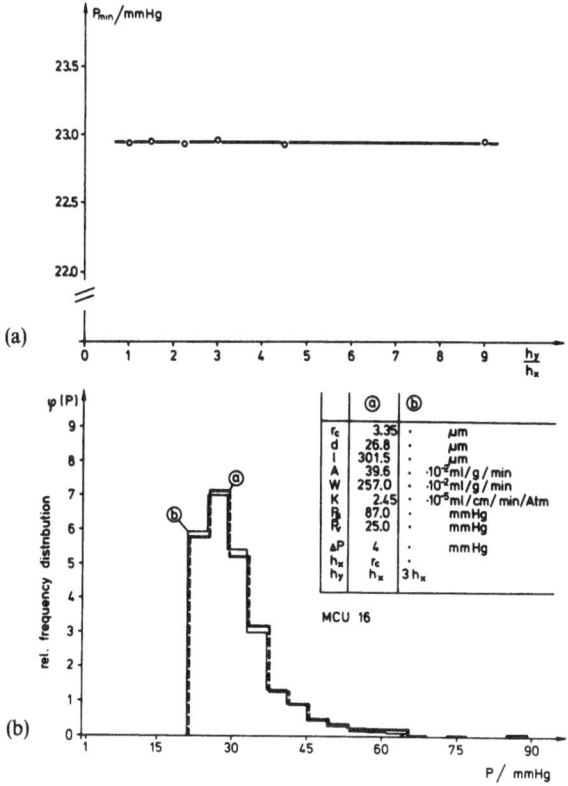

Fig. 28a and b. a) P_{min} of MCU 16 dependent upon the relation $h_y:h_x$ with $h_x=r_c$. P_{min} practically does not change if interstice length h_y is 10 times h_x. b) PO_2 frequency distribution $\varphi(P)$ of MCU 16 at $h_y:h_x=1:1$ (a) and at $h_y:h_x=3:1$ (b) with $h_x=r_c$. Both distributions show practically no difference

Appendix 2

The PO_2 frequency $w=w(P, \Delta P, h)$ is dependent upon the values of P from the P matrix, upon the P class ΔP and upon the distance h between the points of the grid which produces the P matrix in the model. The PO_2 frequency w is normed for the area under the frequency distribution, to allow for a comparison of frequency distributions with different P classes ΔP and different grid point distances h (accordingly, there is a different total number N for all P values in the P matrix). The area F under a frequency distribution curve results in:

$$F=N \cdot \Delta P. \tag{40}$$

Since the P classes ΔP of a distribution should be equal to one another, the relative distribution thus results in the following:

$$\varphi_h(P)=\frac{w(P, \Delta P, h)}{N \cdot \Delta P}. \tag{41}^9$$

---

[9] The relative frequency distribution in %, referring to the total number N of the P values results from $\varphi_h(P)$ by multiplication with ΔP. This is true for all further distributions.

The smaller the grid distance h, so much greater the number N of the values in the P matrix, so much better the P matrix represents the PO_2 field.

Thus, with a decreasing h, the relative frequency distribution describes the PO_2 field more precisely (Fig. 29). During PO_2 measurements, the P classes are between 3–5 mm Hg, a ΔP smaller than 3 mm Hg is therefore no longer advisable. If ΔP lies between 3 and 5 mm Hg and $h=r_c$, a further reduction of the grid distance h results in practically no change in the PO_2 distribution (GRUNEWALD, 1971). We can assume for the most part that h is chosen so small that it has no influence upon the form of the frequency distribution. Such a relative frequency distribution should be called *ideal* and is annotated with $\varphi(P)$.

Each element in the P matrix of a PO_2 field is regarded only once in an ideal frequency distribution. The differing frequencies in the individual P classes result from the fact that there are elements in the P matrix which are equal to one another. The ideal frequency distribution of a PO_2 field could also be produced if, in a random process, a large enough number of matrix elements are chosen and the frequency distribution of the resulting P values were thus formed. The ideal frequency distribution results then, when, in such a process, an *infinite* number of matrix elements are chosen, for only then would each element of the

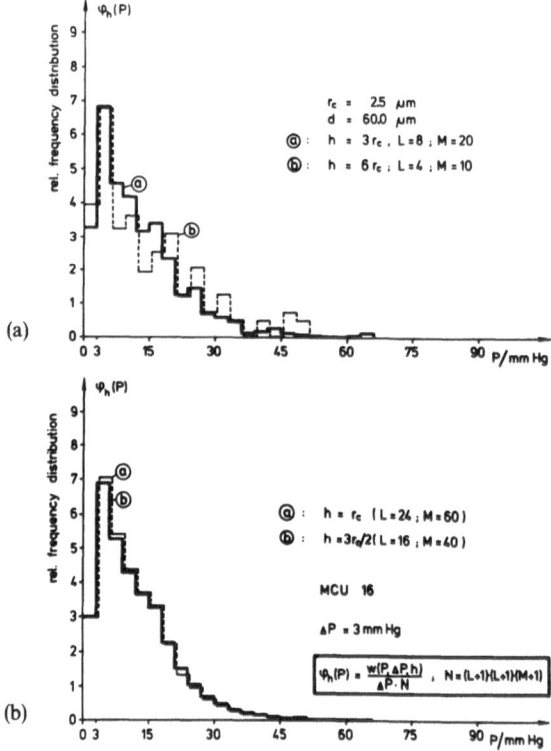

Fig. 29a and b. Frequency distribution $\varphi_h(P)$ of values of P matrix of MCU 16 at different interstice lengths $h(=h_x=h_y)$. $\varphi_h(P)$ is normed such that the area under distribution for all N and ΔP is the same $(=100\,\%)$ (*encircled* Eq.). For interstice length $h=r_c$, PO_2 distribution $\varphi_h(P)$ results from P matrix of $N=75.625$ elements. In figure above, $\varphi_h(P)$ are compared for $h=6r_c$ and $h=3r_c$. In figure below, distributions $\varphi_h(P)$ are compared for insterstice lengths $h=\frac{3}{2}r_c$ and $h=r_c$

matrix be chosen with the same frequency. By a PO_2 measurement, the only thing that occurs is that a *finite* number of PO_2 values from the PO_2 field is chosen. This procedure can be simulated in the diffusion model in that only a finite number of matrix elements is chosen and from these elements a relative frequency distribution is formed. If N^+ is the number of the chosen matrix elements and $w^+ = w^+ (P, \Delta P, h)$ corresponding to w in Equation 41 is the number of P values in a P class ΔP, then analogous to this equation the following results:

$$\varphi_h^+ (P) = \frac{w^+ (P, \Delta P, h)}{N^+ \cdot \Delta P}. \tag{42}$$

This distribution is called a *statistical* frequency distribution. It is normed according to Equation 41.

We must investigate how large the number N^+ of the chosen matrix element must be in order to have a *statistical* frequency distribution that well enough approximates an *ideal* frequency distribution.

In order to produce a statistical frequency distribution $\varphi_h^+ (P)$, the choice of elements from a P matrix of a PO_2 field occurs through a random process. Similar to the point grid system used to calculate the PO_2 field, the three-dimensional P matrix has in the x and z directions, coordinate values between 0 and N and in the y direction, coordinate values between 0 and $2 \cdot M$. From these three values, a random value is produced that has the function of a coordinate. The P value with the three so formed coordinates in the P matrix is singled out and used to form the statistical frequency distribution. This process is repeated N^+ times. The larger the random sample N^+, the better it approximates the ideal distribution $\varphi_h (P)$ (Fig. 30). Figure 30 shows that approximately 100–200 measurements in the tissue by a PO_2 electrode are necessary for a class ΔP between 3–5 mm Hg in order to reach the sufficient approximation of the (ideal) frequency distribution of the PO_2 field in the tissue.

It is necessary to emphasize that with the *ideal* and the *statistical* distribution there are two different frequency distributions to be considered: with a small enough h, the ideal distribution is singly dependent upon the form of the PO_2 field; the statistical distribution can be considered as an ideal distribution which, because of the random process, is superimposed by the statistical variances that become smaller as the number of samples increases. If interpreted in a statistical sense, the *ideal* distribution gives us the expected values for the individual P classes. The distribution of the expected values onto the P classes is not even, as in the case with dice. Each element in the P matrix has the same chance of being singled out, but since there are equally like elements amongst one another, their expected value is larger than those which exist singly for themselves. These differences in the expected values in the P classes are finally the criterion which makes a P matrix of a PO_2 field *characteristic* for a particular MCU and allows for a discrimination between the P matrices of other MCUs.

The above considerations were carried out with the MCU 16. However, this process is transferable onto other MCUs. If one no longer differentiates between an ideal and a statistical distribution and only PO_2 distributions are discussed, then these distributions must have a sampling number N^+ large enough to disregard their differences.

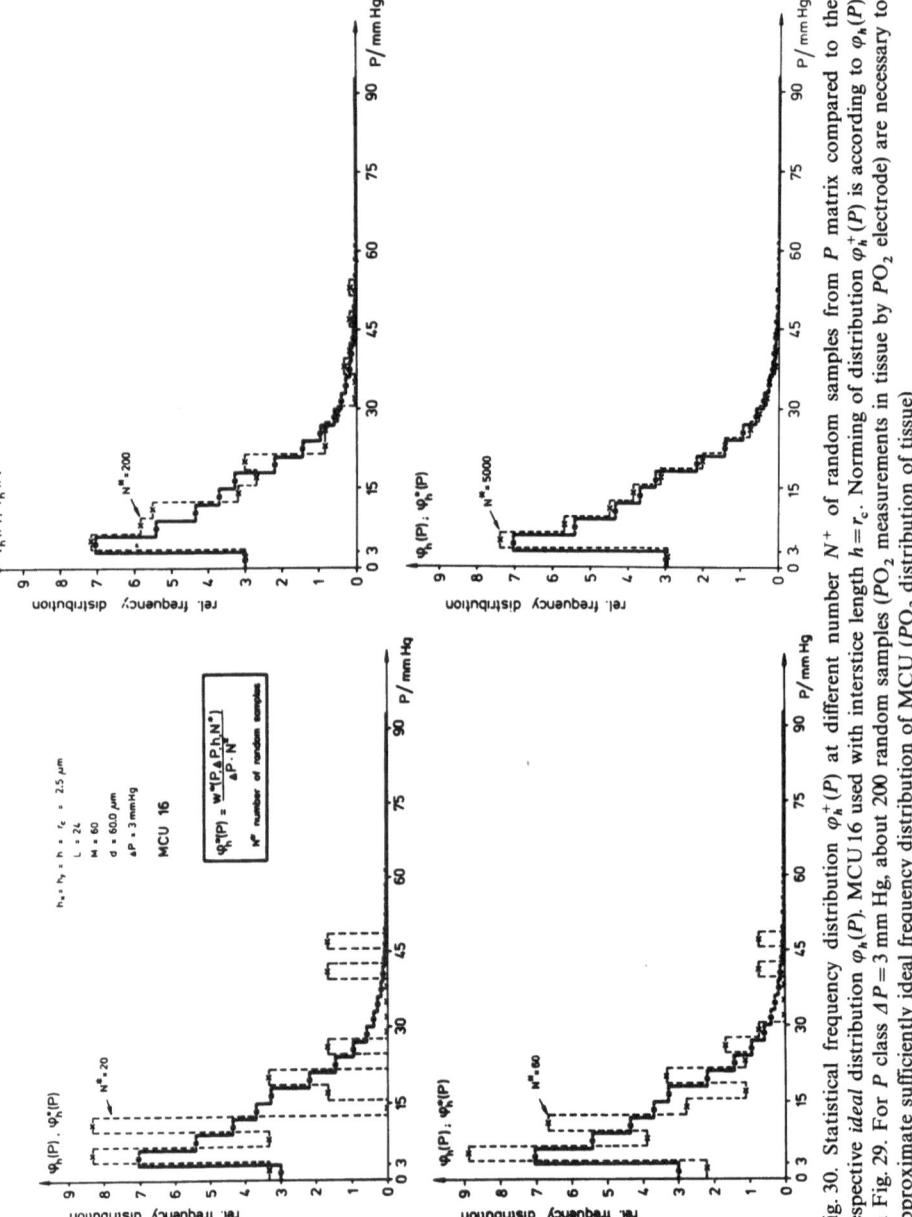

Fig. 30. Statistical frequency distribution $\varphi_h^+(P)$ at different number N^+ of random samples from P matrix compared to the respective *ideal* distribution $\varphi_h(P)$. MCU 16 used with interstice length $h = r_c$. Norming of distribution $\varphi_h^+(P)$ is according to $\varphi_h(P)$ in Fig. 29. For P class $\Delta P = 3$ mm Hg, about 200 random samples (PO_2 measurements in tissue by PO_2 electrode) are necessary to approximate sufficiently ideal frequency distribution of MCU (PO_2 distribution of tissue)

List of Symbols

a	Arterial inflow of a capillary
$A(P)$	O_2 consumption dependent upon PO_2
A	Maximal O_2 consumption
a_i	Weighting factor of basic distribution $\varphi_i(P)$ for different MCUs and different structure and supply parameters
$AVDO_2$	Arterio-venous O_2 difference
$AVDO_2^{(i)}$	Arterio-venous O_2 difference of the i-th capillary
c_j	Weighting factor of structure parameters
c_{Hb}	Hemoglobin concentration in blood
d	Capillary distance
d_0	Anatomical capillary distance
d_m	Functional capillary distance of step m of the mobilization and demobilization of capillary reserve
D	Tissue diffusion coefficient
D_K	Krogh diffusion constant
F	Area under a frequency distribution
$F_j^{(i)}$	Supply area of the i-th capillary in the j-th section
h_x	Interstice length between two grid points in x and z direction
h_y	Interstice length between two grid points in y direction
h	Interstice length between two grid points if $h_x = h_y$
i, j, k	Coordinates of a grid point
K	Tissue diffusion conductivity
l	Capillary length
Δl	Length of a capillary section ($= l/2M$)
δl	Length by which the capillary ends are displaced in order to obtain a superimposed frequency distribution
L	Number of discretization steps for directions x and z
$2 \cdot M$	Number of discretization steps for direction y
MCU	Microcirculatory unit
$MA^{(i)}$	Location of arterial capillary end of the i-th capillary in a MCU
m	Number of steps in mobilization (and demobilization) of the capillary reserve
N	Total number of P values in a P matrix
N^+	Number of random samples of a P matrix
n	Number of iterations
PO_2	Oxygen partial pressure
P $P(x, y, z),$ $P(r)$ $P(i, j, k)$	PO_2 dependent upon location; PO_2 field
$P_j^{(i)}(r)$	PO_2 of the i-th capillary on lateral surface area of the j-th section
P_a	Arterial PO_2
$\bar{P}_v$	Mean venous PO_2 (of a MCU)
$P_v^{(i)}$	Capillary venous PO_2 of i-th capillary
P_m	Maximum PO_2 of the measured PO_2 distribution
$P(r_c)$	PO_2 at the capillary wall
$P(r)$	PO_2 at distance r from the capillary center in a tissue cylinder
$P^{(n)}(i, j, k)$	Approximated PO_2 at the n-th step of the iteration process
P_{50}	PO_2 where $A(P) = A/2$
P_{min}	Minimum (tissue) PO_2 of a MCU
$\hat{P}_{min}$	Minimum (tissue) PO_2 of different MCUs
$P_c^{(i)}$	Intracapillary PO_2 (i.c. PO_2)
ΔP	P class value of a PO_2 frequency distribution
R	Radius of the Krogh tissue cylinder
r_c	Capillary radius

$R_j^{(i)}$	i-th supply boundary on the lateral surface area of the j-th section
$RS^{(n-1)}$	Difference between approximated PO_2 of the $(n-1)$-th and n-th iteration step $(RS^{(n-1)} = P^{(n-1)} - P^{(n)}$ at point $(i, j, k))$
r	Coordinate of the Krogh tissue cylinder
$\mathbf{r}$	Location vector
$s(P)$	Relative HbO_2 saturation of blood (HbO_2 dissociation curve)
$\bar{s}_v$	Mean venous HbO_2 saturation
s_a	Arterial HbO_2 saturation
$s_j^{(i)}$	i.c. HbO_2 saturation of the j-th section
Δs	Arterio-venous HbO_2 saturation difference
$\Delta s_j^{(i)}$	i.c. HbO_2 saturation difference along the capillary section Δl of the i-th capillary of the j-th section $(\Delta s_j^{(i)} = s_j^{(i)} - s_{j-1}^{(i)})$
v	Venous outflow of a capillary
$\mathbf{v}$	Diffusion flow density of the O_2 molecules
$V_c^{(i)}$	Capillary supply volume of the i-th capillary
V	Volume of tissue fragment of a MCU
V_{c_0}, V_{c_m}	Capillary supply volume of step 0 and m of mobilization and demobilization of capillary reserve
ΔV	Volume of the $2M$ sections $(\Delta V = V/2M)$
$\Delta V_j^{(i)}$	Capillary supply volume of the i-th capillary of the capillary section Δl of the j-th section
W	(Local) blood flow of a MCU
$W_c^{(i)}$	Blood flow of supply volume of the i-th capillary
x, y, z	Cartesian coordinates
$w = w(P, \Delta P, h)$	Frequency of all P values in a P matrix
$w^+ = w^+(P, \Delta P, h)$	Frequency of N^+ P values in a P matrix
α	Bunsen solubility coefficient
δ	Diameter of the smallest sphere which includes all grid points
ε	Truncation error of the relaxation process
ε_c	Truncation error of the iterative calculation of i.c. PO_2
κ	Number of MCUs with a displacement of the capillary ends of δl
κ'	Number of nonidentical MCUs
κ''	Number of different (statistically distributed) parameter values
ν	Total number of frequency distributions of an analysis of a measured PO_2 distribution
$\varphi(P)$	Relative PO_2 frequency distribution
$\varphi_h(P)$	Frequency distribution of N P matrix elements (ideal PO_2 frequency distribution)
$\varphi_h^+(P)$	Frequency distribution of N^+ P matrix elements (statistical PO_2 frequency distribution)
$\varphi_i(P)$	PO_2 distribution of i-th MCU
ϕ	Capillary blood flow of a MCU
$\phi_c^{(i)}$	Capillary blood flow of i-th capillary in a MCU
ϕ_{c_0}, ϕ_{c_m}	Capillary blood flow of step 0 and m of mobilization and demobilization of capillary reserve
$\psi(P)$	Superimposed PO_2 frequency distribution
$\psi_m(P)$	Measured PO_2 frequency distribution
ω	Overrelaxation factor
$\nabla^2 \equiv \dfrac{\partial^2}{\partial x^2} + \dfrac{\partial^2}{\partial y^2} + \dfrac{\partial^2}{\partial z^2}$	Laplace operator
grad	Gradient (operator)

References

Acker, H.: Der lokale Sauerstoffdruck im Glomus caroticum der Katze und seine Bedeutung für die nervöse chemorezeptive Aktivität. Habilitationsschrift, Bochum, 1975

Acker, H., Lübbers, D. W., Purves, M.: O$_2$-Transfer in der Bindegewebskapsel des Glomus caroticum der Katze und seine funktionelle Bedeutung. Pflügers Arch. ges. Physiol. **316**, R 30 (1970)

Acker, H., Lübbers, D. W., Purves, M.: Local oxygen tension field in the *Glomus caroticum* of the cat and its change at changing arterial PO_2. Pflügers Arch. ges. Physiol. **329**, 136–155 (1971)

Bänder, A., Kiese, M.: Die Wirkung des sauerstoffübertragenden Fermentes in Mitochondrien aus Rattenlebern bei niedrigen O$_2$-Drucken. Nauyn-Schmiedeberg's Arch. exp. Path. Pharmak. **224**, 312–321 (1955)

Bailey, H. R.: Oxygen exchange between capillary and tissue: some equations describing countercurrent and non-linear transport. In: Physical bases of circulatory transport: Regulation and Exchange. Philadelphia and London: Saunders, 1967

Barcroft, J., Bock, A. V., Hill, A. V., Parsons, T. R., Parsons, W., Shoji, R.: Of the hydrogen-ion concentration and some related properties of normal human blood. J. Physiol. (Lond.) **56**, 157–178 (1922)

Bartels, H., Harms, H.: Sauerstoffdissoziationskurven des Blutes von Säugetieren. Pflügers Arch. ges. Physiol. **268**, 334–365 (1959)

Batschelet, E.: Über die numerische Auflösung von Randwert-Problemen bei elliptischen partiellen Differentialgleichungen. Z. angew. Math. Phys. **3**, 165 (1952)

Baumgärtl, H., Grunewald, W. A., Lübbers, D. W.: Polarographic determination of the oxygen pressure field by Pt-electrodes using the PO_2-field in front of a Pt-macroelectrode as a model. Pflügers Arch. ges. Physiol. **347**, 49–61 (1974)

Baumgärtl, H., Lübbers, D. W.: Platinum needle electrode for polarographic measurement of oxygen and hydrogen. In: Oxygen supply. Theoretical and practical aspects of oxygen supply and microcirculation of tissue. Kessler, M., Bruley, D. F., Clark, L. C., Jr., Lübbers, D. W., Silver, I. A., Strauss, J. (eds.). Munich-Berlin-Vienna: Urban and Schwarzenberg, 1973, p. 130

Bicher, H. I., Bruley, D. F.: Oxygen Transport to Tissue. Advances in Experimental Medicine and Biology, 37A and 37B, New York and London: Plenum Press, 1973

Bicher, H. I., Bruley, D., Reneau, D. W., Knisely, M.: Regulatory mechanisms of brain oxygen supply. In: Oxygen supply. Theoretical and practical aspects of oxygen supply and microcirculation of tissue. Kessler, M., Bruley, D. F., Clark, L. C., Jr., Lübbers, D. W., Silver, I. A., Strauss, J. (eds.). Munich-Berlin-Vienna: Urban and Schwarzenberg, 1973, p. 180

Bicher, H. I., Knisely, M. H.: Brain tissue reoxygenation, demonstrated with a new ultramicro oxygen electrode. J. appl. Physiol. **28**, 387–390 (1970)

Blum, J. J.: Concentration profiles in and around capillaries. Amer. J. Physiol. **198**, 991–998 (1960)

Bork, R., Vaupel, P., Thews, G.: Atemgas-pH-Nomogramme für das Rattenblut bei 37 °C. Anaesthesist **24**, 84–90 (1975)

Bruley, D. F., Knisely, M. H.: Hybrid simulation—oxygen transport in the microcirculation. Chem. Eng. Progr. **66**, 22–32 (1970)

Caligara, F., Rooth, G.: Measurements of the oxygen diffusion coefficient in the subcutis of man. Acta physiol. scand. **53**, 114–127 (1961)

Cobourn, R. F., Mayers, L. B.: Myoglobin O$_2$ tension determined from measurements of carboxymyoglobin in skeletal muscle. Amer. J. Physiol. **220**, 66–74 (1971)

Collatz, L.: Numerische Behandlung von Differentialgleichungen. Berlin-Göttingen-Heidelberg: Springer, 1955

Courant, R., Friedrichs, K., Lewy, H.: Über die partiellen Differenzengleichungen der mathematischen Physik. Math. Ann. **100**, 32–74 (1928)

Diemer, K.: Eine verbesserte Modellvorstellung zur O$_2$-Versorgung des Gehirns. Naturwissenschaften **50**, 617–618 (1963)

Diemer, K.: Über die Sauerstoffdiffusion im Gehirn. I. Mitt.: Räumliche Vorstellung und Berechnung der Sauerstoffdiffusion. II. Mitt.: Die Sauerstoffdiffusion bei O$_2$-Mangelzuständen. Pflügers Arch. ges. Physiol. **285**, 99–108, 109–118 (1965)

Doll, E.: Oxygen pressure and content in the blood during physical exercise and hypoxia. In: Limiting Factors of Physical Performance. Stuttgart: Thieme, 1973

Erdmann, W., Heidenreich, J., Metzger, H.: H_2-Clearance und PO_2-Messung am Ratten- und Katzenhirn mit derselben Pt-Mikroelektrode. Pflügers Arch. ges. Physiol. **307**, R 51 (1969)

Evans, N. T., Naylor, P. F. D.: An electrical analogue for the study of tissue saturation processes with oxygen and inert gases. Phys. in Med. Biol. **9**, 43 (1964)

Fatt, J.: An ultramicro oxygen electrode. J. appl. Physiol. **19**, 326–329 (1964)

Forsythe, G. I., Wasow, W. R.: Finite difference methods for partial differential equations. New York: John Wiley and Sons, 1967

Gonzales-Fernandez, J. M., Atta, S. E.: Transport and consumption of oxygen in capillary-tissue structures. Math. Biosci. **2**, 225–262 (1968)

Grote, J.: Der Einfluß der Temperatur und des CO_2-Druckes auf die O_2-Affinität und den pH-Wert des menschlichen Blutes. Pflügers Arch. ges. Physiol. **300**, R 5 (1968)

Grote, J., Thews, G.: Die Bedingungen für die O_2-Versorgung des Herzmuskelgewebes. Pflügers Arch. ges. Physiol. **276**, 142–165 (1962)

Grunewald, W. A.: Theoretical analysis of the oxygen supply in tissue. In: Oxygen transport in blood and tissue. Stuttgart: Thieme, 1968, pp. 100–114

Grunewald, W. A.: Die Beeinflussung der PO_2-Verteilung im Gewebe durch die PO_2-Messung mit der Pt-Elektrode. Pflügers Arch. ges. Physiol. **312**, R 144–145 (1969a)

Grunewald, W. A.: Digitale Simulation eines räumlichen Diffusionsmodelles der O_2-Versorgung biologischer Gewebe. Pflügers Arch. ges. Physiol. **309**, 266–284 (1969b)

Grunewald, W. A.: Diffusionsfehler und Eigenverbrauch der Pt-Elektrode bei PO_2-Messungen im steady-state. Pflügers Arch. ges. Physiol. **320**, 24–44 (1970)

Grunewald, W. A.: Bedeutung der Kapillarstrukturen für die O_2-Versorgung der Organe und ihre Analyse anhand digital simulierter Modelle. Habilitationsschrift, Bochum, 1971

Grunewald, W. A., Lübbers, D. W.: Die Bedeutung asymmetrischer Capillarstrukturen für die Sauerstoffversorgung der Organe. Pflügers Arch. ges. Physiol. **289**, R 98 (1966)

Grunewald, W. A.: Lübbers, D. W.: Quantitative Beurteilung der "O_2-Diffusions-Kurzschlußgefährdung" bei der O_2-Versorgung des Gewebes. Pflügers Arch. ges. Physiol. **300**, R 20 (1968)

Grunewald, W. A., Sowa, W.: The influence of capillary structures, O_2-supply parameters and O_2-shunt on the frequency distribution of intracapillary HbO_2-saturation. Pflügers Arch. ges. Physiol. **355**, R 3 (1975)

Günther, H., Aumüller, G., Kunke, S., Vaupel, P., Thews, G.: Die Sauerstoffversorgung der Niere. I. Verteilung der O_2-Drucke in der Rattenniere unter Normbedingungen. Res. exp. Med. **163**, 251–264 (1974)

Günther, H., Vaupel, P., Metzger, H., Thews, G.: Stationäre Verteilung der O_2-Drucke im Tumorgewebe (DS-Carcinosarkom). II. Messungen in vivo unter Verwendung von Goldmikroelektroden. Z. Krebsforsch. **77**, 26–39 (1972)

Heidenreich, J., Metzger, H., Erdmann, W.: Änderung des regionalen Sauerstoffpartialdruckes nach plötzlichem Wechsel der inspiratorischen O_2-Konzentration. Pflügers Arch. ges. Physiol. **312**, R 64 (1969)

Henquell, L., Honig, C. R.: Intercapillary distance and capillary reserve in right and left ventricles. Significance for control of tissue PO_2. Microvasc. Res. **12**, 35–41 (1976)

Henquell, L., LaCelle, P. L., Honig, C. R.: Capillary diameter in rat heart in situ, relation to erythrocyte deformability, O_2-transport, and transmural O_2-gradient. Microvasc. Res. In press (1976)

Hill, A. V.: The diffusion of oxygen and lactic acid through tissue. Proc. roy. Soc. B **104**, 39–96 (1929)

Honig, C. R., Bourdeau-Martini, J.: O_2 and the number and arrangement of coronary capillaries; effect on calculated tissue PO_2. In: Oxygen transport to tissue. Instrumentation, methods, and physiology. Advance exp. Med. Biol. Bicher, H. I., Bruley, D. F. (eds.). New York and London, Plenum Press, 1973, Vol. XXXVII A, pp. 519–524

Hudson, J. A., Cater, D. B.: An analysis of factors affecting tissue oxygen tension. Proc. roy. Soc. B **101**, 247–274 (1964)

Iwanow, K. P., Kisliakow, Yn. Ya.: The oxygen available in a neuron and surrounding tissue. Sechenow Physiol. J. USSR, LX, **6**, 900–905 (1974)

Jacobs, M. H.: Diffusion processes. Ergebn. Biol. **12**, 1 (1935)

Kadatz, R.: Sauerstoffdruck und Durchblutung im gesunden und coronarinsuffizierten Myocard des Hundes und ihre Beeinflussung durch coronarerweiternde Pharmaka. Habilitationsschrift, Tübingen, 1967

Kantorowitsch, L. W., Krylow, W. J.: Näherungsmethoden der höheren Analyse. Berlin: Deutscher Verlag der Wissenschaften, 1956

Kessler, M.: Normale und kritische O$_2$-Versorgung der Leber bei Normo- und Hypothermie. Habilitationsschrift, Marburg, 1967

Kessler, M., Bruley, D.F., Clark, L.C., Jr., Lübbers, D.W., Silver, I.A., Strauss, J.: Oxygen supply. Theoretical and practical aspects of oxygen transport and microcirculation of tissue. Munich-Berlin-Vienna: Urban and Schwarzenberg, 1973

Keul, J.: Die Wirkung des sportlichen Trainings auf Substratumsatz, Sauerstoffverbrauch und Durchblutung des Skelettmuskels. In: Probleme der Skelettmuskeldurchblutung. Arzneimittel-Forsch. **21**, 366–376 (1970)

Kisliakow, Yn. Ya., Iwanow, K.P.: PO$_2$-distribution within neurons and cerebral capillaries with regard to the blood flow speed in normal and in hypoxia. Sechenow Physiol. J. USSR, LX, **8**, 1216–1222 (1974)

Knisely, M.H., Reneau, D.D., Bruley, D.F.: The development and use of equations for predicting the limits on the rates of oxygen supply to the cells of living tissues and organs. Angiology (Suppl.) **20**, 1–56 (1969)

Krogh, A.: The number and the distribution of capillaries in muscles with the calculation of the oxygen pressure head necessary for supplying the tissue. J. Physiol. (Lond.) **52**, 409–415 (1918/19 a)

Krogh, A.: The rate of diffusion of gases through animal tissue, with some remarks on the coefficient of invasion. J. Physiol. (Lond.) **52**, 391–409 (1918/19 b)

Kunze, K.: Das Sauerstoffdruckfeld im normalen und pathologisch veränderten Muskel. Schriftenreihe "Neurologie". Berlin-Heidelberg-New York: Springer, 1969, Vol. III

Kunze, K.: Die lokale kontinuierliche Sauerstoffdruckmessung in der menschlichen Muskulatur. Pflügers Arch. ges. Physiol. **282**, 151–160 (1966)

Leichtweiss, H.P., Lübbers, D.W., Weiss, Ch., Baumgärtl, H., Reschke, W.: The oxygen supply of the rat kidney: measurements of intrarenal PO$_2$. Pflügers Arch. ges. Physiol. **309**, 328–349 (1969)

Leonard, E.F., Jørgensen, S.B.: The analysis of convection and diffusion in capillary beds. Ann. Rev. Biophys. Bioeng. **3**, 293–339 (1974)

Liebmann, H.: Die angenäherte Ermittlung harmonischer Funktionen und konformer Abbildungen. S. math.-phys. Kl. Bayer. Akad. Wiss., Munich, 1918

Longmuir, J.S.: Respiration rate of the rat-liver cells at low oxygen concentrations. Biochem. J. **65**, 378–382 (1957)

Lübbers, D.W.: Kritische O$_2$-Versorgung und Mikrozirkulation. Marburger Jahrb. Marburg: Elwerth, 1967, pp. 305–319

Lübbers, D.W., Baumgärtl. H.: Herstellungstechnik von palladinierten Pt-Stichelektroden (1–5 µ Außendurchmesser) zur polarographischen Messung des Wasserstoffdruckes für die Bestimmung der Mikrozirkulation. Pflügers Arch. ges. Physiol. **294**, R 39 (1967)

Lübbers, D.W., Baumgärtl, H., Fabel, H., Huch, A., Kessler, M., Kunze, K., Riemann, H., Seiler, D., Schuchhardt, S.: Principle of construction and application of various platinum electrodes. In: Oxygen pressure recording in gases, fluids, and tissue. Progr. resp. Res. Kreuzer, F., Herzog, H. (eds.). Basel: Karger, 1969, Vol. III, pp. 136–146

Lübbers, D.W., Luft, U.C., Thews, G., Witzleb, E.: Oxygen transport in blood and tissue. Stuttgart: Thieme, 1968

Martini, J., Honig, C.R.: Direct measurement of intercapillary distance in beating rat heart in situ under various conditions of O$_2$-supply. Microvasc. Res. **1**, 244–256 (1969)

Mendler, N., Schuchhardt, S., Sebening, F.: Measurements of intromyocardial oxygen tension during cardiac surgery in man. Res. exp. Med. **159**, 231–238 (1973)

Metzger, H.: Der Sauerstoffaustausch zwischen Blutkapillaren und Gewebe. Dissertation, Darmstadt, 1967

Metzger, H.: Verteilung des O$_2$-Partialdruckes im Mikrobereich des Gehirngewebes. Polarographische Messung und mathematische Analyse. Habilitationsschrift, Mainz, 1972

Metzger, H., Erdmann, W., Thews, G.: Vergleichende Untersuchung der O$_2$- und H$_2$-Übergangsfunktionen im Gehirngewebe nach inspiratorischem Konzentrationswechsel. Pflügers Arch. ges. Physiol. **316**, R 4 (1970)

Niesel, W., Thews, G.: Ein elektronisches Analogrechenverfahren zur Lösung physiologischer Diffusionsprobleme. 1. Mitt.: Die allgemeinen Analogiebeziehungen und ihre Anwendung auf Diffusionsprozesse mit konstanten Nebenbedingungen. Pflügers Arch. ges. Physiol. **269**, 262–305 (1959)

Niesel, W., Thews, G.: Ein elektronisches Analogrechenverfahren zur Lösung physiologischer Diffu-
  sionsprobleme. 2. Mitt.: Anwendung des Verfahrens auf Diffusionsprozesse mit konstanten und
  periodisch veränderlichen Verbrauchsgrößen. Pflügers Arch. ges. Physiol. **276**, 182–191 (1962)
Opitz, E.: Über die Sauerstoffversorgung des Zentralnervensystems. Naturwissenschaften **35**, A 80
  (1948)
Opitz, E., Bartels, H.: Gasanalyse. In: Hoppe-Seyler/Thierfelder Hdb. d. Physiolog. und Patholog.
  Chemie. Analyse, 10th ed. Berlin-Göttingen-Heidelberg: Springer, 1955, Vol. II, pp. 183–311
Opitz, E., Schneider, M.: Über die Sauerstoffversorgung des Gehirns und den Mechanismus von
  Mangelwirkungen. Ergebn. Physiol. **46**, 126–260 (1950)
Opitz, E., Thews, G.: Einfluß von Frequenz und Faserdicke auf die Sauerstoffversorgung des mensch-
  lichen Herzmuskels. Arch. Kreisl.-Forsch. **18**, 137–152 (1952)
Reneau, D.D., Jr., Bicher, H.I., Bruley, D.F., Knisely, M.H.: A mathematical analysis predicting
  cerebral tissue reoxygenation time as a function of the rate of change of effective cerebral blood
  flow. In: Blood Oxygenation. New York: Plenum Press, 1970
Reneau, D.D., Jr., Bruley, D.F., Knisely, M.H.: A digital simulation of transient oxygen transport
  in capillary tissue systems (cerebral grey matter). Aiche J. **15**, 916–925 (1969)
Reneau, D.D., Jr., Bruley, D.F., Knisely, M.H.: A mathematical simulation of oxygen release, diffusion, and consumption in the capillaries and tissue of the human brain. In: Chemical engineering
  in medicine and biology. New York: Plenum Press, 1967
Reneau, D.D., Knisely, M.H.: A mathematical simulation of oxygen transport in the human brain
  under conditions of countercurrent capillary blood flow. Chem. Engng. Progr. **67**, 18–27 (1971)
Renkin, E.M.: Transcapillary exchange in skeletal muscle. In: Probleme der Skelettmuskeldurchblutung. Arzneimittel-Forsch. **21**, 366–376 (1970)
Rodenhäuser, J.H., Baumgärtl, H., Lübbers, D.W., Briggs, D.: Behavior of the oxygen partial pressure
  in the vitreous body under various oxygen conditions. Experimental studies in cats. Exc. Med.
  Intern. Congr. Ser. **222**, Ophthalmology, 1624–1628 (1970)
Roughton, F.R.S.: Diffusion and chemical reaction in cylindrical and spherical systems of physio-
  logical interest. Proc. roy. Soc. B. **140**, 203–230 (1952)
Schuchhardt, S.: Die Sauerstoffdruckverteilung im hämoglobinfrei perfundierten Meerschweinchen-
  herzen bei Ruhe und Tätigkeit. Pflügers Arch. ges. Physiol. **322**, 131–151 (1971a)
Schuchhardt, S.: PO_2-Messung im Myocard des schlagenden Herzens. Pflügers Arch. ges. Physiol.
  322, 83–94 (1971b)
Schuchhardt, S.: Statik und Dynamik der Sauerstoffversorgung des Herzens. Habilitationsschrift,
  Bochum, 1971c
Silver, J.A.: Measurement of oxygen tension in tissues. In: Oxygen measurements in blood and tissues.
  London: Churchill, 1966, p. 135
Silver, J.A.: Some observations on the cerebral cortex with an ultramicro membrane-covered oxygen
  electrode. Med. Electron. biol. Eng. **3**, 377–387 (1965)
Stainsby, W.N., Otis, A.B.: Blood flow, blood oxygen tension, oxygen uptake, and oxygen transport
  in skeletal muscle. Amer. J. Physiol. **206**, 858–866 (1964)
Starlinger, H., Lübbers, D.W.: Polarographic measurement of oxygen pressure performed simul-
  taneously with optical measurements of the redox state of the respiratory chain in suspensions of
  mitochondria under steady-state conditions at low oxygen pressure. Pflügers Arch. ges. Physiol.
  341, 15–22 (1973)
Tang, P.S.: Quart Rev. Biol **8**, 260 (1958)
Thews, G.: Über die mathematische Behandlung physiologischer Diffusionsprozesse in cylinderförmi-
  gen Objekten. Acta biotheor. (Leiden) **10**, 105–136 (1953)
Thews, G.: Ein Verfahren zur Berechnung des O_2-Diffusionskoeffizienten aus Messungen der Sauer-
  stoffdiffusion in Hämoglobin- und Myoglobinlösungen. Pflügers Arch. ges. Physiol. **265**, 138–153
  (1957)
Thews, G.: Die Sauerstoffdiffusion im Gehirn. Pflügers Arch. ges. Physiol. **271**, 197–226 (1960)
Thews, G.: Die Sauerstoffdrucke im Herzmuskelgewebe. Pflügers Arch. ges. Physiol. **276**, 166–181
  (1962)
Thews, G., Niesel, W.: Zur Theorie der Sauerstoffdiffusion im Erythrozyten. Pflügers Arch. ges.
  Physiol. **268**, 318–333 (1959)
Thomas, L.J., Jr.: Algorithmus for selected blood acid-base and blood gas calculations. J. appl.
  Physiol. **33**, 154–158 (1972)

Vaupel, P.: Atemgaswechsel und Glucosestoffwechsel von Implantationstumoren (DS-Carcino-sarkom) in vivo. Funktionsanalyse Biologischer Systeme 1, 1–138 (1974)

Vaupel, P., Braunbeck, W., Thews, G.: Respiratory gas exchange and PO_2-distribution in splenic tissue. In: Oxygen transport to tissue. Instrumentation, methods, and physiology. Advance exp. Med. Biol. Bicher, H. I., Bruley, D. F. (eds.). New York and London: Plenum Press, 1973, Vol. XXXVII A, pp. 401–406

Vaupel, P., Günther, H., Erdmann, W., Kunke, S., Thews, G.: Dreidimensionale Registrierung statio-närer PO_2-Werte im Tumorgewebe unter in vivo-Bedingungen bei Verwendung von Multi-Gold-mikroelectroden. Verh. dtsch. Ges. inn. Med. **78**, 133–136 (1972)

Warburg, O.: Versuche an überlebendem Carcinomgewebe. Biochem. Z. **142**, 317–333 (1923)

Whalen, W. J., Nair, P.: Microelectrode measurements of tissue PO_2 in the carotid body of the cat. In: Oxygen Supply. Theoretical and Practical Aspects of Oxygen Supply and Microcirculation of Tissue. Munich-Berlin-Vienna: Urban and Schwarzenberg, 1973

Whalen, W. J., Riley, J., Nair, P.: A microelectrode for measuring intracellular PO_2. J. appl. Physiol. **23**, 798–801 (1967)

Acknowledgment. I wish to thank the Deutsche Forschungsgemeinschaft for the generous grant of personnel and material without which the present paper would not have been possible. The extensive calculations were carried out in the computer centers of the Max-Planck-Institut für Systemphysiologie in Dortmund (IBM 360/44), the University of Regensburg (Siemens 4004) and the University of Erlangen-Nürnberg (CDC 3300). I am deeply indebted to Mr. TECKHAUS (Dortmund) and Mrs. A. BIRKMANN (Dortmund and Regensburg). Mr. KRAUSENBERGER helped us by quickly and effi-ciently running our computer programs in Erlangen-Nürnberg. Additionally, I want to thank Mrs. A. BIRKMANN for the exquisite preparation of the drawings.

Particular thanks is owed to Mrs. T. GIROUX and Mr. VON DER MOSEL, cand. med., for the translation of this linguistically pretentious paper.

Author Index

Page numbers in *italics* refer to the bibliography

Subject Index

Other Reviews of Interest

BLOUGH, H.A., TIFFANY, J.M.: Theoretical Aspects of Structure and Assembly of Viral Envelopes. Curr. Top. Microbiol. Immunol. **70**, 1–30 (1975)

DOEFLER, W.: Integration of Viral DNA into the Host Genome. Curr. Top. Microbiol. Immunol. **71**, 1–78 (1975)

DZIARSKI, R.: Teichoic Acids. Curr. Top. Microbiol. Immunol. **74**, 113–136 (1976)

EMERSON, S.U.: Vesicular Stomatitis Virus: Structure and Function of Virion Components. Curr. Top. Microbiol. Immunol. **73**, 1–34 (1976)

GEIDER, K.: Molecular Aspects of DNA Replication in *Escherichia coli* Systems. Curr. Top. Microbiol. Immunol. **74**, 55–112 (1976)

HEHLMANN, R.: RNA Tumor Viruses and Human Cancer. Curr. Top. Microbiol. Immunol. **73**, 141–215 (1976)

LEVINE, A.J., VAN DER VLIET, P.C., SUSSENBACH, J.S.: The Replication of Papovavirus and Adenovirus DNA. Curr. Top. Microbiol. Immunol. **73**, 67–124 (1976)

MACARIO, A.J.L., CONWAY DE MACARIO, E.: Antigen-Binding Properties of Antibody Molecules: Time-Course Dynamics and Biological Significance. Curr. Top. Microbiol. Immunol. **71**, 125–170 (1975)

MARTIN, M.A., KHOURY, G.: Integration of DNA Tumor Virus Genomes. Curr. Top. Microbiol. Immunol. **73**, 35–65 (1976)

MOSCOVICI, C.: Leukemic Transformation with Avian Myeloblastosis Virus: Present Status. Curr. Top. Microbiol. Immunol. **71**, 79–101 (1975)

PIRROTTA, V.: The λ Repressor and its Action. Curr. Top. Microbiol. Immunol. **74**, 21–54 (1976)

PURCHASE, H.G., WITTER, R.L.: The Reticuloendotheliosis Viruses. Curr. Top. Microbiol. Immunol. **71**, 103–124 (1975)

RIMON, A.: The Chemical and Immunochemical Identitiy of Amyloid. Curr. Top. Microbiol. Immunol. **74**, 1–20 (1976)

SCHAFFER, P.A.: Temperature-Sensitive Mutants of Herpesviruses. Curr. Top. Microbiol. Immunol. **70**, 51–100 (1975)

SCHOLTISSEK, C.: Inhibition of the Multiplication of Enveloped Viruses by Glucose Derivatives. Curr. Top. Microbiol. Immunol. **70**, 101–119 (1975)

SHARP, P.A., FLINT, S.J.: Adenovirus Transcription. Curr. Top. Microbiol. Immunol. **74**, 137–166 (1976)

STEVENS, J.G.: Latent Herpes Simplex Virus and the Nervous System. Curr. Top. Microbiol. Immunol. **70**, 31–50 (1975)

WESTPHAL, H.: In vitro Translation of Adenovirus Messenger RNA. Curr. Top. Microbiol. Immunol. **73**, 125–139 (1976)

Springer-Verlag Berlin Heidelberg New York

Other Reviews of Interest in this Series

BLOOM, F.E.: The Role of Cyclic Nucleotides in Central Synaptic Function. Rev. Physiol. Biochem. Pharmacol. **74**, 1–103 (1975)

BREW, K., HILL, R.L.: Lactose Biosynthesis. Rev. Physiol. Biochem. Pharmacol. **72**, 105–158 (1975)

DE ROBERTIS, E.: Synaptic Receptor Proteins. Isolation and Reconstruction in Artificial Membranes. Rev. Physiol. Biochem. Pharmacol. **73**, 9–38 (1975)

ELLENDORF, F.: Evaluation of Extrahypothalamic Control of Reproductive Physiology. Rev. Physiol. Biochem. Pharmacol. **76**, 103–127 (1976)

GRUNICKE, H., PUSCHENDORF, B., WERCHAU, H.: Mechanism of Action of Distamycin A and Other Antibiotics with Antiviral Activity. Rev. Physiol. Biochem. Pharmacol. **75**, 69–96 (1976)

HAASE, J., CLEVELAND, S., ROSS, H.-G.: Problems of Postsynaptic Autogenous and Recurrent Inhibition in the Mammalian Spinal Cord. Rev. Physiol. Biochem. Pharmacol. **73**, 73–129 (1975)

HILZ, H.: Poly (ADP-Ribose) and ADP-Ribosylation of Proteins. Rev. Physiol. Biochem. Pharmacol. **76**, 1–58 (1976)

HOFMANN, E.: The Significance of Phosphofructokinase to the Regulation of Carbohydrate Metabolism. Rev. Physiol. Biochem. Pharmacol. **75**, 1–68 (1976)

KATUNUMA, N.: Regulation of Intracellular Enzyme Levels by Limited Proteolysis. Rev. Physiol. Biochem. Pharmacol. **72**, 83–104 (1975)

KULAEV, I.S.: Biochemistry of Inorganic Polyphosphates. Rev. Physiol. Biochem. Pharmacol. **73**, 131–158 (1975)

LAMBERT, A.E.: The Regulation of Insulin Secretion. Rev. Physiol. Biochem. Pharmacol. **75**, 97–159 (1976)

MELANDER, A., ERICSON, L.E., SUNDLER, F., WESTGREN, U.: Intrathyroidal Amines in the Regulation of Thyroid Activity. Rev. Physiol. Biochem. Pharmacol. **73**, 39–71 (1975)

MOE, G.K.: Evidence for Reentry as a Mechanism of Cardiac Arrhythmias. Rev. Physiol. Biochem. Pharmacol. **72**, 55–81 (1975)

RAPPAPORT, A.M., SCHNEIDERMAN, J.H.: The Function of the Hepatic Artery. Rev. Physiol. Biochem. Pharmacol. **76**, 129–175 (1976)

SHAPOVALOV, A.I.: Neuronal Organization and Synaptic Mechanisms of Supraspinal Motor Control in Vertebrates. Rev. Physiol. Biochem. Pharmacol. **72**, 1–54 (1975)

SILBERNAGEL, S., FOULKES, E.C., DEETJEN, P.: Renal Transport of Amino Acids. Rev. Physiol. Biochem. Pharmacol. **74**, 105–167 (1975)

WUTTKE, W.: Neuroendocrine Mechanisms in Reproductive Physiology. Rev. Physiol. Biochem. Pharmacol. **76**, 59–102 (1976)

Springer-Verlag Berlin Heidelberg New York